Advances in Science, Technology & Innovation

IEREK Interdisciplinary Series for Sustainable Development

Advances in Science, Technology & Innovation (ASTI) is a series of peer-reviewed books based on important emerging research that redefines the current disciplinary boundaries in science, technology and innovation (STI) in order to develop integrated concepts for sustainable development. It not only discusses the progress made towards securing more resources, allocating smarter solutions, and rebalancing the relationship between nature and people, but also provides in-depth insights from comprehensive research that addresses the **17 sustainable development goals (SDGs)** as set out by the UN for 2030.

The series draws on the best research papers from various IEREK and other international conferences to promote the creation and development of viable solutions for a **sustainable future and a positive societal** transformation with the help of integrated and innovative science-based approaches. Including interdisciplinary contributions, it presents innovative approaches and highlights how they can best support both economic and sustainable development, through better use of data, more effective institutions, and global, local and individual action, for the welfare of all societies.

The series particularly features conceptual and empirical contributions from various interrelated fields of science, technology and innovation, with an emphasis on digital transformation, that focus on providing practical solutions to **ensure food, water and energy security to achieve the SDGs.** It also presents new case studies offering concrete examples of how to resolve sustainable urbanization and environmental issues in different regions of the world.

The series is intended for professionals in research and teaching, consultancies and industry, and government and international organizations. Published in collaboration with IEREK, the Springer ASTI series will acquaint readers with essential new studies in STI for sustainable development.

ASTI series has now been accepted for Scopus (September 2020). All content published in this series will start appearing on the Scopus site in early 2021.

More information about this series at http://www.springer.com/series/15883

Pardeep Singh • Rishikesh Singh •
Pramit Verma • Rahul Bhadouria •
Ajay Kumar • Mahima Kaushik
Editors

Plant-Microbes-Engineered Nano-particles (PM-ENPs) Nexus in Agro-Ecosystems

Understanding the Interaction of Plant, Microbes and Engineered Nano-particles (ENPS)

Springer

Editors
Pardeep Singh
Department of Environmental Studies
PGDAV College
University of Delhi
New Delhi, Delhi, India

Rishikesh Singh ⓘ
Institute of Environment & Sustainable
Development
Banaras Hindu University
Varanasi, Uttar Pradesh, India

Pramit Verma
Integrative Ecology Laboratory, Institute
of Environment and Sustainable Development
Banaras Hindu University
Varanasi, India

Rahul Bhadouria
Natural Resource Management Laboratory
Department of Botany
University of Delhi
New Delhi, India

Ajay Kumar
Volcani Center
Agriculture Research Organization (ARO)
Rishon LeZion, Israel

Mahima Kaushik
Cluster Innovation Centre (CIC)
University of Delhi
New Delhi, India

ISSN 2522-8714   ISSN 2522-8722 (electronic)
Advances in Science, Technology & Innovation
IEREK Interdisciplinary Series for Sustainable Development
ISBN 978-3-030-66958-4   ISBN 978-3-030-66956-0 (eBook)
https://doi.org/10.1007/978-3-030-66956-0

This Springer imprint is published by the registered company Springer Nature Switzerland AG
The registered company address is: Gewerbestrasse 11, 6330 Cham, Switzerland

Preface

Applications of nanotechnology have quite frequently been explored in varied fields like biomedical, food, energy, defence, textiles, paints and household products, etc. Similarly, nanotechnology has acted as a boon in agriculture sector, because of its utilization in nano-priming for quick seed growth, enhancing crop production due to the use of nano-fertilizers, nano-pesticides and nano-weedicides, etc. In recent times, the nanoparticles are preferred to be synthesized using biological means such as plants or microbes, and hence, these are termed as plant-microbe-engineered nanoparticles (PM-ENPs) nexus. These PM-ENPs nexus have good antimicrobial activity and are comparatively much more efficient, lesser toxic and cost-effective. The applications of these nanomaterials or ENPs in agriculture have helped in using eco-friendly chemicals for crop production, which in turn has also resulted in the decrease of environmental pollution, which otherwise used to happen because of immense use of harmful chemicals, fertilizers, pesticides and insecticides in earlier times. Thus, exploring PM-ENPs nexus approach may help in holistic understanding of the role of ENPs in the agro-ecosystem.

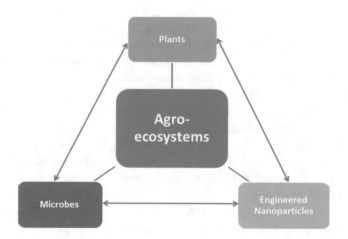

This book is an attempt to summarize the overall and comprehensive work related to the improvement of agriculture production via nanotechnology-aided resources, which includes frequent but quite cautious use of ENPs, as the toxicity, distribution and their fate afterwards in environment are still in question and need to be studied further. The book consists of total 14 chapters divided in six different parts, viz. (i) Engineered Nanoparticles in Agro-ecosystems: General Approach; (ii) Engineered Nanoparticles and Plant Interaction; (iii) Engineered Nanoparticles and Soil Health; (iv) Engineered Nanoparticles as Nano-Fertilizers and Biosensors; (v) Engineered Nanoparticles and Microbial Interaction; and (vi) Plant–Microbe–Soil Health-Engineered Nanoparticles Nexus: Conclusion. Total 49 authors belonging to 7 countries have contributed in this book. Detailed description of the content of individual chapters has been outlined in the following paragraphs.

Chapter "Engineered Nanoparticles in Smart Agricultural Revolution: An Enticing Domain to Move Carefully" of Part One written by Srivastava et al. summarizes the significant role of ENPs in agricultural revolution by discussing their interactions with plants and soil microorganisms. Potential benefits of using ENPs in agriculture include their use in nano-fertilizers and nano-pesticides for their targeted controlled release, without overdosing and leach outs. However, these ENPs need to be studied further for their harmful effects due to their toxicological effects, acquired biological activity and their unknown fate in environment. Chapter "Nanotechnology: Advancement for Agricultural Sustainability" by Upinder and Kumar reviews various advancements in nanotechnology for sustainable agriculture. For increase in agricultural production, various types of nanomaterials have been utilized, but all of them also pose an indirect threat to environment, which is still under investigation. A summary of already utilized nanomaterials in agricultural field has been given. Both these chapters cover the introductory remarks of the book.

Chapters "Nanotechnology for Sustainable Crop Production: Recent Development and Strategies"–"Plant Physiological Responses to Engineered Nanoparticles" come under Part Two in which authors have explored the role of ENPs in crop development and improvement. Chapter "Nanotechnology for Sustainable Crop Production: Recent Development and Strategies" by Abhishek Kumar et al. summarizes a lot of recent developments and strategies utilizing nanotechnology for sustainable crop production. Use of nanomaterials like nano-pesticides, nano-insecticide and nano-weedicides has been discussed at length for exploring applications like increase in seed germination, growth in crop production, identifying pest attack and disease prevalence, etc.

Chapter "Interaction of Titanium Dioxide Nanoparticles with Plants in Agro-ecosystems" by Ranjana Singh et al. discusses about the interaction of titanium dioxide nanoparticles (TiO_2-NPs) with plants in agro-ecosystems. The interactions of TiO_2-NPs in terms of their uptake, translocation and accumulation in plants with an overview of their effect on morphology, physiology, biochemical and molecular pathways have been described in detail. Application of TiO_2-NPs in agriculture as growth boosters and protecting agents against various biotic and abiotic stresses has also been explored. Chapter "Interaction of Nano-TiO_2 with Plants: Preparation and Translocation" by Moodley and Arumugam elaborated on the interaction of nano-titanium dioxide (TiO_2) with plants, their preparation and translocation. These TiO_2 nanoparticles have been used in various products of domestic as well as industrial use. Later, these products end up in soil or water for becoming a cause of pollution, which may let them enter the food chain. Therefore, studying their interaction with various plants becomes important and has been reviewed here. Chapter "Plant Physiological Responses to Engineered Nanoparticles" by Ahmed Abdul Haleem Khan has discussed about the plant physiological responses to ENPs. Potential of ENPs as plant growth stimulants, fungicides, pesticides, weedicides and fertilizers has been explored for agriculture production.

Part Three of the book contains two chapters (Chaps. "Engineered Nanoparticles in Agro-ecosystems: Implications on the Soil Health" and "Effect of Engineered Nanoparticles on Soil Attributes and Potential in Reclamation of Degraded Lands") dealing with the nanoparticle interaction with the soil and regulation of soil health. Chapter "Engineered Nanoparticles in Agro-ecosystems: Implications on the Soil Health" by Mishra et al. has summarized the role of ENPs in agro-ecosystems and their implications on the soil health. Since the delivery of the ENPs in the form of nano-fertilizers, nano-pesticides and nano-weedicides, etc., lead to various transformations in soil environment, which directly affects the soil microbial communities. To tackle these challenges, various risk assessment and mitigation strategies should also be worked out, some of which have been reviewed in this chapter. Chapter "Effect of Engineered Nanoparticles on Soil Attributes and Potential in Reclamation of Degraded Lands" by Vipin Kumar Singh et al. elaborates on the effect of ENPs on soil attributes and potential in reclamation of degraded lands. This chapter has reviewed the applications of ENPs for improving agricultural productivity. Also, the important techniques for nanoparticle quantification, impact of ENPs on soil characteristics and their potential in management of degraded lands have been discussed.

Role of ENPs as nano-fertilizers, nano-pesticides and bio- nanosensors has been explored in the next three chapters (Chaps. "Advances of Engineered Nanofertilizers for Modern Agriculture"–"Bio-nanosensors: Synthesis and Their Substantial Role in Agriculture") under Part Four. Chapter "Advances of Engineered Nanofertilizers for Modern Agriculture" by Theivasanthi Thirugnanasambandan has explored the advances of engineered nano-fertilizers for modern agriculture. Zeolite nano-fertilizers, carbon nanomaterials based fertilizers and slow or controlled release fertilizers (such as hydroxyapatite nanoparticles coated urea, polymer-coated fertilizers) have been elaborated. Also, use of coating bio-polymers (such as chitosan and thermoplastic starch) on nano-fertilizers has been reviewed. Chapter "Nano-fertilizers and Nano-pesticides as Promoters of Plant Growth in Agriculture" by Niloy Sarkar et al. elucidates the applications of nano-fertilizers and nano-pesticides as promoters of plant growth in agriculture. Both positive as well as negative aspects of utilization of nanomaterials in agriculture field have been discussed in detail. This chapter elaborates upon various types, mechanisms, benefits and potential applications of novel nano-agrochemicals, which are necessary to attain more sustainable agriculture practices. Chapter "Bio-nanosensors: Synthesis and Their Substantial Role in Agriculture" by Shailja Dhiman et al. has exploited quite significant role of bio-nanosensors in agriculture. Synthesis of nanomaterials using plants and microbes via bio-nanotechnology has become quite important nowadays. This chapter discusses about the synthesis and further applications of the nanomaterials as bio-nanosensors in pathogen detection, sensing food quality, adulterants, dyes, vitamins, fertilizers and pesticides.

Part Five (based on interaction of ENPs and microbes) consists of two chapters (Chaps. "Interaction of Nanoparticles with Microbes" and "Nano-toxicity and Aquatic Food Chain"). Chapter "Interaction of Nanoparticles with Microbes" by Sudhir S. Shende et al. discusses about the interaction of nanomaterials with microbes, as its crucial role to understand the toxicity mechanisms, impacting the aquatic and soil health. Further, antimicrobial and anti-fungal properties of the nanoparticles are elaborated, for understanding the complex nature of the interaction between the microorganisms and nanoparticles. Chapter "Nano-toxicity and Aquatic Food Chain" by Krishna and Sachan talks about the nanotoxicity, especially in aquatic food chain. The ENPs interact with aquatic organisms on both upper and lower trophic levels throughout the aquatic food chain. Impact of these ENPs on aquatic ecosystem further needs to be studied more carefully, for their effect on food chain resulting in bioaccumulation affecting aquatic animal's well-being, development, reproduction and physiology.

The ultimate part (Part Six) of the book deals with the overall PM-ENPs nexus approach in a single chapter. The ultimate chapter of the book, i.e. Chap. "Impact of Engineered Nanoparticles on Microbial Communities, Soil Health and Plants", by Akhilesh Kumar et al. has elaborated upon the impacts of ENPs on microbial communities, soil health and plants. Sometimes, the use of ENPs in agriculture results in toxicity in economically important crops and trigger severe oxidative stress in plants leading to cell death. This toxicity can later pose potential threats to human and other animal's health, which might consume these plants in due course of time.

We sincerely do expect that the insights from this book on various aspects related to PM-ENPs nexus approach for sustainable agriculture would be quite useful for scientists, researchers, graduate students and policymakers working in this field all across the globe.

We are very much thankful to all the learned contributors for their valuable time and contribution. The completion of this volume would have not been possible without the persistent efforts of eminent reviewers and colleagues. Finally, we extend our sincere thanks to technical staffs of Springer Nature for typesetting and efficient production of the book.

New Delhi, India   Pardeep Singh
Varanasi, India   Rishikesh Singh
Varanasi, India   Pramit Verma
New Delhi, India   Rahul Bhadouria
Rishon LeZion, Israel   Ajay Kumar
New Delhi, India   Mahima Kaushik

Contents

Engineered Nanoparticles in Agro-ecosystems: General Approach

Engineered Nanoparticles in Smart Agricultural Revolution: An Enticing Domain to Move Carefully

Pratap Srivastava, Rishikesh Singh, Rahul Bhadouria, Dan Bahadur Pal, Pardeep Singh, and Sachchidanand Tripathi

Abstract

Nanotechnology may potentially benefit our agro-ecosystems in multiple ways, primarily via reduction in agricultural inputs without yield penalty and enhanced absorption of nutrients by the plants. In this regard, nano-fertilizers (such as engineered metal oxide or carbon-based nano-materials, nano-coated fertilizers, and nano-sized nutrients), and nano-pesticides (inorganic nano-materials or nano-formulations of active ingredients), might bring targeted as well as controlled release of agrochemicals in order to tap the fullest biological efficacy in already stressed agro-ecosystems, without over-dosages and leach-outs. Therefore, such nano-tools may multiply the agricultural yield, providing protection against various pests and diseases, without polluting our soil and water ecosystems at the same time. Though nanotechnology may provide potential solutions on such critical and persistent issues in agricultural management and activities; however, new environmental and human health hazards from their applications itself may pose unforeseen challenges to the humankind. For example, the biosafety, adversity, unknown fate, and acquired biological reactivity/toxicity of these nano-materials once dispersed in environment after application are still an unknown and threatening area, which needs to be investigated carefully and scientifically, before its open field use in our agro-ecosystems. Among other potential benefits, nano-tools may also be utilized for the rapid disease diagnostic in field crops and monitoring of the packaged food quality and contaminations. Similarly, the quality and health of soils and plants can be regularly monitored in real-time manner with the help of sensors based on highly sensitive nano-materials. However, a responsible regulatory consensus on nanotechnology application in agriculture needs to be developed, based upon profound scientific foundations. This chapter explores the area of nanotechnology in revolutionizing agriculture in a smart way via its known interactions with plants and soil microorganisms so far in the literature.

Keywords

Agrochemicals • Carbon nanotubes • Nano-fertilizers • Nano-pesticides • Nanopolymer • Quantum dots • Sustainable agriculture

1 Introduction

Agriculture is fundamental to human civilization, which, therefore, also primarily associates with the sustainability of our system and human health (Srivastava et al. 2016; Mishra et al. 2018). The primary objective of nano-materials, for which they are being explored in agriculture domain, is economy and efficiency (i.e., to reduce agrochemicals, minimize nutrient leach-out with an increase in yield it provides, in a cost- and time-effective manner) (Marchiol et al. 2020; Pirzadah et al. 2020). Agriculture produces and

P. Srivastava (✉)
Shyama Prasad Mukherjee Government PG College, University of Allahabad, Prayagraj, 211013, India
e-mail: prataps103@gmail.com

R. Singh
Institute of Environment and Sustainable Development (IESD), Banaras Hindu University, Varanasi, Uttar Pradesh 221005, India

R. Bhadouria
Department of Botany, University of Delhi, New Delhi, 110007, India

D. Bahadur Pal
Department of Chemical Engineering, Birla Institute of Technology, Mesra, Ranchi, 835215, Jharkhand, India

P. Singh
Department of Environmental Studies, PGDAV College, University of Delhi, New Delhi, Delhi 110065, India

S. Tripathi
Deen Dayal Upadhyaya College, University of Delhi, New Delhi, 110078, India

© Springer Nature Switzerland AG 2021
P. Singh et al. (eds.), *Plant-Microbes-Engineered Nano-particles (PM-ENPs) Nexus in Agro-Ecosystems*,
Advances in Science, Technology & Innovation,
https://doi.org/10.1007/978-3-030-66956-0_1

provides raw materials as human food as well as feed for various industries (Srivastava et al. 2016). The constantly growing human population with limited land, water, and soil availability prompts the agricultural development to keep pace with it and become increasingly more viable economically as well as efficient with time, but safe environmentally for sure. This alteration in agriculture would also be vital for bringing people back in the agricultural business, to opt them out of poverty and hunger (for socio-economic improvement), which is prevalent in most parts of the developing world (Mukhopadhyay 2014). In this regard, new and innovative technology providing better agricultural production in cost- and time-effective manner is need of the hour, and nanotechnology holds a great promise to fill up that space and produce qualitatively and quantitatively better food with lower cost, energy, and waste production in a smart manner (Hossain et al. 2020; Marchiol et al. 2020).

In recent years, a diverse spectrum of potential applications of nanotechnology has been observed in the agriculture, prompting intensive researches across the globe (Chen and Yada 2011; Dasgupta et al. 2015; Parisi et al. 2015). Initially, the term nanotechnology was first coined by Professor Norio Tanaguchi in 1974 (Bulovic et al. 2004), for a domain wherein unique changes in physicochemical properties of materials happen in their nano-size, in sharp contrast to their bulk counterpart (Burman and Kumar 2018). However, it was Eric Drexler who formally introduced the term nanotechnology in his book "Engines of Creation" to the world. Nanotechnology holds a great promise in providing efficiency and economy to the system, particularly in agro-ecosystems. This area of nano-size world (termed as nano-science), with magical properties, evolved gradually, but greatly in last decade, as can be observed by the growing scientific publications and higher captured market size in short time, which also enabled us today to develop cutting-edge applications in most of the important sectors/domains of human life, along with improved instrumental ability to synthesize and isolate engineered nano-materials (ENMs), precisely (Gibney 2015).

Though, nanotechnology in material sciences and electronics has relatively higher dynamics, its potential use in agriculture and food supply chain segment has evolved quite recently. Many engineered nanoparticles (ENPs) have also been synthesized in recent years for a large number of nano-materials based products. Particularly in agriculture, nano-materials are being specially tailored as nanopesticide, nanofertilizer, and nano-biosensor for improving agriculture. However, in-depth scientific studies are being done to understand the impact of ENPs on plant growth, metabolism and physiological processes, and agro-ecosystems productivity/management in order to develop smart nanotechnology applications for revolutionizing agriculture to a next level in a smart manner.

Products that are synthesized via nanotechnology using specialized techniques are known as nano-materials (NMs). It is estimated that over 800 nano-material products are currently available in the market, worldwide. Generally, NMs refer to colloidal particulates with size range lying between 1 and 100 nm, in at least one of their dimension. These NMs reveals size-dependent characteristics, including large surface area/volume ratio and unique optical properties specifically, which lies somewhere intermediate to individual molecule and bulk material. The main categories of NMs include metal oxides, zero-valent metals, quantum dots, carbonaceous, semiconductor, lipids, nanopolymer and dendrimers featuring distinct and diverse characteristics. Additionally, fullerenes and carbon nanotubes are defined as most widely used organic NMs. The change in property of NMs, in sharp contrast to their bulk counterparts and distinct magnetic property in nano-size, owes to the alteration in atoms and larger surface area (due to smaller size of NMs), resulting in high reactivity (Burman and Kumar 2018). The altered property of NMs is specifically related with the change in electronic energy level, specifically due to the alteration in surface area/volume ratio (Prasad et al. 2016). Chemically synthesized nano-materials, being toxic and mostly costly in nature, are now being synthesized alternatively from plant as well in a domain called green nanotechnology. The later is a safe process and is cost- and energy-efficient, but with reduced waste (also because it is mostly produced from waste) and greenhouse gaseous production (Prasad 2014). The recent shift toward the green nanotechnology is at a faster pace, as it is environmentally sustainable. In spite of this green transition, various issues with NMs use in the agricultural field remain open ended, which hopefully would resolve with scientific advancement in the concerned field (Kandasamy and Prema 2015). Quite recently, the biocompatibility, cost-effective synthesis, and enhanced sensitivity to external stimuli have accentuated interest of scientific communities in polymeric NMs, as compared to chemically synthesized counterparts (Baskar et al. 2018).

In modern agriculture, it is quite difficult task to produce crops without pesticides, fertilizers, despite knowing the potential hazardous implications these chemicals unleash upon organisms, not intentioned to affect (including plants, mesofauna, macrofauna, and soil microbiota), human health and environment (Kah 2015; Abbas et al. 2019; Pérez-Hernández et al. 2020). Researches reveal that the primary mechanism through which ENPs cause toxicity is reactive oxygen species (ROS)-mediated oxidative stress, either via physical direct damage or release of toxic ions after nanoparticle dissolution process (Abbas et al. 2019). However, the impact of ENPs on soil microorganisms and plants differs considerably depending upon NMs and soil used. Moreover, the species of microorganism and plant used in the study also considerably affects the results (Khan and Akram 2020). The calibrated use of engineered nanoparticles may drive high-tech agricultural system bringing second revolution in agricultural diaspora. It may thus enhance the quality and quantity of agricultural yield with reduction and/or elimination of the detrimental influence of modern agriculture on environment (Liu and Lal 2015; Shang et al. 2019). In recent years, cost- and time-effective systems are being favored for detection, monitoring, and diagnosis of biological host molecules standing crops in agriculture (Sagadevan and Periasamy 2014). In this regard, NMs can improve the sensitivity, performance, and handiness of the biosensors, in detecting nutritional health status of soil and plant health as well as disease status in real-time manner (Fraceto et al. 2016). Similarly, processed and packaged foods can also be sensed for mycotoxins rapidly with use of NMs biosensors (Sertova 2015). A brief description of major ENPs, potential role of available nano-tools in agriculture via their interface with plant metabolism and soil microorganisms, including eco-toxicity, as well as their potential role in revolutionizing agriculture is discussed in this chapter.

2 A Brief Note on Widely Used Engineered Nanoparticles (ENPs)

2.1 Carbon Nanotubes (CNTs)

Carbon nanotube is equivalent to two-dimensional graphene sheet, which is rolled into a tube shape. Single-walled (SWNTs) and multi-walled (MWNTs) nanotubes are the two distinct forms of carbon nanotubes. The mixing of σ and α bonds as well as rehybridization properties of electron orbital of CNTs confers unique properties (i.e., conductive, optical, and thermal) for nano-device applications to achieve sustainable agricultural conditions (Raliya et al. 2013). CNT-based targeted delivery of agrochemicals to hosts might help control the surplus chemicals, which might bring severe damage to plants and environment after their release in the surrounding (Raliya et al. 2013; Hajirostamlo et al. 2015).

2.2 Quantum Dots (QDs)

Semiconductor QDs possess excellent fluorescence and show size tunable band energy (Androvitsaneas et al. 2016) and unique spectral properties, therefore are generally used in bioimaging and bio-sensing (Bakalova et al. 2004). Therefore, it has been utilized in live imaging of plant root systems as dyes to verify known physiological processes (Hu et al. 2010; Das et al. 2015). It has been found that QDs

at low concentration show no detectable cytotoxicity for seed germination and seedling growth.

2.3 Nano-encapsulation, Nano-rods and Nano-emulsion

Encapsulation protects substances from adverse environments and helps in their controlled delivery with precision in targeting (Ozdemir and Kemerli 2016). Nano-encapsulation term is used as per the size range it achieves after encapsulation. Nano-capsules, which consists of an liquid core ensheathed by a polymeric membrane (Couvreur et al. 1995), have considerable application in drug delivery, enhanced bioavailability of nutrients/nutraceuticals, fortification of food, self-healing of materials, and in the area of plant science research (Ozdemir and Kemerli 2016). Nano-emulsion is a multifunctional material of plasmonic nature, which remarkably couples the sensing phenomenon (Bulovic et al. 2004). Nano-emulsion is nano-scale oil/water droplet, which exhibits size lower than 100 nm (Anton and Vandamme 2011). It appears optically transparent and is particularly advantageous, when incorporated into drinks. It has been observed that the nano-emulsion formation requires very high energy. Nano-rods are nano-sized materials, having standard aspect ratio between 3 and 5, having their wide use in display technologies, as they change their reflectance under electromagnetic field, owing to their change in orientation; however, it has harmful impact on plant processes. For example, the gold nano-rod at high concentration brings lethal physiological change in watermelon plant (Wan et al. 2014) and also considerably affects the transport of auxins in tobacco (Nima et al. 2014).

3 Nanotechnology in Sustainable Agriculture

The nanotechnology might help in improved agricultural productivity, primarily via enhanced nutrient control on release for synchronized availability and monitoring of pesticide's efficient use and water quality (Gruère 2012; Prasad et al. 2014). In this regard, the increased applications of fullerenes, nanotubes, biosensors, controlled and targeted delivery systems, nanofiltration, etc., in the agriculture and associated supply chains are being observed widely in recent years (Ion et al. 2010; Sabir et al. 2014). This emerging technology is efficient in agricultural management of natural resources (nutrient and water), drug delivery mechanisms in plants, and in maintenance of the soil's health (Fig. 1). However, its potential use in agricultural biomass and waste management as well as in the food industry is also being observed (Floros et al. 2010). Recently, nanosensors (e.g.,

Fig. 1 Potential use of nanotechnology in movement towards 4th agricultural revolution

EFFICIENT NUTRIENT MANAGEMENT

- Controlled & synchronized nutrient availability
- No offsite impacts via leaching and run-off

IMPROVED DISEASE AND PEST CONTROL

- Targeted delivery to control disease and pest
- No overdose and harm to non-targeted organisms

SMART AGRICULTURE PRACTICES

- Real time monitoring of biotic and abiotic stresses
- Smart food supply chain (packaging and processing

electrochemical, optical) have been used for monitoring of soil and water contamination for detecting the traces of heavy metals (Ion et al. 2010). Similarly, nano–nano interaction is being tapped to remove the toxic elements in agricultural soils for obtaining healthy foods (Ion et al. 2010; Dixit et al. 2015). NMs catalyze degradation of waste and toxic materials directly as well as indirectly (via improving the efficiency of microorganisms), helping in bio-remediation of the polluted agro-ecosystems. A general assessment of the risks of ENPs is difficult, owing to their diverse inherent and acquired activity under varied set of environmental conditions (Prasad et al. 2014). ENPs may affect the chemical composition, shape, surface properties, extent of particle aggregation (clumping), or disaggregation of other particles, depending on their sizes variability, which may lead to their toxic effects (Ion et al. 2010).

3.1 Engineered Nanoparticles (ENPs) in Agriculture

In recent years, new engineered NMs, using inorganic, polymeric, and lipid nanoparticles, have been synthesized, via techniques called emulsification, ionic gelation, polymerization, oxido-reduction, etc., in order to sustainably increase the agricultural productivity. Such ENPs, which are engineered for distinct physical (shape, size), and associated electrical properties (such as surface properties), further bring a distinct catalytic activity, enhancement in strength and conductivity (thermal and electrical), and controlled delivery of host molecules. Using these remarkably unique

nano-systems, bringing nutrient immobilization and their controlled real-time release in soils, as per plant needs, may bring efficiency and economy in resource use in agro-ecosystems. As an effect, it minimizes nutrient leaching and eutrophication and improves the nutrient uptake by plants (Liu and Lal 2015). Similarly, improvement in pesticides characteristics such as enhancing their solubility potential and resistance against the activity loss, and ability of a highly specific and controlled delivery toward targeted organisms in recent years, may have considerably made the agricultural practices safe, without any off-site repercussion (Mishra and Singh 2015; Grillo et al. 2016; Nuruzzaman et al. 2016). Similarly, the use of hydrogels, nano-clays, and nano-zeolites to improve water holding capacity and capacity of soils to slowly release the water during dry seasons has also been explored. This might help in agricultural sustainability as well as in the most required reforestation programs of degraded lands, limited mostly due to water scarcity. In this regard, organic (polymer and carbon nanotubes) as well as inorganic (nano-metals and metal oxides) NMs have also shown great promise, due to their great capability in quick absorption of the contaminants present in the environment (Khin et al. 2012), helping to remediate soils in cost- and time-effective manner (Sarkar et al 2019).

Quite recently, nanoparticles are also being explored to revolutionize plant genetic engineering aspects in order to develop plants with improved resistance and qualities, easily. Most such studies on how NMs can be used effectively in plant genetic engineering have been observed via plant tissue culture. Recently, carbon nanotubes scaffolds have been applied in plants to successfully deliver linear, DNA plasmid and siRNA in *Nicotiana benthamiana*. Similarly, silicon carbide-based transformation has been observed as a successful method to deliver DNA in various plants such as tobacco, maize, rice, soybean, and cotton (Asad and Arsh 2012). In a similar way, stable genetic transformation in cotton plants via magnetic nanoparticles (MNPs) has also been achieved successfully (Zhao et al. 2017). Moreover, genome editing via mesoporous silica nanoparticles (MSNs) is being tested as a promising approach in recent scientific studies (Valenstein et al. 2013). All these novel approaches are intended to bring novelty and easiness in agricultural production in cost-effective manner.

3.2 Engineered Nano-materials as Stimulant of Plant Growth

Over the last two decades, ENPs research in medicine and pharmacology has been significant, especially for diagnostic or therapeutic purposes (Perrault et al. 2009). Recently, these NMs are receiving an increased interest in the field of crop sciences/agronomy, particularly in the application of NMs as vehicles of agrochemicals or bio-molecules in plants to enhance crop productivity (Khan et al. 2017). Generally, ENPs are applied to roots or vegetative part of plants, preferably to the leaves. Generally, its uptake has been observed a little more complicated in the soils, as compared to the aerial parts of the plants (Sanzari et al. 2019). The uptake, mobilization mechanisms, and biological effects of these NMs with plant are still in infancy, and it is not a wise opinion to move with imperfection in field applications, without knowing their intricate interactions with plants, soil microorganisms and environment, completely and scientifically. In several studies, specific (low dose) concentrations of ENPs, foliar spray/irrigation, and carbon nanotubes have significantly improved plant growth, physiological aspects (chlorophyll *a*, *b*, carotenoid content, photosynthesis, carbohydrates), antioxidants, and plant tolerance against biotic and abiotic stress (Nafees et al. 2020).

In recent studies, ENPs (particularly, based on carbon, metal, and metal oxides) influence on plant physiology and growth showed that it considerably affects seed germination in higher concentration. For example, zinc (Zn) and copper (Cu) oxide nanoparticles, being essential micronutrients, have been observed to act as a significant plant growth promoting complex (Priyanka et al. 2019). Surprisingly, it has been noted that various kinds of ENPs affect the plants ability and behavior, in a differential and sometimes in a contrasting manner. Some plants are even capable of uptake and accumulation of ENPs. Carbon nanotubes and Au, SiO_2, ZnO, and TiO_2 nanoparticles have shown potential to expedite growth of plants, by increasing the uptake of

elements and improved nutrient utilization (Khot et al. 2012). Ag-NPs at low concentrations have shown enhanced shoot and root growth enhancing chlorophyll production and antioxidant enzyme activity, limiting production of reactive oxygen species (ROS) in the plant tissues (Sami et al. 2020). However, the impact of nanoparticles on plant behavior depends largely on the size, surface charge, composition, concentration, and physicochemical properties of the nanoparticle used, besides the susceptibility of the concerned plant species (Ma et al. 2010; Lambreva et al. 2015).

Notably, studies show that nanoparticles might be efficient stimulator of plant growth irrespective of their nature. However, comprehensive experimentations are needed to optimize their application conditions and identifying their specific impact on plant's physiology (Fincheira et al 2020). The plant cell–ENP interaction leads to a change in plant's genetic expression and associated metabolic pathways as well, which affect plant growth and developments as a consequence, in a remarkable manner (Ghormade et al. 2011). For example, a pronounced increase in germination rate of rice and wheat has been observed under carbon nano-materials, especially CNTs (Wang et al. 2012). The beneficial impacts of accumulation of nano-materials in plants, particularly in MWCNTs, ZnO, and Zn, have also been observed (Hussein et al. 2002). Similarly, TiO_2 nanoparticles have been observed to promote nitrate reductase activity in soybean (*Glycine max*), enhance water and nutrient absorption/use, and induce the antioxidant machinery to favor plant's growth. In a similar research, TiO_2-treated seeds have shown 73% higher plant dry weight, due to thrice higher photosynthetic rates and a considerable rise (around 45%) in chlorophyll (Mingfang et al. 2013). Also, it has been found to promote the growth in spinach via improving nitrogen assimilation and photosynthetic rate. In a study, Zn nano-materials have shown to promote chlorophyll production, fertilization, pollen function, and germination and reduce the susceptibility of plants to drought stress. However, contrasting findings with other species have also been observed, signaling more studies to be conducted to understand ENPs-plant interaction. The influence of ENPs on various plants differs greatly depending on growth stage, method, and period of exposure (Khiew et al. 2011). Additionally, symbiotic bacteria and fungi in the soil, associated with plant roots, have shown controversial interactions in relation to ENPs. These microscopic entities increase the heavy metal NPs accumulation in turf grasses, however reduce the uptake of nano-Ag and nano-FeO in legumes (Guo and Chi 2014). Therefore, to better understand the interaction of these ENPs with plants and associated microflora, new and improved protocols and techniques (such as magnetic resonance imaging (MRI), microscopy, and fluorescence spectroscopy) might help in reaching appropriate scientific conclusion (Srivastava et al. 2019).

In recent years, the potential use of polymeric soft NMs in delivery of bio-molecules in a smart manner and for developing new mythologies of genetic engineering in plants to enhance their defense mechanisms and induction of growth and development is being actively pursued, worldwide (Sanzari et al. 2019).

There are some major bottlenecks in use of ENPs, which are primarily checking the progress of NMs application in plant growth are: (i) design and synthesis of safe NMs; (ii) understanding mechanisms of NMs uptake and mobilization in plants, and, (iii) the lack of global multidisciplinary collaboration for adequate development and controlled use of nano-applications in plants (Sanzari et al. 2019). Despite, these obvious hurdles to be resolve in years to come, we have multiple nano-applications to boost agricultural development indirectly via controlled release of agrochemicals and smart monitoring systems, to manage agricultural production, cost effectively and environmentally sound manner. Nanotechnology has shown promising observations in laboratory tests in controlling the overuse of agricultural inputs and causing negligible impact on the environment. In this respect, metal oxide nanoparticles offer promising perspective for the development of effective nano-scale formulations of fertilizers/pesticides for their controlled release capacity and targeted delivery, in sharp contrast to the conventional fertilizers and pesticides.

3.2.1 Nano-Fertilizers

Quite recently, nano-fertilizers have been recognized as novel nutrient delivery tools, utilizing nanoparticles of C, Mn, Fe, and ZnO (Liu and Lal 2015). Researchers across the globe have shown that some engineered NMs can increase plant growth in certain concentration ranges, mostly at smaller concentrations. Several studies showed that nanoparticles of essential minerals affected plant growth, depending on their size, concentration, composition, and mode of application. It was reported to enhance increasing crop yields promoting germination, seedling growth, affecting photosynthetic activity, N metabolism, and changes in gene expression (Tapan and Sivakoti 2019). Also, their use in nano-fertilizers can increase the agronomic yields many fold with minimum environmental pollution. Specifically, developing nitrogen and phosphorous macronutrient nano-fertilizers are being given a high research and development priority in current times, both for food production and environmental protection. For example, hydroxyapatite nanoparticles, being used as phosphorous nano-fertilizers today, have been reported to enhance the soybean growth rate and yield considerably, as compared to the ordinary phosphorous fertilizers (Liu and Lal 2015). Also, the slower release of phosphate from the nano-fertilizer contributes to maintain the soil fertility along with eutrophication,

nullifying the runoff or leaching (Liu and Lal 2015). Similarly, Zn deficiency, a key factor limiting agricultural yield, particularly in alkaline soils (Sadeghzadeh 2013), can easily be ameliorated with the use of Zn nanoparticles, in a cost-effective manner. Different nano-fertilizers and nano-pesticides such as Ag, Zn, Fe, Ti, P, Mo, and polymer nanoparticles have shown significant potential as plant growth promoting and pest control agent. Similarly, different kinds of nano-technological tools such as (materials, formulations, composites, emulsions, and encapsulations) have all shown promising result in providing increased nutrition to plants and targeted toxins to the concerned pests in a precise and controlled manner.

Recent studies stated that nanoparticles, made up of essential minerals and non-essential elements, affect plant physiological processes and growth considerably, which primarily depends on size, composition, concentration, and type of application (via foliar and soil routes). Nano-fertilizers may contain nano-zinc, iron, silica and titanium dioxide, InP/ZnS core shell QDs, ZnCdSe/ZnS core shell QDs, Mn/ZnSe QDs, core shell QDs, gold nano-rods, etc. However, comprehensive studies on uptake, fate in biological systems, and toxic influence of several metal oxide NPs (viz., Al_2O_3, CeO_2, TiO_2, FeO, and ZnO) were studied intensively in agricultural production, which equally lauds for their cautious use, as well (Dimkpa 2014; Zhang et al. 2016; Parada et al. 2019a, b).

3.2.2 Nano-pesticides

The potential role of NMs in plant protection and food production is still in infancy. Insect pests, affecting plants as well as stored foods, may be controlled with the use of ENPs (Khot et al. 2012). It has been observed that nano-encapsulated pesticides are released slowly in the applied system and shows greater solubility, permeability, specificity, and stability (i.e., long-lasting pest control efficacy) (Bhattacharyya et al. 2016). Use of such nano-encapsulated agricultural tool leads to the development of non-toxic and promising pesticide delivery systems for better control of such pests with reduced dose and no associated off-site harm to human life and environmental health (Bhattacharyya et al. 2016; Grillo et al. 2016; Nuruzzaman et al. 2016). Nano-encapsulation is designed for desired chemicals delivery to the target biological process. Some products such as Karate ZEON, Ospray's Chyella, Subdue MAXX, Penncap-M, Banner MAXX, Primo MAXX, Subdue MAXX, etc., are available in market as micro-suspensions. Organic and polymeric ENPs as nano-capsules/nanospheres have been used in agro-ecosystem as nano-carriers for application of herbicides. For example, polymeric ENP is highly biocompatible and is being largely used for atrazine encapsulation, a potent herbicide. Similarly, triazine-coated chitosan nanoparticles have shown lower environmental impact and genotoxicity in *Allium cepa* (Grillo et al. 2016).

4 Engineered Nanoparticles Impact on Soil Microbial Processes

Having diverse range of nanotechnology products around us, its presence in air, water, and soil is unavoidable, owing to no strict regulation and monitoring placed in this regard. Similar to pollution, sources of ENPs into these three systems can also be described as point (production and storage units, research laboratories) or non-point sources. Also, ENPs stand a better change to mobilize to other places via air and water owing to their small sizes. Soil is known to be the highest recipient of ENPs, owing to their extreme resistance and tendency to accumulate. As soil microbial biomass and diversity is crucial for the sustainable use of soils, using ecological subsidies in the form of ecological processes (Torsvik and Øvreås 2002), the nanoparticles may have considerable influences on this ecosystem, mediating a change in soil microbial community characteristics. Metal/metal oxide nanoparticles have been identified as most toxic to soil microbial community which support important ecosystem processes such as nutrient cycling (Fig. 2), thus threatening soil health and fertility (Parada et al. 2019a; b). TiO_2 nanoparticles impact on nitrification process and ammonia-oxidizing bacteria has been observed strongly negative, triggering a cascading negative effect on denitrification activity and considerable change in bacterial community structure (Simonin et al. 2016).

However, contradictory report has also been observed (Chavan and Nadanathangam 2020). ENPs have been observed to affect soil humic acid content, influencing soil bacterial community characteristics (including diversity) affecting decomposition process (Kumar et al. 2012; Ben-Moshe et al. 2013). Soil contaminations of ENPs persist in the soil for long, or they may contaminate ground water (Tripathi et al. 2012).

Among the nano-applications, widely used paints, coatings, and pigments have the highest possibility of getting released into water and soil systems. Owing to close linkage of soil and plant system, ENPs in soil may harm microorganisms and plants, and thus animals and human beings as a consequence, present down the line in trophic food chain. They may also affect soil rhizospheric and phyllospheric microbial community to indirectly affect the plant functioning/metabolism. The presence and persistence of ENPs into the natural environment (such as agro-ecosystems) owing to their widespread use may threaten the favorable microbial communities (bacteria and fungi). Nanoparticles accumulate in our natural systems via soil and water remediation technologies, use as nano-fertilizers and nano-pesticides, and their unintentional emission through water, air, sludge, and sewage (Tourinho et al. 2012; Tripathi et al. 2012; Shandilya et al. 2015; Coll et al. 2016). The measurement of soil CO_2 efflux/respiration and enzyme activity is often used to observe how the ENPs affect soil microbial activity (Simonin and Richaume 2015).

In some recent studies, TiO_2 and CuO ENPs have been found to decrease soil microbial biomass and enzymatic activities, in addition to microbial community structure in

Fig. 2 Harmful aspects of engineered nanoparticles (ENPs) application in agriculture

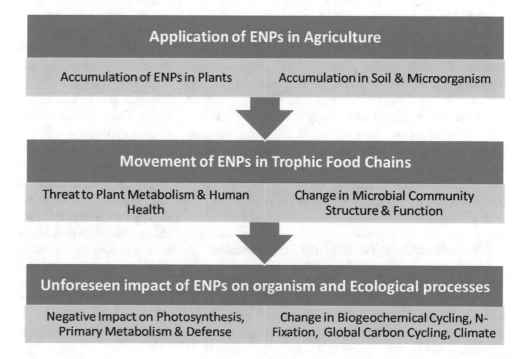

Application of ENPs in Agriculture

Accumulation of ENPs in Plants | Accumulation in Soil & Microorganism

Movement of ENPs in Trophic Food Chains

Threat to Plant Metabolism & Human Health | Change in Microbial Community Structure & Function

Unforeseen impact of ENPs on organism and Ecological processes

Negative Impact on Photosynthesis, Primary Metabolism & Defense | Change in Biogeochemical Cycling, N-Fixation, Global Carbon Cycling, Climate flooded paddy soils (Xu et al. 2015). Similarly, You et al. (2017) studied the effect of inorganic ENPs on soil enzyme activities (such as phosphatase and urease) and microbial communities of alkaline soils. The study observed a considerable change in abovementioned properties along with harmful impact on biological nitrogen fixation. In another study, Fe_3O_4 ENPs at higher concentration significantly decreased the bacterial count in soil (Jiling et al. 2016). Similarly, zinc oxide and CeO_2 ENPs considerably affected various bacterial groups (such as azotobacter, phosphorous, and potassium solubilizing bacteria) and inhibited various enzymatic activities (Chai et al. 2015). TiO_2 has shown to rapidly decline the soil active bacteria and enzymatic activity, affecting soil microbial characteristics such as activity, abundance, and diversity (Buzea et al. 2007). In a similar study, Concha-Guerrero et al. (2014) observed that CuO ENPs unleashed similar, but a relatively more toxic impact on soil microbial community. It has been generally observed that ENPs of inorganic nature have a relatively greater toxicity than organic ENPs on soil microbial characteristics (Frenk et al. 2013).

In a functional study, CuO and Ag ENPs have shown reduction in decomposition of leaf (Pradhan et al. 2011). Ag ENPs, used in a variety of consumer products due to its antimicrobial properties, significantly impact soil microbial functional and genomic diversity (Samarajeewa et al. 2017). However, contrasting studies also exists in the literature (de Oca-Vásquez et al. 2020). The soil enzymatic activities have also shown a drastic reduction at high concentrations of ENPs (Josko et al. 2014; Asadishad et al. 2018). The impact of ENPs show significant variation with type and dose of NPs as well as soil properties (Xin et al. 2020). Moreover, these ENPs have shown negative impact on self-cleaning ability as well as nutrient providing capacity of soil systems, which determines the level of plant nutrition and soil fertility (Suresh et al. 2013). In a manner, soil properties also determine the toxicity of ENPs. For example, soil pH, textural composition, structure/aggregation, and organic content affect the soil microbial community and the capability of these ENPs to unleash toxic effects on soil microorganisms (Fierer and Jackson 2006; Simonin and Richaume 2015; Read et al 2016). On the contrary, nanoparticles have also been termed as "remediation of the future" owing to their significant role in soil remediation (Sarkar et al. 2019).

5 Nanoparticle's Toxicity on Environment

The invisible pollution due to ENPs is considered as the most complicated type of pollutant to control, owing to its size. The ever-increasing applications and concentrations of ENPs pose enormous threat of their release into the environment, whose risk assessments are very difficult to quantify and understand at present stage (Servin and White 2016). The existing literature on eco-toxicological impact of nanoparticles is somewhat contradictory; however, in general, low to moderate toxicity of these nanoparticles on terrestrial plants has been observed in most of the scientific studies. A large number of research studies have focused on the toxicity assessment of the ENPs used in industries (Du et al. 2017; Tripathi et al. 2017a, b, c). Generally, the effect of ENPs on crops (such as spinach, onion, coriander, rice, wheat, soybean, lettuce, radish, barley, cucumber) has shown considerable inhibition of seed germination, reduction in shoot and root growth, toxicological effects, decreased photosynthesis and chlorophyll concentration (Tripathi et al. 2017a; b, c, d, e). The toxicity level of a nanoparticle primarily depends upon its solubility and specificity in binding to the biological site. ENPs of metallic nature are primarily antimicrobial in nature (Aziz et al. 2016; Patra and Baek 2017) and show toxicity on the plant cells, depending on surface charge at the membrane (electrostatic interaction), which follows the order: mold > yeast > Gram-negative > Gram-positive. Thus, it may unleash an entirely unknown cascade of change in microbial community dynamics in the concerned ecosystems, which may turn lethal on humans in return (Fig. 2).

Carbon-based nano-materials (nanotubes and fullerenes) can be degraded easily under a wide range of conditions; however, fullerene is preferably absorbed by wood decaying fungi and metabolized. As an effect, fullerene nanoparticles accumulate in microbial cells and are transferred across the food chain further, owing to feeding relationships (Warheit et al. 2004). In case of no acute toxicity, bioaccumulation and long-term exposure to these ENPs may have unforeseen effects on food chain/web. Similarly, the uptake, accumulation, and build-up of nanoparticles vary in plants, depending on its type and size, as well as the plant composition. Among the metal-based NMs studied in this regard (e.g., TiO_2, Fe_3O_4, CeO_2, ZnO, Ag, Au, Fe, and Cu), only fullerene and fullerols show a ready uptake tendency in plants. These NMs enter plant cells variously via aquaporins (a carrier protein), ion channels, endocytosis, and formations of entirely new pores across the plant cells, following apoplastic and symplastic movement and via xylem and phloem. Remarkably, seed, flower, and fruit strongly import fluid from the phloem (i.e., sink activity) and have greater tendency to accumulate ENPs, in relatively higher concentration. Besides toxicological impact on the plant, it raises issue of safety in human and animal consumption of such plant organs (Pérez-de-Luque 2017). In all these cases, they might enter the food chain to unleash unforeseen consequences. Similarly, the excess Fe_3O_4 nanoparticles produce some oxidative stress in plant system, affecting photosynthesis, leading to decline in metabolic process rates. ZnO

NMs are hazardous in nature and may affect the chromosomal and cellular traits.

Several ENPs such as TiO_2, ZnO, SiO_2 are photo-chemically active and generate superoxide radicals under light in oxygenic condition by direct transfer of electrons (Hoffmann et al. 2007). Studies demonstrate that in cultivated plants (such as tomato and wheat), metal-based ENPs triggering an oxidative burst, mediating electron transport chain and impairing ROS detoxifying mechanisms, bring enormous genotoxicity in the plants as a consequence (Pakrashi et al. 2014; Pagano et al. 2016). Moreover, this eco-toxicity is multiplied under simultaneous exposure of ENPs and UV light. The consequent generation of ROS as a response is exploited in determination of toxicity (Sayes et al. 2004). However, their protective effect against oxidative stress has also been observed in some studies (Venkatachalam et al. 2017). Therefore, mechanistic understanding of ENPs metabolism in organism and specific cell need investigation to clarify this ambiguity. Also, delayed impacts of environmental exposure to ENPs need exploration to determine potential mechanisms of adaptation (Cox et al. 2017; Singh et al. 2017). Studies on bioaccumulation of ENPs in food chain and their interaction with other environmental pollutants needs investigation as well, as it may affect major plant processes, compromising agricultural sustainability, detrimentally (Rana and Kalaichelvan 2013; Du et al. 2017).

The introduction of chemical or green ENPs in the fields must be monitored carefully and closely. The nanoparticles, having no harmful NMs, should only be allowed in agriculture for any improvement in yield and other critical issues. The uses of polymeric ENPs in the agriculture having plant-based insecticides coating are unique in itself and are increasingly being permeated (Chakravarthy et al. 2012; Perlatti et al. 2013). As soil health, ecosystem, and crop productivity are primarily determined by soil microorganisms (Mishra and Kumar 2009), the impact of NMs on such organisms also needs through assessment to avoid unseen consequences due to microbial community change across ecosystems. Accumulation of these ENPs in treated/applied soils may threaten soil microbial communities along with associated organisms in food chain (Simonin et al. 2016), which may impair the ecosystem functioning at large in an unpredictable way, owing to their crucial importance.

6 Nano-Biosensor Technology: A Path to Smart Agriculture

In the era of changing climate, smart agriculture to achieve the long-term goal of climate resilient development is need of the hour (Helar and Chavan 2015). Diminishing the material size to nano-scale brings radical change in physicochemical properties (i.e., quantum size effect) and good transduction properties owing to huge surface area/volume ratio, which can be utilized for analytical purpose in agricultural products (Kandasamy and Prema 2015). The gold ENPs (AuNPs) may be used as transducers for several improvements of agricultural products, such as bio-sensing devices. Biological tests measuring the presence or activity of selected analytics of key importance become highly sensitive and fast with its use (Vidotti et al. 2011; Kandasamy and Prema 2015). The use of nano-biosensors for detection of phyto-regulators and secondary metabolite may help in real-time monitoring of plant growth and development and understanding its environmental interactions in limiting growth conditions (Sanzari et al. 2019). It indicates that the application of nano-scale particles may provide numerous advantages over traditional procedures, which can revolutionize the present-day agriculture in a more smart way.

Nanotubes, nanocrystals, or nanoparticles and nanowires are mostly used in optimizing signal transduction, which are derived by the sensing elements in response to exposure to biological and chemical analytes, having similar size. The surface chemistry and other distinct properties of ENPs (such as thermal, electrical, and optical) help enhance the sensitivity, thereby reducing response time along with improvement in detection limit, which can, therefore, be utilized in multiplexed systems (Aragay et al. 2010; Yao et al. 2014). The distinct physicochemical properties of materials in nano-scale size have been exploited in development of biosensors, as signals are improved remarkably with its use (Sagadevan and Periasamy 2014). It enables us to develop rapid, sensitive, and cost-effective nano-biosensor systems in agriculture, food processing industries, and environmental monitoring. Currently, the sensors based on nanotechnology are at initial phase of development (Fogel and Limson 2016). Metal ENPs (such as silver, gold, and cobalt), CNT, magnetic ENPs, and QDs are some chief candidates which have been actively used in biosensor (device combining biological recognition element with physical/ chemical principles) development. Therefore, biosensor converts the biological response (such as an enzyme, a protein, an antibody, or a nucleic acid) into an electrical signal.

Recently, different natural and artificial bio-receptors have been identified and used widely, such as thin films, enzymes, dendrimers (Rai et al. 2012). The progress in nanofabrication and other techniques (such as mass spectrometry, chromatography, surface plasmon resonance, electrophoresis chips) may stimulate sensor development. Considerable scientific efforts in nanosensor development to supplement decision-making in crop monitoring, in order to achieve precise nutrients and pesticides application and higher water use efficiency via its easy testing in soils for smart agricultural development, are already in action. In the context of smart agriculture revolution, nanosensors may potentially manage the food supply chain right from crop cultivation to distribution (such as harvesting, food processing, transportation, packaging) (Scognamiglio et al. 2014). The regular monitoring of soil pH and nutrients, residual pesticides in soil and crops tissues, soil humidity, pathogens detection, and prediction of nitrogen uptake using nanosensor can give way to a more sustainable and smart farming system (Bellingham 2011). Also, the presence of pests, pathogens, or pesticides with use of biosensor tools may help us tune the amount of chemicals to use, utilizing the high sensitivity of nanosensors. A network of nanosensors installed across cultivated fields may help in comprehensive monitoring of crop growth in real-time manner, providing quality data for scientific analysis and interpretation (El Beyrouthya and El Azzi 2014). Similarly, bringing automation in the irrigation systems using nanosensors technology under changing climate conditions toward water scarcity may potentially maximize the efficiency of water use in agro-ecosystems in a simple way (de Medeiros et al. 2001).

6.1 Nanotechnology in Food Industry and Supply Chain

Nanotechnology may help in developing analytical devices dedicated specifically to the control of quality, safety, and bio-security from agricultural field to throughout the food supply chain (Valdes et al. 2009). Nanotechnology has multiple uses in food industry. For example, it can help in pathogen detection and diagnosis (via nano-scale biosensors), supply bioactive ingredients in foodstuffs, texture, and color modification in food (via nano-scale filtration system) (Martirosyan and Schneider 2014). Nano-printed, intelligent packaging (Ghaani et al. 2016), nano-coding of paper and plastics materials (Bhushani and Anandharamakrishnan 2014), and nano-additives (Khond and Kriplani 2016) have already been used for authentication and identification purposes in supply food chains. In food quality testing, monitoring, and control of food quality (e.g., smell, taste, color, texture), sensing ability of label and package and nutraceutical delivery can be monitored by using nanotechnology tools.

6.2 Food Processing

In food processing, use of nano-carriers for the delivery of nutrients/supplements, nano-sized organic additives, supplements, and animal feed is in limited use in recent times. Recently, vitamins are being encapsulated and delivered into human blood efficiently via foods through digestion system. Further, many foods and drinks have also been fortified with ENPs adding benefits to the product, without affecting the appearance/texture and taste. For example, nanoparticle emulsions are added in ice creams, which improve their texture and uniformity (Berekaa 2015). For example, KD Pharma BEXBACH GMBH (Germany) is known to provide encapsulated Omega-3 fatty acids in suspension and powder forms in nano- as well as micro-sizes, which is gaining higher market with time over the conventional one.

6.3 Food Packaging and Labeling

Nanosensors used in recent times in supply food chain ensure food authenticity, quality, freshness, safety, and traceability across food supply chain via faster, highly sensitive, and cost-effective detection of various target molecules. Currently, the assessment of food quality and safety is best using nanosensors, providing smart monitoring of chief food ingredients (sugar, vitamin, amino acid and mineral) and contaminants (heavy metals, pesticides, toxins, etc.). Such kind of intelligent and smart packaging of foods to monitor integrity and freshness of food during transportation and storage is also a trademark of nano-sensor technology (Vanderroost et al. 2014). In them, nanosensors observe the physical parameters (such as pH, humidity, and temperature), to identify gas mixtures (e.g., O_2 and CO_2) in order to detect toxins and pathogens and to control freshness (via ethanol, acetic acid, lactic acid) and decomposition (via cadaverine, putrescine).

Recently, some packaging materials incorporated with "nanosensors" have been used in food industry to detect the oxidation process in milk and meat (Bumbudsanpharoke and Ko 2015). NP-based sensors indicate the color change in case of oxidation/deterioration. ENPs being good barriers for gaseous diffusion, which can be exploited in drink industry (beer, soda waters) to increase in shelf life. Similarly, ENPs in packaging, nano-coating over plastic polymers, slow down processes, such as oxidation and microbial degradation (owing to antibacterial property) further extending the shelf life of food products (Berekaa 2015). Therefore, nanotechnology is a forward-looking technique in agricultural bio-security (Bumbudsanpharoke and Ko 2015).

Engineered nanoparticles show broad-spectrum antibacterial properties against Gram-positive and Gram-negative bacteria. For example, ZnO-NPs have been observed to suppress *Staphylococcus aureus* (Liu et al. 2009). Similarly, Ag-NPs show antimicrobial activity against *Escherichia coli*, *Aeromonas hydrophila*, and *Klebsiella pneumoniae* in a concentration-dependent manner (Aziz et al. 2016). According to recent studies, the major processes through which ENPs unleash their antibacterial effects: (1) bacterial cell membrane disruption; (2) ROS production; (3) induction of intracellular antibacterial effects following entry into the cell variously (including impact on DNA replication as well as inhibition of protein synthesis) (Aziz et al. 2015; Wang et al. 2017).

7 Future Perspectives: Identification of Gaps and Obstacles

Despite immense smart applications of nanotechnology in agriculture, multiple issues, critical to human and environmental health and sustainability, remain to be resolved with advancement in nanotechnology applications in the area of agriculture. Some key areas requiring critical attention are: (i) hybrid carriers development for delivering nutrients, pesticides and fertilizers to maximize their efficiency in agricultural production (De Oliveira et al. 2014); (ii) risk and life-cycle assessment of NMs (i.e., phytotoxicity) on non-target microorganisms, plants and pollinators insects; and (iii) strict regulations for the use of NMs based on fundamental scientific findings.

The implementation of nanotechnology in agriculture requires even higher technical advancement, enabling ENPs quantification at lowest possible concentrations, present in different environmental compartments for its life-cycle assessment (Kookana et al. 2014; Sadik et al. 2014; Parisi et al. 2015). ENPs interaction with organisms (target as well as non-target) and the presence of synergistic effects are undeniable. Therefore, infrastructure and methodologies to characterize, localize, and quantify ENPs in the environments should be developed beforehand, mobilizing knowledge exchange and co-ordination between scientists across research fields throughout the world (Malysheva et al. 2015). In time to come, these ENPs would provide us enormous potential in identifying cutting edge and cost- and time-effective development routes to achieve smart human civilization across the globe.

8 Conclusion

It is a ripe time to take a modern knowledge and tools in agricultural management to prepare ourselves self-sufficient to feed the growing population in a sustainable manner, under changing climate conditions, without damaging our environment any further. The emergence of engineered nano-materials application for achieving sustainable agriculture has revolutionized world agriculture to meet global food demand in environmentally sound and resource efficient manner, with reduced farming risks at the same time. These nanotechnology applications take us forward to efficiently use the natural resources, via nano-scale carriers and compounds to avoid loss and overdose of pesticides and fertilizers, causing pollution. Similar smart applications can be found today across the food supply chain, starting from agricultural production, animal feed, food processing, and additives, with ever-growing importance. Despite having plenty of information available on individual nano-materials in relation to agricultural benefits, theirs unpredictable course of eco-toxicity level, once they reach in our environment, is still challenging, which can be largely attributed to the scanty understanding of risk assessment, particularly in relation to human and environmental health. Therefore, we need to strike a balance between nanotechnology applications and implications in agriculture and food production, as this smart technology stands a better place to promote social and economic equity as well. Also, we have to thoroughly perform a reliable risk–benefit assessment, and full cost accounting evaluation before open field applications. Likewise, reliable methods to characterize and quantify these NMs in different environmental compartments, and evaluation of their interaction with bio-macromolecules present in living systems and environments must be given top priority. At the same time, development of comprehensive database and alarm system with multidisciplinary collaborative mindset, as well as international cooperation in regulation and legislation are necessary for potential exploitation of this ENP technology. Furthermore, engaging all stakeholders including non-governmental (NGOs) and consumer associations in an open dialogue to acquire consumer acceptance and public support for this technology is also critically required.

Author Contributions

PS* developed the idea in major consolation with RS and RB, which was revised with the help of RS, RB, DBP, PS, and SNT. All authors have proofread and approved the final draft of the chapter.

Conflict of Interest Statement

The authors declare that the research was conducted in the absence of any commercial or financial relationships that could be construed as a potential conflict of interest.

Acknowledgements The authors would like to thank University Grants Commission (UGC), New Delhi, India, for providing funding support as Start-up Grant (BSR): No. F 30-461/2019 (PS), JRF/SRF (RS) and DS Kothari fellowship (RB). Also, the corresponding author would like to acknowledge the University of Allahabad and Shyama Prasad Mukherjee Government PG college for their infrastructural and other supports in developing a research facility.

References

Abbas Q, Yousaf B, Ullah H, Ali MU, Ok YS, Rinklebe J (2019) Environmental transformation and nano-toxicity of engineered nano-particles (ENPs) in aquatic and terrestrial organisms. Crit Rev Env Sci Tec 1–59

Androvitsaneas P, Young AB, Schneider C, Maier S, Kamp M, Höfling S et al (2016) Charged quantum dot micropillar system for deterministic light-matter interactions. Phys Rev B 93:241409. https://doi.org/10.1103/physrevb.93.241409

Anton N, Vandamme TF (2011) Nano-emulsions and micro-emulsions: clarifications of the critical differences. Pharm Res 28:978–985. https://doi.org/10.1007/s11095-010-0309-1

Aragay G, Pons J, Ros J, Merkoci A (2010) Aminopyrazole-based ligand induces gold nanoparticle formation and remains available for heavy metal ions sensing. A simple "mix and detect" approach. Langmuir 26:10165–10170. https://doi.org/10.1021/la100288s

Asad S, Arsh M (2012) Silicon carbide whisker-mediated plant transformation. In: Gerhardt R (ed) Properties and applications of silicon carbide. BoD-Books on Deman, Rijeka, pp 1–16. https://doi.org/10.5772/15721

Asadishad B, Chahal S, Akbari A, Cianciarelli V, Azodi M, Ghoshal S, Tufenkji N (2018) Amendment of agricultural soil with metal nanoparticles: effects on soil enzyme activity and microbial community composition. Environ Sci Technol 52 (4):1908–1918

Aziz N, Faraz M, Pandey R, Sakir M, Fatma T, Varma A et al (2015) Facile algae-derived route to biogenic silver nanoparticles: synthesis, antibacterial and photocatalytic properties. Langmuir 3:111605–111612. https://doi.org/10.1021/acs.langmuir.5b03081

Aziz N, Pandey R, Barman I, Prasad R (2016) Leveraging the attributes of Mucor hiemalis-derived silver nanoparticles for a synergistic broad-spectrum antimicrobial platform. Front Microbiol 7:1984. https://doi.org/10.3389/fmicb.2016.01984

Bakalova R, Zhelev Z, Ohba H, Ishikawa M, Baba Y (2004) Quantum dots as photosensitizers? Nat Biotechnol 22:1360–1361. https://doi.org/10.1038/nbt1104-1360

Baskar V, Meeran S, Shabeer STK, Sruthi S, Ali J (2018) Historic review on modern herbal nanogel formulation and delivery methods. Int J Pharm Pharm Sci 10:1–10. https://doi.org/10.22159/ijpps.2018v10i10.23071

Bellingham BK (2011) Proximal soil sensing. Vadose Zone J 10:1342–1342. https://doi.org/10.2136/vzj2011.0105br

Ben-Moshe T, Frenk S, Dror I, Minz D, Berkowitz B (2013) Effects of metal oxide nanoparticles on soil properties. Chemosphere 90 (2):640–646

Berekaa MM (2015) Nanotechnology in food industry; advances in food processing, packaging and food Safety. Int J Curr Microbiol App Sci 4:345–357

Bhattacharyya A, Duraisamy P, Govindarajan M, Buhroo AA, Prasad R (2016) Nano-biofungicides: emerging trend in insect pest control. In: Prasad R (ed) Advances and applications through fungal nanobiotechnology. Springer International Publishing, Cham, pp 307–319. https://doi.org/10.1007/978-3-319-42990-8_15

Bhushani JA, Anandharamakrishnan C (2014) Electrospinning and electrospraying techniques: potential food based applications. Trends Food Sci Technol 38:21–33. https://doi.org/10.1016/j.tifs.2014.03.004

HusseinMZ B, Zainal Z, Yahaya AH, Foo DWV (2002) Controlled release of a plant growth regulator, α-naphthaleneacetate from the lamella of Zn-Al-layered double hydroxide nanocomposite. J Contr Rel 82(2–3):417–427

Bulovic V, Mandell A, Perlman A (2004) Molecular memory device. US 20050116256, A1

Bumbudsanpharoke N, Ko S (2015) Nano-food packaging: an overview of market, migration research, and safety regulations. J Food Sci 80:R910–R923. https://doi.org/10.1111/1750-3841.12861

Burman U, Kumar P (2018) Plant response to engineered nanoparticles. In: Nanomaterials in plants, algae, and microorganisms. Academic Press, pp 103–118

Buzea C, Pacheco II, Robbie K (2007) Nanomaterials and nanoparticles: sources and toxicity. Biointerphases 2, MR17–MR71

Chai H, Yao J, Sun J, Zhang C, Liu W, Zhu M, Ceccanti B (2015) The effect of metal oxide nanoparticles on functional bacteria and metabolic profiles in agricultural soil. Bull Environ Contam Toxicol 94:490–495

Chakravarthy AK, Bhattacharyya A, Shashank PR, EpidiTT DB, Mandal SK (2012) DNA-tagged nano gold: a new tool for the control of the armyworm, Spodoptera litura Fab. (Lepidoptera: Noctuidae). Afr J Biotechnol 11:9295–9301. https://doi.org/10.5897/AJB11.883

Chavan S, Nadanathangam V (2020) Shifts in metabolic patterns of soil bacterial communities on exposure to metal engineered nanomaterials. Ecotoxicol Environ Saf 189:110012

Chen HD, Yada R (2011) Nanotechnologies in agriculture: new tools for sustainable development. Trends Food Sci Technol 22:585–594. https://doi.org/10.1016/j.tifs.2011.09.004

Coll C, Notter D, Gottschalk F, Sun T, Som C, Nowack B (2016) Probabilistic environmental risk assessment of five nanomaterials (nano-TiO$_2$, nano-Ag, nano-ZnO, CNT, and fullerenes). Nanotoxicology 10:4

Concha-Guerrero SI, Brito EMS, Piñón-Castillo HA et al (2014) Effect of CuO nanoparticles over isolated bacterial strains from agricultural soil. J Nanomater 2014:13

Couvreur P, Dubernet C, Puisieux F (1995) Controlled drug delivery with nanoparticles: current possibilities and future trends. Eur J Pharm Biopharm 41:2–13

Cox A, Venkatachalam P, Sahi S, Sharma N (2017) Reprint of: silver and titanium dioxide nanoparticle toxicity in plants: a review of current research. Plant Physiol Biochem 110:33–49. https://doi.org/10.1016/j.plaphy.2016.08.007

Das S, Wolfson BP, Tetard L, Tharkur J, Bazata J, Santra S (2015) Effect of N-acetyl cysteine coated CdS:Mn/ZnS quantum dots on seed germination and seedling growth of snow pea (Pisum sativum L.): imaging and spectroscopic studies. Environ Sci 2:203–212. https://doi.org/10.1039/c4en00198b

Dasgupta N, Ranjan S, Mundekkad D, Ramalingam C, Shanker R, Kumar A (2015) Nanotechnology in agro-food: from field to plate. Food Res Int 69:381–400. https://doi.org/10.1016/j.foodres.2015.01.005

de Medeiros GA, Arruda FB, SakaiE FM (2001) The influence of crop canopy on evapotranspiration and crop coefficient of beans (Phaseolusvulgaris L.). Agric Water Manage 49:211–224. https://doi.org/10.1016/S0378-3774(00)00150-5

de Oca-Vásquez GM, Solano-Campos F, Vega-Baudrit JR, López-Mondéjar R, Odriozola I, Vera A, Moreno JL, Bastida F (2020) Environmentally relevant concentrations of silver nanoparticles diminish soil microbial biomass but do not alter enzyme activities or microbial diversity. J Hazard Mat 391:122224

De Oliveira JL, Campos EVR, Bakshi M, Abhilash PC, Fraceto LF (2014) Application of nanotechnology for the encapsulation of botanical insecticides for sustainable agriculture: prospects and promises. Biotechnol Adv 32:1550–1561. https://doi.org/10.1016/j.biotechadv.2014.10.010

Dimkpa CO (2014) Can nanotechnology deliver the promised benefits without negatively impacting soil microbial life? J Basic Microbiol 54:889–904. https://doi.org/10.1002/jobm.201400298

Dixit R, Wasiullah MD, Pandiyan K, Singh UB, Sahu A et al (2015) Bioremediation of heavy metals from soil and aquatic environment:

an overview of principles and criteria of fundamental processes. Sustainability 7:2189–2212. https://doi.org/10.3390/su7022189

Du W, Tan W, Peralta-Videa JR, Gardea-Torresdey JL, Ji R, Yin Y et al (2017) Interaction of metal oxide nanoparticles with higher terrestrial plants: physiological and biochemical aspects. Plant Physiol Biochem 110:210–225. https://doi.org/10.1016/j.plaphy.2016.04.024

El Beyrouthya M, El Azzi D (2014) Nanotechnologies: novel solutions for sustainable agriculture. Adv Crop Sci Technol 2:e118. https://doi.org/10.4172/2329-8863.1000e118

Fierer N, Jackson RB (2006) The diversity and biogeography of soil bacterial communities. PNAS 103(3):626–631

Fincheira P, Tortella G, Duran N, Seabra AB, Rubilar O (2020) Current applications of nanotechnology to develop plant growth inducer agents as an innovation strategy. Crit Rev Biotechnol 40(1):15–30

Floros JD, Newsome R, Fisher W, Barbosa-Cánovas GV, Chen H, Dunne CP et al (2010) Feeding the world today and tomorrow: the importance of food science and technology. Compr Rev Food Sci Food Saf 9:572–599. https://doi.org/10.1111/j.1541-4337.2010.00127.x

Fogel R, Limson J (2016) Developing biosensors in developing countries: South Africa as a case study. Biosensors 6:5. https://doi.org/10.3390/bios6010005

Fraceto LF, Grillo R, de Medeiros GA, Scognamiglio V, Rea G, Bartolucci C (2016) Nanotechnology in agriculture: which innovation potential does it have? Front Environ Sci 4:20. https://doi.org/10.3389/fenvs.2016.00020

Frenk S, Ben-Moshe T, Dror I, Berkowitz B, Minz D (2013) Effect of metal oxide nanoparticles on microbial community structure and function in two different soil types. PLoS ONE 8:84441

Ghaani M, Cozzolino CA, Castelli G, Farris S (2016) An overview of the intelligent packaging technologies in the food sector. Trends Food Sci Tech 51:1–11. https://doi.org/10.1016/j.tifs.2016.02.008

Ghormade V, Deshpande MV, Paknikar KM (2011)Perspectives for nano-biotechnology enabled protection and nutrition of plants. Biotech Adv 29(6):792–803

Gibney E (2015) Buckyballs in space solve 100-year-old riddle. Nat News. https://doi.org/10.1038/nature.2015.17987

GrilloR APC, Fraceto LF (2016) Nanotechnology applied to bio-encapsulation of pesticides. J Nanosci Nanotechnol 16:1231–1234. https://doi.org/10.1016/j.tifs.2003.10.005

Gruère GP (2012) Implications of nanotechnology growth in food and agriculture in OECD countries. Food Policy 37:191–198. https://doi.org/10.1016/j.jhazmat.2014.05.079

Guo J, Chi J (2014) Effect of Cd-tolerant plant growth-promoting rhizobium on plant growth and Cd uptake by Loliummultiflorum Lam. and Glycinemax (L.) Merr.in Cd-contaminated soil. Plant Soil 375:205–214. https://doi.org/10.1007/s11104-013-1952-1

Hajirostamlo B, Mirsaeedghazi N, Arefnia M, Shariati MA, Fard EA (2015) The role of research and development in agriculture and its dependent concepts in agriculture. Asian J Appl Sci Eng 4:79–81. https://doi.org/10.1016/j.cocis.2008.01.005

Helar G, Chavan A (2015) Synthesis, characterization and stability of gold nanoparticles using the fungus Fusarium oxysporum and its impact on seed. Int J Recent Sci Res 6:3181–3318

Hoffmann M, Holtze EM, Wiesner MR (2007) Reactive oxygen species generation on nanoparticulate material. In: Wiesner MR, Bottero JY (eds) Environmental nanotechnology. Applications and impacts of nanomaterials. McGraw Hill, New York, pp 155–203

Hossain K, Abbas SZ, Ahmad A, Rafatullah M, Ismail N, Pant G, Avasn M (2020) Nanotechnology: a boost for the urgently needed second green revolution in Indian agriculture. In: Nanobiotechnology in agriculture. Springer, Cham, pp 15–33

Hu Y, Li J, Ma L, PengQ FW, Zhang L et al (2010) High efficiency transport of quantum dots into plant roots with the aid of silwet

L-77. Plant Physiol Biochem 48:703–709. https://doi.org/10.1016/j.plaphy.2010.04.001

Ion AC, Ion I, Culetu A (2010) Carbon-based nanomaterials: environmental applications. Univ Politehn Bucharest 38:129–132

Jiling C, Youzhi F, Xiangui L, Junhua W (2016) Arbuscular mycorrhizal fungi alleviate the negative effects of iron oxide nanoparticles on bacterial community in rhizospheric soils. Front Environ Sci 4:10

Josko I, Oleszczuk P, Futa B (2014) The effect of inorganic nanoparticles (ZnO, Cr_2O_3, CuO and Ni) and their bulk counterparts on enzyme activities in different soils. Geoderma 232:528–537

Kah M (2015) Nanopesticides and nanofertilizers: emerging contaminants or opportunities for risk mitigation? Front Chem 3:64. https://doi.org/10.3389/fchem.2015.00064

Kandasamy S, Prema RS (2015) Methods of synthesis of nano particles and its applications. J Chem Pharm Res 7:278–285

Khan MN, Mobin M, Abbas ZK, AlMutairi KA, Siddiqui ZH (2017) Role of nanomaterials in plants under challenging environments. Plant Physiol Biochem 110:194–209. https://doi.org/10.1016/j.plaphy.2016.05.038

Khan MR, Akram M (2020) Nanoparticles and their fate in soil ecosystem. In: Biogenic nano-particles and their use in agro-ecosystems. Springer, Singapore, pp 221–245

Khan ST (2020) Interaction of engineered nanomaterials with soil microbiome and plants: their impact on plant and soil health. In: Hayat S, Pichtel J, Faizan M, Fariduddin Q (eds) Sustainable agriculture reviews, vol 41. Springer, Cham

Khiew P, Chiu W, Tan T, Radiman S, Abd-Shukor R, Chia CH (2011) Capping effect of palm-oil based organometallic ligand towards the production of highly monodispersed nanostructured material. In: Palm oil: nutrition, uses and impacts, pp 189–219, Nova Science

Khin MM, NairAS BVJ, Murugan R, Ramakrishna S (2012) A review on nanomaterials for environmental remediation. Energy Environ Sci 5:8075–8109. https://doi.org/10.1039/c2ee21818f

Khond VW, Kriplani VM (2016) Effect of nanofluid additives on performances and emissions of emulsified diesel and biodiesel fueled stationary CI engine: a comprehensive review. Renew Sustain Energy Rev 59:1338–1348. https://doi.org/10.1016/j.rser.2016.01.051

Khot LR, Sankaran S, Maja JM, Ehsani R, Schuster EW (2012) Applications of nanomaterials in agricultural production and crop protection: a review. Crop Prot 35:64–70. https://doi.org/10.1016/j.cropro.2012.01.007

Kookana RS, Boxall AB, Reeves PT, Ashauer R, Beulke S, Chaudhry Q et al (2014) Nanopesticides: guiding principles for regulatory evaluation of environmental risks. J Agric Food Chem 62:4227–4240. https://doi.org/10.1021/jf500232f

Kumar N, Shah V, Walker VK (2012) Influence of a nanoparticle mixture on an arctic soil community. Environ Toxicol Chem 31:131–135

Lambreva MD, Lavecchia T, Tyystjärvi E, Antal TK, Orlanducci S, Margonelli A et al (2015) Potential of carbon nanotubes in algal biotechnology. Photosyn Res 125:451–471. https://doi.org/10.1007/s11120-015-0168-z

Liu RQ, Lal R (2015) Potentials of engineered nanoparticles as fertilizers for increasing agronomic productions. Sci Total Environ 514:131–139. https://doi.org/10.1016/j.scitotenv.2015.01.104

Liu Y, He L, Mustapha A, Li H, Hu ZQ, Lin M (2009) Antibacterial activities of zinc oxide nanoparticles against Escherichia coli O157: H7. J Appl Microbiol 107:1193–1201. https://doi.org/10.1111/j.1365-2672.2009.04303.x

Ma X, Geiser-Lee J, Deng Y, Kolmakov A (2010) Interactions between engineered nanoparticles (ENPs) and plants: phytotoxicity, uptake and accumulation. Sci Total Environ 408:3053–3061. https://doi.org/10.1016/j.scitotenv.2010.03.031

Malysheva A, Lombi E, Voelcker NH (2015) Bridging the divide between human and environmental nanotoxicology. Nat Nanotechnol 1:0835–0844. https://doi.org/10.1038/nnano.2015.224

Marchiol L, Iafisco M, Fellet G, Adamiano A (2020) Nanotechnology support the next agricultural revolution: perspectives to enhancement of nutrient use efficiency. In: Advances in agronomy, vol 161. Academic Press, pp 27–116

Martirosyan A, Schneider YJ (2014) Engineered nanomaterials in food: implications for food safety and consumer health. Int J Environ Res Public Health 11:5720–5750. https://doi.org/10.3390/ijerph110605720

MingfangQ YL, Tianlai L (2013) Nano-TiO$_2$ improve the photosynthesis of tomato leaves under mild heat stress, biological trace element research. Biol Trace Element Res 156(1):323–328

Mishra S, Fraceto LF, Yang X, Singh HB. (2018) Rewinding the history of agriculture and emergence of nanotechnology in agriculture. In: Emerging trends in agri-nanotechnology: fundamental and applied aspects. CABI UK, p 1

Mishra S, Singh HB (2015) Biosynthesized silver nanoparticles as a nanoweapon against phytopathogens: exploring their scope and potential in agriculture. Appl Microbiol Biotechnol 99:1097–1107. https://doi.org/10.1007/s00253-014-6296-0

Mishra VK, Kumar A (2009) Impact of metal nanoparticles on the plant growth promoting rhizobacteria. Dig J Nanomater Biostruct 4:587–592

Mukhopadhyay SS (2014) Nanotechnology in agriculture: prospects and constraints. Nanotechnol Sci Appl 7:63–71. https://doi.org/10.2147/NSA.S39409

Nafees M, Ali S, Rizwan M, Aziz A, Adrees M, Hussain SM, Ali Q, Junaid M (2020) Effect of nanoparticles on plant growth and physiology and on soil microbes. In: Nanomaterials and environmental biotechnology. Springer, Cham, pp 65–85

Nima AZ, Lahiani MH, Watanabe F, Xu Y, Khodakovskaya MV, Biris AS (2014) Plasmonically active nanorods for delivery of bio-active agents and high-sensitivity SERS detection in planta. RSC Adv 4:64985–64993. https://doi.org/10.1039/C4RA10358K

Nuruzzaman M, Rahman MM, Liu Y, Naidu R (2016) Nanoencapsulation, nano-guard for pesticides: a new window for safe application. J Agric Food Chem 64:1447–1483. https://doi.org/10.1021/acs.jafc.5b05214

Ozdemir M, Kemerli T (2016) Innovative applications of micro and nanoencapsulation in food packaging. In: Lakkis JM (ed) Encapsulation and controlled release technologies in food systems. Wiley, Chichester

Pagano L, Servin AD, De La Torre-Roche R, Mukherjee A, Majumdar S, Hawthorne J et al (2016) Molecular response of crop plants to engineered nanomaterials. Environ Sci Technol 50:7198–7207. https://doi.org/10.1021/acs.est.6b01816

Pakrashi S, Jain N, Dalai S, Jayakumar J, Chandrasekaran PT, Raichur AM et al (2014) In vivo genotoxicity assessment of titanium dioxide nanoparticles by Allium cepa root tip assay at high exposure concentrations. PLoS ONE 9:e98828. https://doi.org/10.1371/journal.pone.0087789

Parada J, Rubilar O, Fernández-Baldo MA, Bertolino FA, Durán N, Seabra AB, Tortella GR (2019) The nanotechnology among US: are metal and metal oxides nanoparticles a nano or mega risk for soil microbial communities? CritRevBiotech 39(2):157–172

Parada J, Rubilar O, Fernández-Baldo MA, Bertolino FA, Durán N, Seabra AB, Tortella GR (2019) The nanotechnology among US: are metal and metal oxides nanoparticles a nano or mega risk for soil microbial communities? Crit Rev Biotechnol 39(2):157–172

Parisi C, Vigani M, Rodriguez-Cerezo E (2015) Agricultural nanotechnologies: what are the current possibilities? Nano Today 1:124–127. https://doi.org/10.1016/j.nantod.2014.09.009

Patra JK, Baek KH (2017) Antibacterial activity and synergistic antibacterial potential of biosynthesized silver nanoparticles against foodborne pathogenic bacteria along with its anticandidal and antioxidant effects. Front Microbiol 8:167. https://doi.org/10.3389/fmicb.2017.00167

Pérez-de-Luque A (2017) Interaction of nanomaterials with plants: what do we need for real applications in agriculture? Front Environ Sci 5:12. https://doi.org/10.3389/fenvs.2017.00012

Pérez-Hernández H, Fernández-Luqueño F, Huerta-Lwanga E, Mendoza-Vega J, Álvarez-Solís José D (2020) Effect of engineered nanoparticles on soil biota: Do they improve the soil quality and crop production or jeopardize them? Land Degrad Dev 31(6):1–25

Perlatti B, Bergo PLS, Silva MFG, Fernandes JB, Forim MR (2013) Polymeric nanoparticle-based insecticides: a controlled release purpose for agrochemicals. In: Trdan S (ed) Insecticides-development of safer and more effective technologies. InTech, Rijeka, pp 523–550. https://doi.org/10.5772/53355

Perrault SD, Walkey C, Jennings T, Fischer HC, Chan WCW (2009) Mediating tumor targeting efficiency of nanoparticles through design—nano letters. Nano Lett 9:1909–1915. https://doi.org/10.1021/nl900031y

Pirzadah B, Pirzadah TB, Jan A, Hakeem KR. (2020) Nanofertilizers: a way forward for green economy. In: Nanobiotechnology in agriculture. Springer, Cham, pp 99–112

Pradhan A, Seena S, Pascoal C, Cássio F (2011) Can metal nanoparticles be a threat to microbial decomposers of plant litter in streams? Microb Ecol 62:58–68

Prasad R (2014) Synthesis of silver nanoparticles in photosynthetic plants. J Nanopart 2014:963961. https://doi.org/10.1155/2014/963961

Prasad R, Kumar V, Prasad KS (2014) Nanotechnology in sustainable agriculture: present concerns and future aspects. Afr J Biotechnol 13:705–713. https://doi.org/10.5897/AJBX2013.13554

Prasad R, PandeyR BI (2016) Engineering tailored nanoparticles with microbes: quo vadis. WIREs Nanomed Nanobiotechnol 8:316–330. https://doi.org/10.1002/wnan.1363

Priyanka N, Geetha N, Ghorbanpour M, Venkatachalam P (2019) Role of engineered zinc and copper oxide nanoparticles in promoting plant growth and yield: present status and future prospects. In: Advances in phytonanotechnology. Academic Press, pp 183–201

Rai V, Acharya S, Dey N (2012) Implications of nanobiosensors in agriculture. J Biomater Nanobiotechnol 3:315–324. https://doi.org/10.4236/jbnb.2012.322039

Raliya R, Tarafdar JC, Gulecha K, Choudhary K, Ram R, Mal P et al (2013) Review article; scope of nanoscience and nanotechnology in agriculture. J Appl Biol Biotechnol 1:041–044

Rana S, Kalaichelvan PT (2013) Ecotoxicity of Nanoparticles. ISRN Toxicol 2013:574648. https://doi.org/10.1155/2013/574648

Read DS, Matzke M, Gweon HS, Newbold LK, Heggelund L, Ortiz MD, Lahive E, Spurgeon D, Svendsen C (2016) Soil pH effects on the interactions between dissolved zinc, non-nano-and nano-ZnO with soil bacterial communities. Environ Sci Poll Res 23(5):4120–4128

Sabir S, Arshad M, Chaudhari SK (2014) Zinc oxide nanoparticles for revolutionizing agriculture: synthesis and applications. Sci World J 2014:8. https://doi.org/10.1155/2014/925494

Sadeghzadeh B (2013) A review of zinc nutrition and plant breeding. J Soil Sci Plant Nutr 13:905–927. https://doi.org/10.4067/S0718-95162013005000072

Sadik OA, Du N, Kariuki V, Okello V, Bushlyar V (2014) Current and emerging technologies for the characterization of nanomaterials. ACS Sustain Chem Eng 2:1707–1716. https://doi.org/10.1021/sc500175v

Sagadevan S, Periasamy M (2014) Recent trends in nanobiosensors and their applications—a review. Rev Adv Mater Sci 36:62–69

Samarajeewa AD, Velicogna JR, Princz JI, Subasinghe RM, Scroggins RP, Beaudette LA (2017) Effect of silver nano-particles on soil microbial growth, activity and community diversity in a sandy loam soil. Environ Poll 220:504–513

Sami F, Siddiqui H, Hayat S. (2020) Impact of silver nanoparticles on plant physiology: a critical review. In: Sustainable agriculture reviews, vol 41. Springer, Cham, pp 111–127

Sanzari I, Leone A, Ambrosone A (2019) Nanotechnology in plant science: to make a long story short. FrontBioeng Biotechnol 7:120. https://doi.org/10.3389/fbioe.2019.00120

Sarkar A, Sengupta S, Sen S (2019) Nanoparticles for soil remediation. InNanoscience and biotechnology for environmental applications. Springer, Cham, pp 249–262

Sayes CM, Fortner JD, Guo W, LyonD BAM, Ausman KD, Tao YJ et al (2004) The differential cytotoxicity of water-soluble fullerenes. Nano Lett 4:1881–1887. https://doi.org/10.1002/btpr.707

Scognamiglio V, Arduini F, Palleschi G, Rea G (2014) Biosensing technology for sustainable food safety. Trac-Trends Anal Chem 62:1–10. https://doi.org/10.1016/j.trac.2014.07.007

Sertova NM (2015) Application of nanotechnology in detection of mycotoxins and in agricultural sector. J Cent Eur Agric 16:117–130. https://doi.org/10.5513/JCEA01/16.2.1597

Servin AD, White JC (2016) Nanotechnology in agriculture: next steps for understanding engineered nanoparticle exposure and risk. NanoImpact 1:9–12

Shandilya N, Le BO, Bressot C, Morgeneyer M (2015) Emission of titanium dioxide nanoparticles from building materials to the environment by wear and weather. Environ Sci Technol 49 (4):2163–2170

Shang Y, Hasan M, Ahammed GJ, Li M, Yin H, Zhou J (2019) Applications of nanotechnology in plant growth and crop protection: a review. Molecules 24(14):2558

Shrivastava M, Srivastav A, Gandhi S, Rao S, Roychoudhury A, Kumar A, Singhal RK, Jha SK, Singh SD (2019) Monitoring of engineered nanoparticles in soil-plant system: a review. Environ Nanotechnol Monitor Manag 11:100218

Simonin M, Richaume A (2015) Impact of engineered nanoparticles on the activity, abundance, and diversity of soil microbial communities: a review. Environ Sci Poll Res 22:13710–13723

Simonin M, Richaume A, Guyonnet J et al (2016) Titanium dioxide nanoparticles strongly impact soil microbial function by affecting archaeal nitrifiers. Sci Rep 6:33643

Singh S, Vishwakarma K, Singh S, Sharma S, Dubey NK, SinghVK et al (2017) Understanding the plant and nanoparticle interface at transcriptomic and proteomic level: a concentric overview. Plant Gene 11:265-272. https://doi.org/10.1016/j.plgene.2017.03.006

Srivastava P, Singh R, Tripathi S, Raghubanshi AS (2016) An urgent need for sustainable thinking in agriculture–An Indian scenario. Ecol Ind 67:611–622

Suresh AK, Pelletier DA, Doktycz MJ (2013) Relating nanomaterial properties and microbial toxicity. NANO 5:463–474

Tapan A, Sivakoti R (2019) Nano fertilizer: its impact on crop growth and soil health. J Res PJTSAU 47(3):1–1

Torsvik V, Øvreås L (2002) Microbial diversity and function in soil: from genes to ecosystems. Curr Opin Microbiol 5:240–245

Tourinho PS, van Gestel CA, Lofts S, Svendsen C, Soares AM, Loureiro S (2012) Metal-based nanoparticles in soil: fate, behavior, and effects on soil invertebrates. Environ Toxicol Chem 31 (8):1679–1692

Tripathi DK, Mishra RK, Singh S, Singh S, Vishwakarma K, Sharma S et al (2017) Nitric oxide ameliorates zinc oxide nanoparticles phytotoxicity in wheat seedlings: implication of the ascorbate-glutathione cycle. Front Plant Sci 8:1. https://doi.org/10.3389/fpls.2017.00001

Tripathi DK, Singh S, Singh S, Srivastava PK, Singh VP, Singh S, Prasad SM, Singh PK, Dubey NK, Pandey AC, Chauhan DK (2017) Nitric oxide alleviates silver nanoparticles (AgNPs)-induced

phytotoxicity in Pisum sativum seedlings. Plant Physiol Biochem 110:167–177

Tripathi DK, Singh S, Singh S, Srivastava PK, Singh VP, Singh S et al (2017) Nitric oxide alleviates silver nanoparticles (AgNps)-induced phytotoxicity in Pisum sativum seedlings. Plant Physiol Biochem 110:167–177. https://doi.org/10.1016/j.plaphy.2016.06.015

Tripathi DK, Singh S, Singh VP, Prasad SM, Dubey NK, Chauhan DK (2017) Silicon nanoparticles more effectively alleviated UV-B stress than silicon in wheat (Triticum aestivum) seedlings. Plant Physiol Biochem 110:70–81. https://doi.org/10.1016/j.plaphy.2016.06.026

Tripathi DK, Tripathi A, Shweta SS, Singh Y, Vishwakarma K, Yadav G et al (2017) Uptake, accumulation and toxicity of silver nanoparticle in autotrophic plants, and heterotrophic microbes: a concentric review. Front Microbiol 8:07. https://doi.org/10.3389/fmicb.2017.00007

Tripathi S, Champagne D, Tufenkji N (2012) Transport behavior of selected nanoparticles with different surface coatings in granular porous media coated with Pseudomonas aeruginosa biofilm. Environ Sci Technol 46(13):6942–6949

Valdes MG, Gonzalez ACV, Calzon JAG, Diaz-Garcia ME (2009) Analytical nanotechnology for food analysis. Microchim Acta 166:1–19. https://doi.org/10.1007/s00604-009-0165-z

Valenstein JS, Lin VSY, Lyznik LA, Martin-Ortigosa S, Wang K, Peterson DJ et al (2013) Mesoporous silica nanoparticle-mediated intracellular cre protein delivery for maize genome editing via loxP site excision. Plant Physiol 164:537–547. https://doi.org/10.1104/pp.113.233650

Vanderroost M, Ragaert P, Devlieghere F, De Meulenaer B (2014) Intelligent food packaging: the next generation. Trends Food Sci Technol 39:47–62. https://doi.org/10.1016/j.tifs.2014.06.009

Venkatachalam P, Jayaraj M, Manikandan R, Geetha N, Rene ER, Sharma NC et al (2017) Zinc oxide nanoparticles (ZnONPs) alleviate heavy metal-induced toxicity in Leucaena leucocephala seedlings: a physiochemical analysis. Plant Physiol Biochem 110:59–69. https://doi.org/10.1016/j.plaphy.2016.08.022

Vidotti M, Carvalhal RF, Mendes RK, Ferreira DCM, Kubota LT (2011) Biosensors based on gold nanostructures. J Braz Chem Soc 22:3–20. https://doi.org/10.1590/S0103-50532011000100002

Wan Y, Li J, Ren H, HuangJ YH (2014) Physiological investigation of gold nanorods toward watermelon. J Nanosci Nanotechnol 14:6089–6094. https://doi.org/10.1166/jnn.2014.8853

Wang L, Hu C, Shao L (2017) The antimicrobial activity of nanoparticles: present situation and prospects for the future. Int J Nanomed 12:1227–1249. https://doi.org/10.2147/IJN.S121956

WangX HH, Liu X, Gu X, Chen K, Lu D (2012) Multi-walled carbon nanotubes can enhance root elongation of wheat (Triticum aestivum) plants. J Nano Res 14(6):841

Warheit DB, Laurence BR, Reed KL, Roach DH, Reynolds GAM, Webb TR (2004) Comparative pulmonary toxicity assessment of single-wall carbon nanotubes in rats. Toxic Sci 77(1):117–125

Xin X, Zhao F, Zhao H, Goodrich SL, Hill MR, Sumerlin BS, Stoffella PJ, Wright AL, He Z (2020) Comparative assessment of polymeric and other nanoparticles impacts on soil microbial and biochemical properties. Geoderma 367:114278

Xu C, Peng C, Sun L, Zhang S, Huang H, Chen Y, Shi J (2015) Distinctive effects of TiO_2 and CuO nanoparticles on soil microbes and their community structures in flooded paddy soil. Soil Biol Biochem 86:24–33

Yao J, Yang M, Duan YX (2014) Chemistry, biology, and medicine of fluorescent nanomaterials and related systems: new insights into biosensing, bioimaging, genomics, diagnostics, and therapy. Chem Rev 114:6130–6178. https://doi.org/10.1021/cr200359p

You T, Liu D, Chen J, Yang Z, Dou R, Gao X, Wang L (2017) Effects of metal oxide nanoparticles on soil enzyme activities and bacterial communities in two different soil types. J Soils Sediments 18:211–221. https://doi.org/10.1007/s11368-017-1716-2

Zhang Q, Han L, Jing H, Blom DA, Lin Y, Xin HL (2016) Facet control of gold nanorods. ACS Nano 10:2960–2974. https://doi.org/10.1021/acsnano.6b00258

Zhao X, Meng Z, Wang Y, Chen W, Sun C, Cui B (2017) Pollen magnetofection for genetic modification with magnetic nanoparticles as gene carriers. Nat Plants 3:956–964. https://doi.org/10.1038/s41477-017-0063-z

Nanotechnology: Advancement for Agricultural Sustainability

Upinder and Rabindra Kumar

Abstract

Nanotechnology plays a vital role in agriculture for food production, security, and safety. Due to sensing, applicability of nanotechnology include the use of fertilizers to enhance food production, pesticides for pest, and disease management for monitoring soil quality and plant health. To improve the sustainability of agricultural practice, there is incorporation of nanomaterials in it as nanopesticides, nonfertilizer, and nanosensors. To suppress crop disease by requiring less input and generating less waste than conventional products, there is the use of nanoscale nutrients (metals, metal oxides, carbon) for subsequent enhancement of plant growth and yield. Directing this enhanced yield not only reduce growth of pathogens but also increase the nutritional value of the nanoparticles themselves, for the essential micronutrients that are necessary for host defense. We also posit that these positive effects are the greater availability of the nutrients in the "nano" form for disease-controlled plant yield. Keeping these points of view, we offer comments on the current regulatory perspective for such applications with the increased demand to agricultural field. This book chapter focused on engineered nanoparticles and naturally occurring nanoparticles for identifying their nanoscale properties in the agriculture fields.

Keywords

Engineered nanoparticles • Nonfertilizer • Nanopesticides • Nanosensors • Sustainable agriculture

Upinder (✉)
Department of Chemistry, D. A. V. College, Bathinda, India
e-mail: kaurreet604@gmail.com

R. Kumar
Central Instrumentation Laboratory, Central University of Punjab, Bathinda, India
e-mail: rabindrabiochem@gmail.com

1 Introduction

In sustainable agriculture, sensing and selective potentiality of nanotechnology plays important role in global food production (Rodrigues et al. 2017), food security (Hosseini et al. 2010), and food safety (Ali et al. 2018). Natural nanomaterials (NMs) followed the specific mechanisms to reduce its harmful effects on living organisms surviving in its ecosystem (Bernhardt et al. 2010). Under the research, advancement to reduce acute toxic effects of potential hazards of engineered NMs in living beings there has put the light over extensive research of nanotoxicology and strict laws of government to identify and avoid toxic nanoparticles (NPs) (Shrader-Frechette 2007). Having these unique properties with enhanced potential, development of engineered nanoparticles (ENPs) have raised considerable concerns with plants, as well as with other components of all ecosystems.

The ENP interaction with plants varies, depending on plant anatomy and its potentiality to transport through the vascular system from root to shoot part of the plant. Based on the literature review, the surface activity of ENPs on phytotoxicity to enhance crop yield and minimize disease-causing factors within the plant species has been thoroughly described. Under such interaction, it is prominent to say that agriculture is not an unseen sector in our life because of its better yield with disease-free products playing an essential role in the living world. On the other hand, agricultural productivity is facing unsolved problems because of insufficient space, disease, and change in agro-climate conditions. Moreover, with the rapid population growth and reducing/deteriorating environmental resources, there is a strong need of producing more by using less (Singh et al. 2019). This demand needs to adopt innovative technologies such as nanotechnology (Tilman et al. 2002; Singh et al. 2019). We are having remarkable properties such as less than 100 nm size of the materials considered as nanomaterial showing the abilities of signaling by fluorescence, optical activities, etc. (Giraldo et al. 2019).

Nanoparticles (NPs) are categorized based on natural nanoparticles and engineered nanoparticles which are present in the environment (Fig. 1). Further NPs are categorized as atmospheric, terrestrial, aquatic, unintentional, and synthesized (Shrivastava et al. 2019). Thus, the nanoformulation has been nowadays recommended for enhancement of agro-production increased utility of life-threatening pesticides and fertilizers (Servin et al. 2015) where the water dissolved fertilizers on the soil surface become unavailable to plant for its proper utility and that same amount leach off into the groundwater as a toxic pollutant nanofertilizers used including metal oxide NPs like Al_2O_3, TiO_2, CeO_2, FeO, ZnO, etc. These metal oxide NPs have been extensively come into the light while studied under the ENP plant interaction (Dimkpa 2014; Thwala et al. 2016).

2 Interaction of Engineered Nanoparticles (ENPs) with Crop Plants

Extensive use of pesticides in agro-production enhancement and disease control factors results in addition to toxic pollutants into the ecosystem. Nano-pesticides in the form of encapsulated pesticides act as environmental-friendly factor with huge potential to control disease-causing pests attack. Encapsulated pesticides synthesis by matrix coated active amount of slowly moving pesticides, which inhibit the accumulation of phytotoxic content in the ecosystem food chain (Chhipa 2017). Nanofertilizers and nanopesticides not only act as an environment-friendly factors which provide a wide array to find and control disease-causing aspects to getting quality yield crops. However, smart sensors developed by nanotechnology include nanomaterials (NMs), magnetic NPs, metal NPs (cobalt (Co), silver (Ag), gold (Au), and quantum dots (QDs)). These metal nanoparticles have been actively demonstrated for their application such as gene transformation, where genes of tobacco tissue found as the best result for tagging with gold nanoparticle (AuNPs) (Martin-Ortigosa et al. 2014).

3 Engineered Nanoparticles as a Smart Sensor

Plant tissue culture is a part of nanotechnology technique. Incorporation of AuNPs into the basal medium results in the improvement of the higher fraction out of total yield of seed germination and seedling growth in *Arabidopsis thaliana* (Kumar et al. 2014). AuNPs and AgNPs individually and in combination enhance the callus proliferation of *Prunella vulgaris* (Fazal et al. 2016). With a high rate of variation in both calli and regenerate shoot of AuNPs, it results in the enhancement of somaclonal variety (Kokina et al. 2013).

Nanomaterials also influence the production of secondary metabolites during plant tissue culture (Kim et al. 2017). Nanofabrication and characterization technology understand plant disease management (Ismail et al. 2017). Recently, existing green technology enables to reduce potential by risks with efficiently detecting the diseases and controlling the pollutants spread by acting as a smart sensor.

4 Detection and Diagnosis of Pathogens by Nanoparticles

Conventionally, when morphological symptoms appear on the plant, then pesticides are sprayed on it delayed situation to protect the plant from pest attack and seek quick remedial actions for control and protection of the plant from nutritional deficiencies, timely in most of the Indian agricultural cases (Singh 2008; Mahlein 2016). Thus, under mild infection condition, detection and plant protection both seem complicated, so under that situation highly sensitive, precise, and accurate detection technology follow to develop disease detection and management strategies that depend on cultural practices, embryo tests, and examination of infected parts visually and microscopically (Been 1995). Biochemical identification of the pathogen is time-consuming method so for following short way solution and nanotechnology has proved good with its high potential in the agriculture field and overcome disease-causing factors. Such advanced initiation will help the agricultural industry to overcome crop pathogens survival in plants, including virus, bacteria, and fungus. Nanoparticles have widespread application in disease diagnostic and its control by introducing nano-forms such as gold (Au), carbon (C), silver (Ag), and silica (Si). Gold nanoparticles (AuNPs) have found widespread applications in disease diagnostics (Jo et al. 2009; Servin et al. 2015) as well as used for early detection of diseases particularly cancer in humans.

Presently, research field of nanotechnology and analyte nanoparticles complex, i.e., DNA nanoparticle and antibody nanoparticle conjugate systems, emerged as a tool for disease diagnosis (Fang and Ramasamy 2015). Nanotechnology and quantum dots (QDs) for detection of plant viral infections for quarantine and indexing of quality planting material are the other areas where nanotechnology can be used in crops in the study of diagnosis host-virus interaction of crop plants (Sanzari et al. 2019). The disease-causing agent, especially phytoplasmas, is a unique group of obligate plant pathogens. The efficacy of tissue penetration of AuNPs depends not only on the plant species used but also on the particles' size and surface charge. Positively charged AuNPs are well absorbed only by plant roots, whereas negatively charged AuNPs also can effectively move from roots to stems and leaves (Li et al. 2016). Some part of this process is

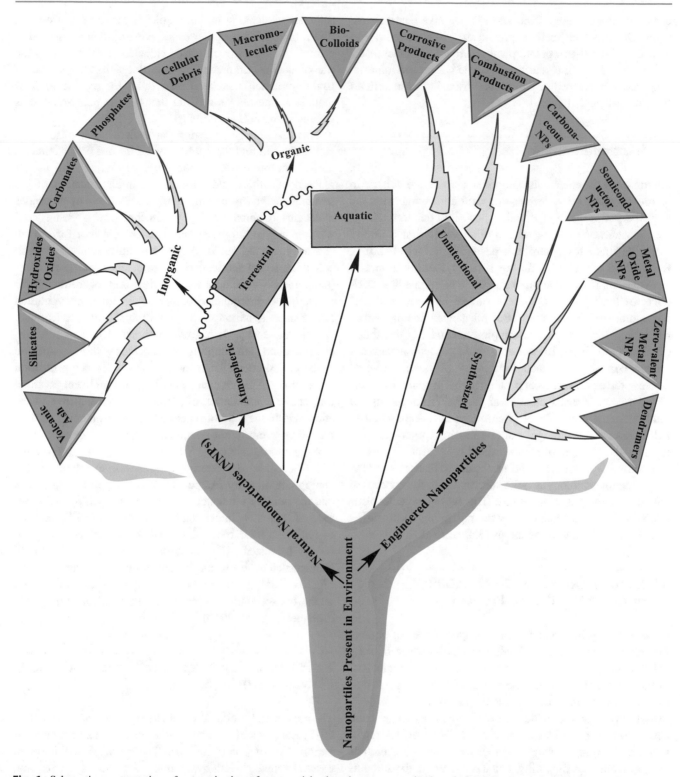

Fig. 1 Schematic representation of categorization of nanoparticles based on presence in the whole environment (Adopted and redrawn from Shrivastava et al. 2019)

played by the plant vascular system, as well as by plasmodesmata (Koelmel et al. 2013). When AuNPs are sprayed on seedlings of watermelon (*Citrullus lanatus* var. Lanatus (Thunb.), it enter leaves via stomata and are translocated from leaves to roots by the phloem transport mechanism (Raliya et al. 2016). It is complicated to detect phytoplasmas diseases by using ELISA and PCR test (Sankaran et al. 2010). Although this technique is highly useful, but it is time consuming and becomes too late to undertake control measures. So, dip-stick method proved as an effective way to detect viral disease in potato plant where plant extract reacts with the stick, and the detection is found within few minutes. The precision and validity improved by using nanoparticles (Koelmel et al. 2013).

5 Nanopheromone

Pheromones are chemicals compounds which are volatile in nature secreted by a species for communicating with the opposite sex of own individuals' species. In this technology, the pheromone compound attracts different sex female insects and kills them for controlling pests (Ragaei et al. 2014). Researchers find the way for the isolation, identification, and synthesis of insect pheromone, which is compelling track trace for management of pest as one of the integrated pest management (IPM) strategy. But till nanotechnology needs refinement for use in mass trapping and mating disruption (Shelton et al. 2006; Brewer et al. 2012). Encapsulate pheromone compounds in nanotechnology plays vital role to increase the effectiveness of prolonged use of semi-chemicals (Poddar et al. 2018; Kumar et al. 2019). Prolonged use of semi-chemical molecule of pheromones is encapsulated to reduce its cost and wastage by slow-releasing compounds, and it solves the problem of photo-instability by protecting these compounds from sunlight and molecular oxidation. Application of nanotechnology is quite successful for the concept of volatile pheromone compounds. The highly volatile pheromone compounds identified with regulating its releasing pattern through nanoformulations (Kumar et al. 2019).

6 Nanopesticides for Sustainability in Agricultural Crop-Production

Nanopesticides in nanometer (nm) range consists of organic as well as inorganic ingredients (e.g., polymers and metal oxides) in various forms (e.g., particles and micelles) (Ragaei and Sabry 2014). The use of (ENPs) for plant protection products are termed as "nanopesticides." "Small-sized nanopesticides are engineered active structure having useful pesticidal properties." Nanopesticides represent an emerging technological development that include increased efficacy, durability, and a reduction in the amounts of active ingredients. For seeking environmental safety measures, use of varied products have been done at different stages in the product development cycle for enhancing the efficacy of existing pesticide active ingredients (Kah and Hofmann 2014).

In some cases, the ENP itself may "drive" the biological effect (e.g., nanosilver when used as a pesticide where the active component is the ionic gold that is released from the ENP). In contrast, in other cases, nanotechnology is used to protect an active ingredient or enhanced its delivery to the site of action (Chinnamuthu and Boopathi 2009). Use of nanoformulation has been also done for enhancing IPM module bio-pesticides self-life that has the poor self-life as well as its pesticidal action.

Nanotechnology following "smart field systems" to detect pathogens and find the application of the use of pesticides is needed for protection of the environment. By the use of reduced quantities and targeted the implementation of the pesticide active ingredient, monitor the effects of pesticides molecules (Chhipa 2017). Nanotechnology is being used with the aim to improve plant disease resistance for growth enhancement and sufficient nutrient utilization, agrochemicals design, and fabrication of nanoplant protection inputs (herbicides, fungicides, insecticides, and pheromone) for protecting the natural environment. Rather than conventional agrochemicals, nanoformulations designed more effectively where nano-encapsulation which shows more benefits, efficient use, and safe handling of pesticides with less exposure to the environment. With the guarantees for eco-protection transformation use of nanotechnology has sufficient potential for genetic manipulation of plants to obtain improved varieties. Within the field of plant pathology, problems and their protection from plant-pathogen interaction found a way through nanotechnology-based précised process and product which capably delivered the nutrients to plant in the appropriate quantity. As per the laboratory results, it shows that the nutrient efficiency of nanofertilizer nitrogen (N) increased from 32% in conventional fertilizers to 72% amount of fertilizers used reduced by half to the amount of grain being harvested (Chinnamuthu and Boopathi 2009). Nanopesticides/fertilizers influence the soil–plant system (Fig. 2) with improvement of an analytical tool for detection of the transportation system of ENPs in plant–soil system (Shrivastava et al. 2019).

7 Effects of Plant Exposure to the Gold Metal Nanoparticle (AuNPs)

The growth and productivity of *Brassica juncea* (Arora et al. 2012) is studied after spraying the plant with suspensions of various AuNP concentrations. The particles within plant tissues were detected by atomic absorption spectroscopy. The effects of AuNP applications were positive, including increased stem length and diameter, increased numbers of leaves and shoots, and improved productivity (Gunjan et al. 2014). The addition of AuNPs to soil used for plant growth enhanced seed germination in *Zea mays* (Mahakam et al. 2016) and *Pennisetum glaucum* (Parveen et al. 2016). Using synchrotron-based X-ray microanalysis and high-resolution

Fig. 2 Schematic representation of nanoparticles release in soil–plant–groundwater system

transmission electron microscopy (TEM), (Sabo-Attwood et al. 2012) showed that 3.5-nm AuNPs entered *Nicotiana xanthi* through the roots and moved into the vascular system. Aggregates of 18-nm AuNPs were detected only in the root cell cytoplasm. Exposure to small particles led to leaf necrosis after 14 days, but with large particles, no differences from the control were observed. At very high AuNP concentrations, no physiological effect has been described in *Glycine max* L. (Falco et al. 2011) and aquatic aquarium plants (Glenn et al. 2012).

Several studies on the mechanisms of nanoparticle entry into plants and on nanoparticle phytotoxicity (Taylor 2011) have been done with *Arabidopsis thaliana*, a classical object in current plant physiology. The addition of 24-nm AuNPs (10–80 mg l^{-1}) to the growth medium led to a threefold increase in total seed yield, as compared with the control. It also markedly increased the length and diameter of the stem and roots. Exciting results came from the study by Taylor et al. (2014), who found that the root length in *A. thaliana* grown on an agar medium with 100 mg l^{-1} of K(AuCl4) was reduced by 75%. But there also was a slight decrease in the expression of genes coding for aquaporins and proteins implicated in the transport metal ions such as copper,

cadmium, iron, and nickel ions. Oxidized gold was found simultaneously in the vascular system of roots and shoots of plant species *A. thaliana*, but AuNPs synthesized in plants were detected only in root tissues. Gold chlorides were much more genotoxic than AuNPs (Taylor et al. 2014). Overall, the toxicity of metal ions was much higher than that of nanoparticles, and AgNPs were more phytotoxic than AuNPs (Notter et al. 2014).

In a study of the toxicity of metallic nanoparticles to callus cultures, Fazal et al. (2016) showed that AuNPs enhance callus proliferation in *Prunella vulgaris* (L.). Cellular entry and toxicity of nanoparticles are often investigated with suspension cultures (Alkilany et al. 2010). Suspension cultures of plant cells are more sensitive to a broad range of compounds and abiotic effects (Rains et al. 1989). Biochemical and physiological responses develop within a short time and are fairly evenly distributed across the population, unlike what is observed in a whole plant or its organs. Additionally, one can expect that the effects of nanoparticles on suspension culture cells will be more significant owing to the absence of specialized protective structures such as cuticles or epidermis. The addition of 20-nm AuNPs to the growth medium of *A. thaliana*

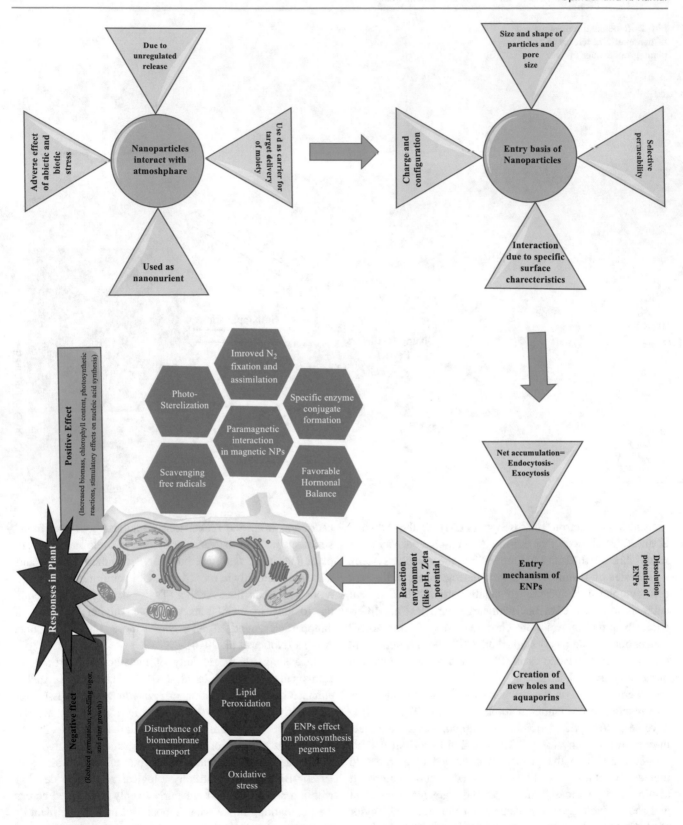

Fig. 3 Schematic representation of different factors influencing interaction of nanoparticles and plants (Adopted and redrawn from Burman and Kumar 2018)

promoted biomass growth in suspension culture cells (Selivanov et al. 2017).

Gold (Au) NPs change the pH of the culture medium and made it alkaline. AuNPs caused a slight but steady decrease in the specific respiratory activity of the *A. thaliana* suspension culture cells, as well as an increase in the intracellular pool of free amino acids (alanine, γ-aminobutyric acid, and valine). Furthermore, the nanoparticles changed the extracellular protein composition and the actin cytoskeleton structure in the *A. thaliana* cell culture (Selivanov et al. 2017). AuNP-induced increases in stem height and diameter, leaf and shoot numbers, and yield were observed in mustard and tobacco and that increases in seed germination and biomass amount were observed in maize, gloriosa lily, mung bean, and pearl millet. Adverse effects of AuNPs on plants were also reported: in tobacco, small (3.5-nm) AuNPs caused leaf necrosis, while the impacts of large (18-nm) nanoparticles were no different from the control (Alkilany et al. 2010). In onion, AuNP toxicity was manifested as an increase in the mitotic index. In barley, wormwood, and rice, the root and shoot length decreased slightly. In soybean and pumpkin, AuNPs did not affect the plants' morphological and functional characteristics (Chhipa 2017).

8 Interaction of Nanoparticles in Plant Cells and Tissue Culture

The effect of AuNPs on plant growth endocytosis is the way for nanoparticles to enter plant cells (Onelli et al. 2008). Other types of nanoparticles, such as gold nanostars (Su et al. 2010), paramagnetic nanoparticles (González-Melendi et al. 2008), nanoparticles of silicon oxide (Torney et al. 2007), magnesium oxide (Wang et al. 2013), and carbon nanotubes (Khodakovskaya et al. 2009), presumably enter plant tissues through endocytosis. Several studies (Koemel et al. 2013) have pointed out that AuNPs are never found in the aerial parts of radish, pumpkin, barley, poplar, and wheat, unlike what is observed in tobacco, tomato, alfalfa, ryegrass, maize, bamboo, and rice. The efficacy of tissue penetration of AuNPs depends not only on the plant species used but also on the particles' size and surface charge. Some part in this process is played by the plant vascular system, as well as by plasmodesmata (Koemel et al. 2013). When AuNPs are sprayed on seedlings of watermelon, they enter leaves via stomata and are translocated from leaves to roots by the phloem transport mechanism (Raliya et al. 2016).

Small nanoparticles penetrate the aerial parts better than large ones; also they are more toxic. With AgNPs, this fact could be explained by the better solubility of small particles and by the toxicity of the metal ions (Ivask et al. 2014). AuNPs were found not only in tobacco leaves but also in the tissues of tobacco hornworm (*Manduca sexta*), which feeds on tobacco leaves. Using an artificial aquatic ecosystem, Ferry et al. (2009) showed that gold nanorods penetrated the tissues of molluscs, shrimp, and fish better than they penetrated the tissues of the water plant *Spartina alterniflora* Loisel. Seedlings of spring barley grown hydroponically for two weeks with 1–10 µg^{-1} ml of 10-nm AuNPs accumulated the nanoparticles both in leaves and in roots. Factors that determine the intracellular penetration of nanoparticles is their chemical nature, size, shape, surface charge, and dose to detect metals in organs, localize, and identify nanoparticles at the cellular and subcellular levels, and assess cytotoxicity. Having both positive and negative charge, NPs make suspension of cell culture that helps in studying the effect of nanoparticle on the plant. There have been different factors involved in influencing NPs and plants, which are shown in Fig. 3 (Burman and Kumar 2018).

9 Conclusion

The existence of innovation always comes with the pros and cons of our ecosystem. Nanotechnology in sustainable agriculture leads to synthesize nanomaterials with a vast range of its applicability to human and environmental health. On behalf of the precautionary principle for keeping the environmental health safety by reducing contamination through toxic pesticide aspects replace with the environmental release of nanoparticles. Although nanopesticides may also create new kinds of contamination in soils and waterways due to enhanced transport, longer persistence, and higher toxicity than the conventional chemicals utilization way outs, that they replace.

- The recent development in the synthesis and characterization of engineered nanoparticle show the significant transformation from hazardous, environmental polluting aspects that lead toward green chemistry.
- Nanosciences include the utility of nanofertilizers and nanopesticides and influence environment friendly synthesis of the metal nanoparticle.
- Metal nanoparticle has drawn much attention because of controlled properties of shape, size, and dispersity which is imperative. Such metal nanoparticle is widely employed as a catalyst for enhancing the reaction kinetics at its nanoscale use.

References

Ali Z, Zhang C, Zhu J, Jin G, Wang Z, Wu Y, Khan MA, Dai J, Tang Y (2018) The role of nanotechnology in food safety: Current status and future perspective. J Nanosci Nanotechnol 18(12):7983–8002

Alkilany AM, Murphy CJ (2010) Toxicity and cellular uptake of gold nanoparticles: what we have learned so far? J Nanopart Res 12 (7):2313–2333

Arora S, Sharma P, Kumar S, Nayan R, Khanna PK, Zaidi MGH (2012) Gold-nanoparticle induced enhancement in growth and seed yield of *Brassica juncea*. Plant Growth Reg 66(3):303–310

Been BO (1995) Integrated pest management for the control of lethal yellowing: quarantine, cultural practices and optimal use of hybrids. In *Lethal yellowing: research and practical aspects* (pp. 101–109). Springer, Dordrecht.

Bernhardt ES, Colman BP, Hochella MF, Cardinale BJ, Nisbet RM, Richardson CJ, Yin L (2010) An ecological perspective on nanomaterial impacts in the environment. J Environ Qual 39 (6):1954–1965

Brewer MJ, Goodell PB (2012) Approaches and incentives to implement integrated pest management that addresses regional and environmental issues. Ann Rev Entomology 57:41–59

Burman U, Kumar P (2018) Plant response to engineered nanoparticles. In *Nanomaterials in Plants, Algae, and Microorganisms* (pp. 103–118). Academic Press.

Chhipa H (2017) Nanofertilizers and nanopesticides for agriculture. Environ Chem Lett 15(1):15–22

Chinnamuthu CR, Boopathi PM (2009) Nanotechnology and Agroecosystem. Madras Agric J 96(1/6):17–31

Dimkpa CO (2014) Can nanotechnology deliver the promised benefits without negatively impacting soil microbial life? J Basic Microb 54 (9):889–904

Falco WF, Botero ER, Falcão EA, Santiago EF, Bagnato VS, Caires ARL (2011) In vivo observation of chlorophyll fluorescence quenching induced by gold nanoparticles. J Photochem Photobiol a: Chemistry 225(1):65–71

Fazal H, Abbasi BH, Ahmad N, Ali M (2016) Elicitation of medicinally important antioxidant secondary metabolites with silver and gold nanoparticles in callus cultures of *Prunella vulgaris* L. Appl Biochem Biotechnol 180(6):1076–1092

FerryJL CP, Hexel C, Sisco P, Frey R, Pennington PL, Murphy CJ (2009) Transfer of gold nanoparticles from the water column to the estuarine food web. Nat Nanotechnol 4(7):441

Giraldo JP, Wu H, Newkirk GM, Kruss S (2019) Nanobiotechnology approaches for engineering smart plant sensors. Nat Nanotech 14 (6):541–553

Glenn JB, White SA, Klaine SJ (2012) Interactions of gold nanoparticles with freshwater aquatic macrophytes are size and species dependent. Environ Toxic Chem 31(1):194–201

González-Melendi P, Fernández-Pacheco R, Coronado MJ, Corredor E, Testillano PS, Risueño MC, Pérez-de-Luque A (2008) Nanoparticles as smart treatment-delivery systems in plants: assessment of different techniques of microscopy for their visualization in plant tissues. Ann Botany 101(1):187–195

Gunjan B, Zaidi MGH (2014) Impact of gold nanoparticles on physiological and biochemical characteristics of *Brassica juncea*. J Plant Biochem Physiol 2:133

Hosseini SJF, Dehyouri S, Mirdamadi SM (2010) The perception of agricultural researchers about the role of nanotechnology in achieving food security. African J Biotechnol 9(37):6152–6157

Ismail M, Prasad R, Ibrahim AI, Ahmed AI (2017) Modern prospects of nanotechnology in plant pathology. In *Nanotechnology* (pp. 305–317). Springer, Singapore.

Ivask A, Kurvet I, Kasemets K, Blinova I, Aruoja V, Suppi S, Visnapuu M (2014) Size-dependent toxicity of silver nanoparticles to bacteria, yeast, algae, crustaceans and mammalian cells in vitro. PLoS ONE 9(7):e102108

Jo YK, Kim BH, Jung G (2009) Antifungal activity of silver ions and nanoparticles on phytopathogenic fungi. Plant Dis 93(10):1037–1043

Fang Y, Ramasamy RP (2015) Current and prospective methods for plant disease detection. Biosensors 5(3):537–561

Kah M, Hofmann T (2014) Nanopesticide research: current trends and future priorities. Environ Int 63:224–235

Khodakovskaya M, Dervishi E, Mahmood M, Xu Y, Li Z, Watanabe F, Biris AS (2009) Carbon nanotubes are able to penetrate plant seed coat and dramatically affect seed germination and plant growth. ACS Nano 3(10):3221–3227

Kim DH, Gopal J, Sivanesan I (2017) Nanomaterials in plant tissue culture: the disclosed and undisclosed. RSC Adv 7(58):36492–36505

Koelmel J, Leland T, Wang H, Amarasiriwardena D, Xing B (2013) Investigation of gold nanoparticles uptake and their tissue level distribution in rice plants by laser ablation-inductively coupled-mass spectrometry. Environ Poll 174:222–228

Kokina I, Gerbreders V, Sledevskis E, Bulanovs A (2013) Penetration of nanoparticles in flax (*Linum usitatissimum* L.) calli and regenerants. J biotechnol 165(2):127–132.

Kumar S, Nehra M, Dilbaghi N, Marrazza G, Hassan AA, Kim KH (2019) Nano-based smart pesticide formulations: Emerging opportunities for agriculture. J Controlled Rel 294:131–153

Kumar V, Kumar V, Som S, Neethling JH, Olivier E, Ntwaeaborwa OM, Swart HC (2014) The role of surface and deep-level defects on the emission of tin oxide quantum dots. Nanotechnology 25(13):135701

Li H, Ye X, Guo X, Geng Z, Wang G (2016) Effects of surface ligands on the uptake and transport of gold nanoparticles in rice and tomato. J Hazard Mat 314:188–196

Mahakham W, Theerakulpisut P, Maensiri S, Phumying S, Sarmah AK (2016) Environmentally benign synthesis of phytochemicals-capped gold nanoparticles as nanopriming agent for promoting maize seed germination. Sci Total Environ 573:1089–1102

Mahlein AK (2016) Plant disease detection by imaging sensors–parallels and specific demands for precision agriculture and plant phenotyping. Plant Dis 100(2):241–251

Martin-Ortigosa S, Peterson DJ, Valenstein JS, Lin VSY, Trewyn BG, Lyznik LA, Wang K (2014) Mesoporous silica nanoparticle-mediated intracellular Cre protein delivery for maize genome editing via loxP site excision. Plant Physiol 164(2):537–547

Notter DA, Mitrano DM, Nowack B (2014) Are nanosized or dissolved metals more toxic in the environment? a Meta-Analysis. Environ Toxic Chem 33(12):2733–2739

Onelli E, Prescianotto-Baschong C, Caccianiga M, Moscatelli A (2008) Clathrin-dependent and independent endocytic pathways in tobacco protoplasts revealed by labelling with charged nanogold. J Exp Botany 59(11):3051–3068

Parveen A, Mazhari BBZ, Rao S (2016) Impact of bio-nanogold on seed germination and seedling growth in *Pennisetum glaucum*. Enz Microb Technol 95:107–111

Poddar K, Vijayan J, Ray S, Adak T (2018) Nanotechnology for sustainable agriculture. In *Biotechnology for Sustainable Agriculture* (pp. 281–303). Woodhead Publishing.

Ragaei M, Sabry AKH (2014) Nanotechnology for insect pest control. Int J Cci, Environ Technol 3(2):528–545

Rains DW (1989) 10 Plant tissue and protoplast culture: applications to stress physiology and. Plants Under Stress: Biochemistry, Physiology and Ecology and Their Application to Plant Improvement 39:181

Raliya R, Franke C, Chavalmane S, Nair R, Reed N, Biswas P (2016) Quantitative understanding of nanoparticle uptake in watermelon plants. Front Plant Sci 7:1288

Rodrigues SM, Demokritou P, Dokoozlian N, Hendren CO, Karn B, Mauter MS, Sadik OA, Safarpour M, Unrine JM, Viers J, Welle P (2017) Nanotechnology for sustainable food production: promising opportunities and scientific challenges. Environ Sci: Nano 4 (4):767–781

Sabo-Attwood T, Unrine JM, Stone JW, Murphy CJ, Ghoshroy S, Blom D, Newman LA (2012) Uptake, distribution and toxicity of gold nanoparticles in tobacco (*Nicotiana xanthi*) seedlings. Nanotoxicology 6(4):353–360

Sankaran S, Mishra A, Ehsani R, Davis C (2010) A review of advanced techniques for detecting plant diseases. Computers Electr Agric 72 (1):1–13

Sanzari I, Leone A, Ambrosone A (2019) Nanotechnology in plant science: to make a long story short. Front Bioeng Biotechnol 7:120

Selivanov NY, Selivanova OG, Sokolov OI, Sokolova MK, Sokolov AO, Bogatyrev VA, Dykman LA (2017) Effect of gold and silver nanoparticles on the growth of the *Arabidopsis thaliana* cell suspension culture. Nanotechnol Russia 12(1–2):116–124

Servin A, Elmer W, Mukherjee A, De la Torre-Roche R, Hamdi H, White JC, Dimkpa C (2015) A review of the use of engineered nanomaterials to suppress plant disease and enhance crop yield. J Nanoparticle Res 17(2):92

Shelton AM, Badenes-Perez FR (2006) Concepts and applications of trap cropping in pest management. Ann Rev Entomol 51:285–308

Shrader-Frechette K (2007) Nanotoxicology and ethical conditions for informed consent. Nanoethics 1(1):47–56

Shrivastava M, Srivastav A, Gandhi S, Rao S, Roychoudhury A, Kumar A, Singh SD (2019) Monitoring of engineered nanoparticles in soil-plant system: A review. Environ Nanotechnol, Mon Manag 11:100218

Singh MV (2008) Micronutrient deficiencies in crops and soils in India. In *Micronutrient deficiencies in global crop production* (pp. 93–125). Springer, Dordrecht.

Singh R, Srivastava P, Singh P, Upadhyay S, Raghubanshi AS (2019) Human overpopulation and food security: challenges for the agriculture sustainability. In *Urban agriculture and food systems: breakthroughs in research and practice* (pp. 439–467). IGI Global.

Su YH, Tu SL, Tseng SW, Chang YC, Chang SH, Zhang WM (2010) Influence of surface plasmon resonance on the emission intermittency of photoluminescence from gold nano-sea-urchins. Nanoscale 2(12):2639–2646

Taylor A (2011) *Gold uptake and tolerance in* Arabidopsis .Doctoral dissertation, University of York.

Taylor AF, Rylott EL, Anderson CW, Bruce NC (2014) Investigating the toxicity, uptake, nanoparticle formation and genetic response of plants to gold. *PLoS One 9*(4): : e93793.

Thwala M, Klaine SJ, Musee N (2016) Interactions of metal-based engineered nanoparticles with aquatic higher plants: A review of the state of current knowledge. Environ Toxicol Chem 35(7):1677–1694

Tilman D, Cassman KG, Matson PA, Naylor R, Polasky S (2002) Agricultural sustainability and intensive production practices. Nature 418(6898):671–677

Torney F, Trewyn BG, Lin VSY, Wang K (2007) Mesoporous silica nanoparticles deliver DNA and chemicals into plants. Nature Nanotechnol 2(5):295–300

Wang WN, Tarafdar JC, Biswas P (2013) Nanoparticle synthesis and delivery by an aerosol route for watermelon plant foliar uptake. J Nanoparticle Res 15(1):1417

Engineered Nanoparticles and Plant Interaction

Nanotechnology for Sustainable Crop Production: Recent Development and Strategies

Abhishek Kumar, Shilpi Nagar, and Shalini Anand

Abstract

The conventional agricultural farming system has adversely affected the natural ecosystem with the heavy use of fertilizers, pesticides and contaminated water irrigation. Although the conventional agricultural system plays a significant role in the feeding of world population, it has also damaged our pristine ecosystem, simultaneously. In order to solve this problem, nanotechnology has gained a lot of popularity in last few decades. This could be because of the fact that the traditional farming techniques are neither able to substantially enhance the crop production nor are sustainable in the long term. The intervention of nanotechnology in the agricultural system has not only improved the crop yield but also restored and improved the quality of this ecosystem. Moreover, nanotechnology-based products like nanofertilizers, nanopesticides, nanoweedicides and nanosensors have improved the crop yield and income of the farmers. They have helped in boosting seed germination, photosynthesis and nutrient levels in soils. Additionally, they have aided in identifying pest attack and disease prevalence. Simultaneously, they have remediated polluted lands and filtered polluted waters. Further, these products have enabled plants to face climate changing scenarios.

Keywords

Climate change • Nanofertilizers • Nanoherbicides • Nanopesticides • Nano-robots • Nanosensors • Nanoweedicides

1 Introduction

Human population has been increasing leaps and bounds since the twenty-first century. With a billion people to feed at the beginning of nineteenth century, exponential population growth had increased this count to 7 billion by 2011. It is expected that the 9 billion mark would be reached by 2045 (Van Bavel 2013). Food and nutrition demand would increase by 70% from its current levels, throughout the planet (Chen and Yada 2011). With such a large number of empty stomachs to feed and meet the nutritional demand, pressure on the natural resources like soil and water would enhance exceedingly, to grow more and more crops. The limited natural resources are currently being exploited at unexpected rates and for the world to sustain in the longer run, natural resources need to be conserved and preserved (Cropper and Griffiths 1994). In such a scenario, growing crops by the utilization of minimal resources become very critical.

Agriculture has been the most crucial sector since the dawn of *Homo sapiens*. It produces crops to feed human hunger, apart from providing raw materials to various industries. However, the current situation of agriculture is not encouraging with prevalence of subsistence farming being practised on a large scale, uncontrolled and improper application of agrochemicals, low productivity and ever-degrading soil quality. The disappointing scenario is worsened by high poverty levels, illiteracy and negligence towards environmental degradation (Szargut et al. 2002). It is the absence of financial resources that pressurize the poor to go for subsistence farming. Lack of awareness has led to

A. Kumar (✉)
Department of Civil & Environmental Engineering, Birla Institute of Technology, Mesra, Ranchi, Jharkhand, 835215, India
e-mail: abhishekkumar1205@gmail.com

S. Nagar (✉)
Department of Environmental Studies, University of Delhi, New Delhi, Delhi 110007, India
e-mail: shilpinagar7507@gmail.com

S. Nagar · S. Anand
Center for Fire, Explosive and Environment Safety, DRDO, New Delhi, Delhi 110054, India

© Springer Nature Switzerland AG 2021
P. Singh et al. (eds.), *Plant-Microbes-Engineered Nano-particles (PM-ENPs) Nexus in Agro-Ecosystems*, Advances in Science, Technology & Innovation, https://doi.org/10.1007/978-3-030-66956-0_3

overexploitation of natural resources. Additionally, the lust for economic growth neglecting the environmental degradation has intensified global sufferings. Therefore, it is critical to incorporate sustainable approach in agriculture sector, so that there is holistic development and not just economic growth (Prasad et al. 2017). Sustainable approach refers to survival of the present generation, keeping in mind the needs and existability of future generations (Lélé 1991). With existential threats of changing climate, food security, depleting non-renewable energy resources and urban sprawl, sustainable agriculture is meant to be our sole saviour. Search for alternatives to strengthen sustainable agriculture has become inevitable. Several innovations and advancements have been made to address food security and sustainable production challenges (Shang et al. 2019). In such conditions, nanotechnology has grown as an encouraging and promising technology to provide efficient solutions to agricultural issues (Dwivedi et al. 2016).

The term nanotechnology was given by Norio Tanaguchi in 1974, referring to the manipulation of matter at level of nanoscale (Prasad et al. 2017). Research has grown exponentially since 1980s, and nanotech advancements have penetrated everywhere, all across the globe (Ahmed et al. 2013; Kah and Hofmann 2014; Servin et al. 2015; Nuruzzaman et al. 2016). A nanometre is one-billionth on a metre scale. The physico-chemical properties of the material change at such a small scale. The alterations at atomic and molecular level, such as surface area enhancement and magnetic power development, contribute in changing the properties at nanoscale and bring about differences in reactivity of atoms (Pokropivny et al. 2007; Sun 2007; Aziz et al. 2015; Prasad et al. 2017). Interestingly, nanotechnology has boomed in the recent past to promote environmentally safe practices for sustainable development (Prasad 2014; Ram et al. 2014).

Green revolution brought about a drastic enhancement in crop yield (Conway and Barbie 1988; Nin-Pratt 2016). Humongous amounts of groundwater were pumped out, more than they could be replenished by rainwater (Gleick 1993; Postel et al. 1996; Presley et al. 2004; Rodell et al. 2009). Such irrigation patterns have been in practice for very long time and have damaged the soil quality by salt accumulation and accelerated weathering of minerals (Österholm and Åström 2004; Mukhopadhyay 2005). Similarly, boosting crop production is nearly impossible today without the use of agrochemicals. The term 'agrochemical' is inclusive of fertilizers, pesticides, insecticides and weedicides. Addition of fertilizers to soil is essential to increase the soil fertility. Sadly, the excessive application of fertilizers has damaged the nutrient composition of soil and given rise to the problem of eutrophication. Further, the damage caused to nutritional composition of soil has resulted in reduction in size of arable land. Additionally, the use of insecticides,

pesticides and weedicides has affected the biotic components of soil and enhanced resistance among pests like insects and pathogens. Correspondingly, health of animals and human beings is put to risk and biodiversity loss could be a major resulting outcome (Prasad et al. 2017; Shang et al. 2019). Incorporation of sustainable agriculture would mean minimal use of these agrochemicals. The cumulative impacts of reckless groundwater exploitation and agrochemical application have eventually led to a rise in abandoned arable lands. The extent of damage, unleashed by these practices, could still be felt in many parts across the globe (such as Latin America and India).

Nanomaterials (NMs) have been reported to facilitate the agricultural input requirements of the soil. They do so by targeted delivery of the nutrients to plants and plant protection from various diseases. Additionally, with persistent problems of global warming and changing climate, it is important that plants adapt to such consistent changes (Vermeulen et al. 2012). Sensors could be developed to monitor the soil conditions, prevalence of diseases, plant health and their growth (Shrivastava and Dash 2009; Giraldo et al. 2014; Chen et al. 2016). Apart from provisioning of agrochemicals, NMs like zeolites and nanotubes have been reported to retain water, which could help in enhancement of crop production (Navrotsky 2000; Manjaiah et al. 2018; Tripathi et al. 2018). Nanomaterial engineering, therefore, has the potential to provide cutting edge technology to boost crop production, eliminate the harms associated with modern agricultural practices, decrease the anthropogenic footprint on environment and enrich the food quality (Singh Sekhon 2014; Liu and Lal 2015; Panpatte et al. 2016; He et al. 2019). Nanotechnology has various agricultural applications which have been discussed in the upcoming sections.

2 Detection and Control of the Plant Diseases

Nanomaterials would help in monitoring the diseases prevalent in food crops. They could target the pathogens, thereby treating the diseases caused because of them (Philip 2011). A large number of nanoparticles (NPs) have been used, mainly focusing on nano-forms of silver, gold, carbon and alumina to control the spreading of diseases (Jo et al. 2009; Sharma et al. 2012). Nano-Ag has been employed on a wide scale by the researchers (Kim et al. 2012; Prasad and Swamy 2013). It has been reported to remove unwanted microbes from the soil, thereby restricting the occurrence of diseases (Bhattacharyya et al. 2010; Singh et al. 2015a, b). They do so by altering the biochemical processes of the microorganisms and preventing the ATP production in them, consequently killing them and preventing plants from getting affected (Yamanaka et al. 2005; Pal et al. 2007).

In conjugation with silver NPs, ZnO and CuO NPs were used for suppressing soil-borne diseases in *Prunus domestica* (Malandrakis et al. 2019). Al_2O_3 NPs were used to control root rot in *Solanum lycopersicum* (Shenashen et al. 2017). Silver NPs were used to protect *Vigna unguiculata* from disease attack (Vanti et al. 2019). CuO NPs were used to protect *Solanum lycopersicum* from late blight disease caused because of *Phytophthora infestans* (Giannousi et al. 2013). MgO NPs were applied to *Solanum lycopersicum* to suppress pathogens like *Ralstonia solanacearum* (Imada et al. 2016). Interestingly, NMs like ZnO and MgO possess antimicrobial properties, smooth and optically transparent and are easily dispensable, which increases their demand for agricultural purposes and preservatives (Aruoja et al. 2009; Sharma et al. 2009). Further, deoxyribonucleic acid (DNA) and chemicals could be easily delivered to the plant cells by silica NPs, modifying the genetic composition of the cells to initiate defence mechanism against any pathogen attack (Torney et al. 2007).

Nanomaterials have been used as nanosensors to measure and monitor disease prevalence in crops. Nanosensors could provide timely detection of pest attack by early identification of symptoms to protect crops from diseases and increase crop yield. Wireless sensors were developed to detect insect attack (Afsharinejad et al. 2016). Nano-Au-based sensors have been reported to be efficient in detection of a fungal disease named Karnal bunt in *Triticum aestivum* (Singh et al. 2010). Interestingly, protection of crops from diseases might help in decreasing the utilization of agrochemicals and enhancing crop yield, thereby boosting the national economy (González-Fernández et al. 2010; Rai and Ingle 2012).

3 Seed Germination and Plant Growth

Germination of a seed could be considered the most critical and sensitive stage in a plant life. It enables seedling growth, the development of which establishes a plant. Seed germination could be altered by various factors including soil fertility, moisture content, genetics and environmental factors (Manjaiah et al. 2018). NMs have been reported to facilitate seed germination, thereby promoting plant growth. The role of NMs in seed germination is not explained very well so far. It has been reported that NMs enable seed coat penetration, activate enzymes and enhance water absorption and usage, which result in improved seed germination and seedling growth (Changmei et al. 2002; Khodakovskaya et al. 2012a; Banerjee and Kole 2016). Additionally, water retention uplifts root growth (Shojaei et al. 2018). However, the mechanism behind water uptake is unclear.

Carbon nanotubes (CNTs) have a beneficial effect on germination of seeds in a number of plants such as *Triticum aestivum* (wheat), *Lycopersicon esculentum* (tomato), *Zea mays* (maize), *Arachis hypogaea* (peanut), *Allium sativum* (garlic), *Glycine max* (soybean) and *Hordeum vulgare* (barley) (Khodakovskaya et al. 2012a; Lahiani et al. 2013; Joshi et al. 2018). Low concentrations of multiwalled CNTs have been reported in augmenting growth by 60% in tobacco plants (Bheemidi 2011; Khodakovskaya et al. 2012b; Suresh et al. 2013; Gottschalk et al. 2015). Zeolite, silicon oxide and titanium oxide NMs have been reported to facilitate seed germination in plants (Changmei et al. 2002; Manjaiah et al. 2018). Iron/silicon oxide NMs were reported to promote seed germination in *Zea mays* and *Hordeum vulgare* (Najafi Disfani et al. 2017). FeS_2 has been noted to enhance germination in chick pea, spinach, mustard and sesame (Srivastava et al. 2014; Das et al. 2016). Fullerenes have been stated to stimulate cell division, thereby increasing the hypocotyl growth in *Arabidopsis* (Gao et al. 2011). Fullerols escalate fruit quality and quantity, double the crop yield and revitalize bioactive components like lycopene as observed in *Momordica charantia* (Kole et al. 2013). Kaolin NPs have been stated to enable seed growth and strengthen roots (Gogos et al. 2012).

Zinc and boron as NMs have shown to enrich fruit quality and quantity, without altering fruit properties (Davarpanah et al. 2016). Hydroxyapatite NM-coated fertilizers facilitate slow release of nutrients for the crops to consume in the longer run (Lateef et al. 2016; Madusanka et al. 2017). NMs have been reported to boost photosynthesis levels. A 2.5% application of nano-titanium oxide increased the photosynthetic activity (Zheng et al. 2005). Nano-iron/silicon oxide boosted shoot length in *Hordeum vulgare* and *Zea mays* seedlings when the application rate was 15 mg/kg (Najafi Disfani et al. 2017). Interestingly, NM application rate is very important with regards to increasing crop production. Application rate of 25 mg/kg had negative impact on *Hordeum vulgare* and *Zea mays* seedlings (Najafi Disfani et al. 2017). Further, mode of NM application is also critical for crop productivity enhancement. For example, foliar application of nano-magnetite is preferable to soil application to boost overall plant growth in *Ocimum basilicum* (Elfeky et al. 2013).

4 Photosynthetic Upgradation

Research has been focussed upon catalyzing photosynthetic upgradation by increasing RuBisCO efficiency, engineering C_3 plants for manoeuvring C_4 pathway, bringing changes in chlorophyll efficacy and enhancing photosynthetic waveband (Hibberd et al. 1996; Amthor 2001; Evans 2013). NMs have been reported to favour photosynthesis. Integration of plants and NMs is referred to as plant nanobionics. TiO_2 NPs were reported to stimulate photocatalytic activity by improving the light absorbed by the leaves. Further, TiO_2

NPs aid in delaying ageing of chloroplasts caused because of the photochemical stress (Hong et al. 2005a; b). TiO_2 NPs initiate RuBisCO carboxylation and favour electron transport chain, thereby pumping photosynthesis (Gao et al. 2006, 2008; Linglan et al. 2008; Qi et al. 2013). Additionally, TiO_2 NPs promote transpiration rate (Lei et al. 2007). Therefore, plant nanobionics could be used to enhance crop growth and crop production (Giraldo et al. 2014).

5 Nano-Agrochemicals and Nanobionics

5.1 Nanofertilizers (NFs)

It is extremely crucial to add fertilizers to soil to obtain a higher yield (Barker and Pilbeam 2015). However, the use of chemicals has a potential to damage the soil health and the environment. Unfortunately, the efficiency of fertilizer utilization by crops is around 35–40% (Dijk van and Meijerink 2014). Nanofertilizers (NFs) could be used to increase nutritional status of soil without damaging the environment (Naderi and Abedi 2012) and could be used to replace use of conventional fertilizers (Naderi and Danesh-Shahraki 2011; Batsmanova et al. 2013). NFs are capable of escalating the nutrient facilitation to the seeds and boosting nourishment to seedlings that holistically increases the shoot and root length. NFs increase the nutrient availability for the leaves and branches of crops, resulting in crop production enhancement (Tapan et al. 2010; Stamp and Visser 2012). NMs are absorbed by the pores available on the roots or the stomata in leaves (Eichert and Goldbach 2008). NFs could be taken up by the plants via ion channels and endocytosis (Rico et al. 2011).

Nanofertilizers could provide nutrients to the plants in multiple ways. It could be encapsulated in NMs (like CNTs), coated with polymers or delivered as emulsions (Derosa et al. 2010). NFs could promote a slow release of nutrients and avoid the subsequent loss of valuable nutrients. Additionally, NFs could release the nutrients when crops could directly use them (Derosa et al. 2010). The advantages related to use of NFs have been shown in Fig. 1.

Nanofertilizers would make the nutrients available to plants in the following ways:

- In the form of NPs or emulsions, such as CNTs, fullerenes, SiO_2 and TiO_2 NPs which could directly bring about changes in the nutritional availability to plants (Millán et al. 2008).
- Controlled release of nutrients on stimulation by environmental factors, changes in pH or magnetic/ultrasonic pulses. Types of controlled release include slow release, quick release or specific release (Aouada and De Moura 2015).

- Complexed with organic polymers (Corradini et al. 2010) such as zeolites, chitosan or polyacrylic acid to deliver the nutritional contents to the plants (Ohlsson 1996; Ditta 2012; Servin et al. 2015).

Ammonium charged zeolites enhance phosphate solubility, thereby increasing its plant availability (Dwivedi et al. 2016). Graphene oxide derived NMs extend KNO_3 release, minimizing its losses (Shalaby et al. 2016). Calcite NMs applied with SiO_2, MgO and Fe_2O_3 NPs improve the phosphorus, calcium, magnesium, manganese, iron and zinc uptake (Sabir et al. 2014). Cationic and anionic nutrients could be delivered in the form of NM emulsions (Subramanian et al. 2015). Application of ZnO NPs along with fertilizers was shown to double the barley production (Kale and Gawade 2016). Controlled release of nourishment by NFs reduces nitrogen loss through leaching by 22% and through runoff by 25% and enhances the yield of crops (Liu et al. 2016). Various NFs and their impact on crop yield have been represented in Table 1.

5.2 Nanosensors (NSs)

Growth of crops is dependent upon adequate climatic conditions and protection from insect and pathogen attack. Nanosensors (NSs) would help in collecting the data related to soil, water, plants and climatic conditions, which could aid in boosting the crop growth (Rai et al. 2012; Alfadul et al. 2017). NSs could increase the crop yield with meagre financial requirements (Rai and Ingle 2012). NSs could be delivered as NPs, nanowires or nanocrystals, by incorporating the physico-chemical properties of NMs (Khiari 2017). NSs could help in the early detection of diseases, identification of nutritional deficiency and real-time control of nutrient provisioning and water delivery. Further, NS-based global positioning system could be installed to monitor the crops and agricultural lands (El Beyrouthya 2014; Mariano et al. 2014). The variety of data generated and collected by NSs has been shown in Fig. 2.

A nanosensor is composed of a biological probe, a transducer and a data recorder. The probe interacts with the target producing signals, which is converted by the transducer into digital signals and the data recording unit captures the signals and stores them (Habibi and Vignon 2008; Espinosa et al. 2016). The data is relayed to the internet for its analyses and further application (Dufresne et al. 2000). These NSs could be placed on the aerial parts of plants by spraying them on these leaves (Marchiol 2018). CNTs have the potential to act as precise NSs which could aid in pest control and crop yield enhancement (Alejandro and Rubiales 2009; De La Torre-Roche et al. 2013). Single-walled CNTs have been reported to augment photo-absorption, increasing

Fig. 1 Advantages of using Nanofertilizers

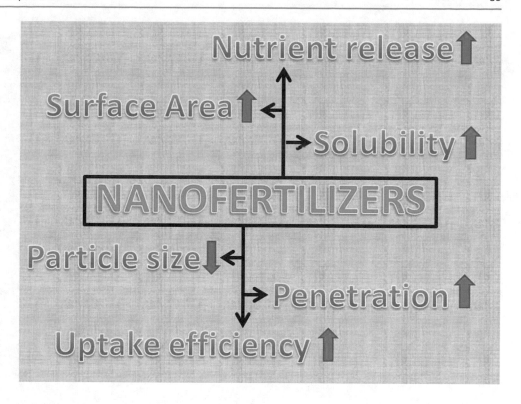

Table 1 Various nanofertilizers used for boosting crop production

Nanomaterial	Used for plants	Effects	References
ZnO	*Arachis hypogaea*	Yield rise by 30%	(Prasad et al. 2012)
Ag	*Zea mays*	23% rise in yield	(Berahmand et al. 2012)
CeO$_2$	*Solanum lycopersicum*	Rise in yield	(Singh et al. 2015a)
Fe$_2$O$_3$	*Glycine max*	48% rise in yield	(Sheykhbaglou et al. 2010)
Zn	*Pennisetum americanum*	38% boost in yield	(Tarafdar et al. 2014)
Hydroxyapatite	*Glycine max*	Growth rise by 33%	(Liu and Lal 2014)
SiO$_2$	*Lycopersicon esculentum*	Growth rise	(Siddiqui and Al-Whaibi 2014)
TiO$_2$	*Triticum aestivum*	Plant growth rise	(Jaberzadeh et al. 2013)

the electron transport, reducing the reactive oxygen species formation in chloroplasts and consequently boosting photosynthesis in the crops (Giraldo et al. 2014). Multi-walled CNTs regulate hormones, like auxin, in the plants, aiding in altering the plants growth (McLamore et al. 2010). A wireless NS was developed to identify the insect attack by differentiating volatiles released by the plant–insect interaction (Afsharinejad et al. 2016). Nano-gold has been used in NSs to identify karnal bunt disease in *Triticum aestivum* (Singh et al. 2010). Potassium niobate-based NSs have been stated to possess moisture detection capability, thereby could assist productivity enhancement (Ganeshkumar et al. 2016). Moreover, NSs could help quantify pollution levels in the soils (Shang et al. 2019).

5.3 Nano-robots

Nano-robots could be used in combination with wireless communication systems to scan plants under consideration. The capillaries present inside the plants could be explored thoroughly for the identification of deficient water and nutrient levels. Additionally, the agricultural soils could be monitored by the nano-robots for the determination of nutritional status and prevalent moisture contents (Pandey 2018). Further, the quality of irrigational water provisioned to the soils could be monitored using the nano-robots (Cavalcanti et al. 2003). The details generated by the nano-robots could be furnished through communicational channels to remote computer centres for further analysis.

Fig. 2 Variety of data generated by nanosensors

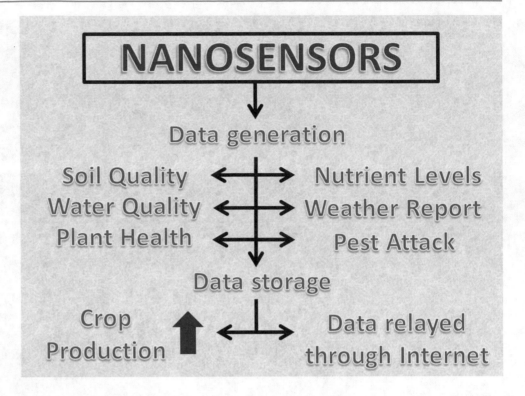

These analyses could turn out to help in enhancement of crop production and crop yield (Lindblade et al. 1999; Ghormade et al. 2011; Parisi et al. 2015). Further, analysis of soil and water could help in the determination of contamination status which could help in prevention and minimization of pollution levels. This could prove essential in decreasing food and crop contamination and supply of pollutant-free crops to the hungry human population (Cavalcanti et al. 2003). Interestingly, nano-robots could also be used for measurement of physico-chemical properties of soil like temperature, salinity levels and pH. Changes in environmental conditions could be picked up by the nano-robots to identify alterations in weather patterns which may help in preventing crop damage. A radio frequency identifier (RFID) tag sensor or complementary metal-oxide semiconductor (CMOS) sensors could be used for transmission of data generated by the nano-robots (Cavalcanti et al. 2003; Collins 2006).

5.4 Nanopesticides (NPCs)

Agrochemicals, like pesticides and weedicides, have been utilized for pest and weed control so that crop productivity is enhanced. Less than 10% of the pesticides reach their targets and help in pest control (Nuruzzaman et al. 2016). Unfortunately, they deplete the health of the soil system, damage the food chain and give birth to agrochemical resistant super-pests. NPCs have been used as an alternative for the conventional pesticides to overcome the related issues (Sasson et al. 2007). The NPCs are pest specific and do not damage essential biota of the soil (Kah et al. 2013; Kah and Hofmann 2014). Additionally, their stiff and crystalline shapes are more stable, soluble, permeable and biodegradable in comparison with the conventional pesticides (Ul Haq and Ijaz 2019). Different materials have been used as NPCs like polymers, surfactants and inorganic NPs (Alfadul et al. 2017). Properties of NPCs and its effect on pests have been shown in Fig. 3.

Technologies like nanoencapsulation and nanoformulation have been used to design a controlled release of pesticides to avoid leaching and the resulting losses without compromising with the efficiency (Scrinis and Lyons 2007). Genetic material could also be delivered to plants to defend them against pest attack (Torney et al. 2007; Torney 2009). Nanoencapsulation refers to coating pesticides with NMs, while nanoformulation describes NMs exploited as active components of pesticides (Nuruzzaman et al. 2016; Ul Haq and Ijaz 2019). The nanocapsules deliver the pesticides via dissolution, diffusion, degradation or osmosis at defined pH (Ding and Shah 2009; Vidhyalakshmi et al. 2009). Polymeric nanocapsules were combined with pyrethroid bifenthrin in the nanoformulation and were reported to be vehicles of pesticide delivery by greater dispersion and diminished runoffs (Petosa et al. 2017).

Nanopesticides could be delivered in the form of emulsions like gel or creams and liquids. NPC products available in the market include Karate ZEON and 'Gutbuster', which

Fig. 3 Properties of NPCs and its impact on pests

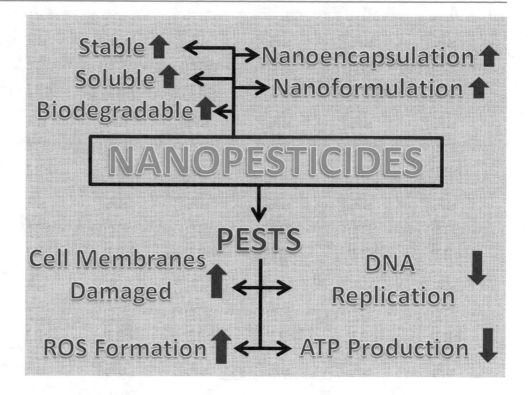

contain lambda-cyhalothrin, that help in pest control (Ram et al. 2014). Porous hollow silica NPs were used to deliver a prolonged release of a pesticide called validamycin (Liu et al. 2006). Oil-in-water emulsions were used as NPCs in a report (Wang et al. 2007). Various NMs have been used as NPCs to tackle a number of pests, which have been given in Table 2. Nano-silica could be used for altering the absorption of cuticular lipids by insects, thereby acting as useful insecticides (Barik et al. 2008). Modified nanosilica was utilized against insects damaging the standing crops (Ulrichs et al. 2006). Polyethylene glycol loaded NPs enriched with garlic essential oil was used as an insecticide against *Tribolium castaneum* with 80% efficacy (Yang et al. 2009). Nano-alumina was used against *Sitophilus oryzae* and *Rhyzopertha dominica* insects with results of very high mortality rate (Stadler et al. 2010). Clay nanotubes like halloysites have been used as a cost-effective delivery agent of pesticides prolonging their release into the soil (Dwivedi et al. 2016). Similarly, nanofiber formulation of insect pheromone and pesticides was used to attract and destroy pests like *Grapholita molesta* (Czarnobai De Jorge et al. 2017). NPCs like silica NMs could penetrate the plants and mix with cell sap to exert systemic effects on insects, and the nanoformulation was resistant to photodegradation as well (Li et al. 2007). Ferbam formulated with gold NPs enhance leaf penetration capability and changes non-systemic property of conventional pesticides (Hou et al. 2016). Latex fabricated gold NPs have been reported to interact with proteins like

trypsin, thereby decreasing their activity in insects, consequently killing them (Patil et al. 2016).

5.5 Nanoherbicides/Nanoweedicides (NHs or NWs)

Weeds survive and proliferate by their deeply seated roots. They are deleterious for the crops because of their property to strongly compete against plants for nourishment, sunlight and water (Shang et al. 2019). They could be removed by ploughing, but could create trouble in uninfected areas. An easy way to remove weeds would be to eliminate their germination by destruction of their seeds (Ram et al. 2014). Conventional herbicides destroy the weeds but they are not helpful in preventing their regrowth because of their inactivity against deeply seated root systems (Shang et al. 2019). The NHs, being miniscule in size, could blend with the soil and eliminate spread of conventional herbicide resistant weeds without causing toxicity or boosting resistance. Interestingly, NHs are helpful in preventing the regrowth of weeds and harmful herbs (Dwivedi et al. 2016). Removing the weeds from the agricultural fields would help in improving the crop yield and reducing the manual labour required for weed removal.

The functioning of NHs is similar to NPCs, i.e. by nanoencapsulation or nanoformulation. Various NMs have been used as NHs to control and diminish weed growth,

Table 2 Nanomaterials used as nanopesticides and nanoweedicides/nanoherbicides

Nanomaterial	Used against	References
Silica	*Spodoptera litura*	(Debnath et al. 2011)
Polyethylene glycol	*Tribolium castaneum*	(Yang et al. 2009)
Alumina	*Sitophilus oryzae*	(Stadler et al. 2010)
Ag	*Erwinia carotovora*	(Al-Askar et al. 2013)
Al_2O_3, TiO_2	*Sitophilus oryzae*	(Sabbour 2012)
CdS, Ag, TiO_2	*Spodoptera litura*	(Chakravarthy 2012)
Ag	*Alternia alternata*	(Al-Askar et al. 2013)
Clay	Pests	(Dwivedi et al. 2016)
Fe	*Meloidogyne incognita*	(Sharma et al. 2017)
Ferbam + Au	Insects	(Hou et al. 2016)
SiO_2	*Caenorhabditis elegans*	(Acosta et al. 2018)
TiO_2, ZnO, Al_2O_3, Ag	*Caenorhabditis elegans*	(Wang et al. 2009)
Ag, SiO_2 and TiO_2	*Meloidogyne incognita*	(Ardakani 2013)
ZnO, MgO, TiO_2, SiO_2, Ag, Cu	Weeds	(Elmer and White 2016)
Cu	*Phytophthora infestans*	(Giannousi et al. 2013)
Cu	*Fusarium graminearum*	(Brunel et al. 2013)
Cu	*Fusarium oxysporum*	(Saharan et al. 2015)
ZnO	*Fusarium graminearum*	(Servin et al. 2015)
CarboxyMethyl Cellulose NPs	Weeds	(Satapanajaru et al. 2008)
SiO_2	Weeds	(Sharifi-Rad et al. 2016)
Ag	*Lemna minor*	(Gubbins et al. 2011)
Ag + chitosan	*Eichhornia crassipes*	(Namasivayam et al. 2014)
CuO	*Lolium perenne*	(Atha et al. 2012)
Metsulfuron methyl + pectin	*Chenopodium*	(Kumar et al. 2017)
Atrazine Nanocapsule	*Amaranthus viridis*	(Sousa et al. 2018)

which have been given in Table 2. NPs like silver, copper, ZnO, MgO, TiO_2 and SiO_2 could help in altering the growth of weeds (Servin et al. 2015; Elmer and White 2016). ZnO NPs are effective to control the spreading of weeds like *Fusarium graminearum*, *Mucor plumbeus* or *Rhizopus stolonifer* (Servin et al. 2015; Vanathi et al. 2016). Cu NPs were reported to successfully remove *Phytophthora infestans* from *Lycopersicon esculentum* plantations (Giannousi et al. 2013). SiO_2 NPs were reported to diminish the germination and growth of weeds by changing photosynthetic activity (Sharifi-Rad et al. 2016). Metsulfuron methyl enriched pectin NPs have been stated to be effective against *Chenopodium* weed growth (Kumar et al. 2017).

6 Water Conservation and Treatment

The importance of water for crops could never be undermined. Provisioning of nutrients is majorly dependent upon water used for irrigation. It is, therefore, essential to treat water before its supply to the crops. Most of the countries across the globe are generally dependent upon seasonal rainwater for growing the crops. However, dependence on irregular rain patterns threatens the crop growth and agricultural productivity (Pramanik and Pramanik 2016). Sustainable water management could be utilized in determining the water level requirements, and detection and prevention of its contamination (Iavicoli et al. 2014; Dasgupta et al. 2015). Various NMs have been used to aid sustainable water management as shown in Fig. 4. Nano-zeolites could be used for enhancing water retention because of it porous property and capability to boost capillary action. It could, exceptionally, be helpful in arid and dry soils and remarkably enhance porous nature of clayey soils. Nutrients could, simultaneously, be made available to the plants. This could help in increasing crop production (Lateef et al. 2016; Manjaiah et al. 2018).

Nanotechnology has been used for a safe and sustainable water treatment and has enabled a clean water supply (Qu et al. 2013). It could be incorporated for desalination and decontamination through nanosorption and nanophotocatalysis. NSs could detect presence of contaminants and the

Fig. 4 Various applications of
nanomaterials in Agriculture

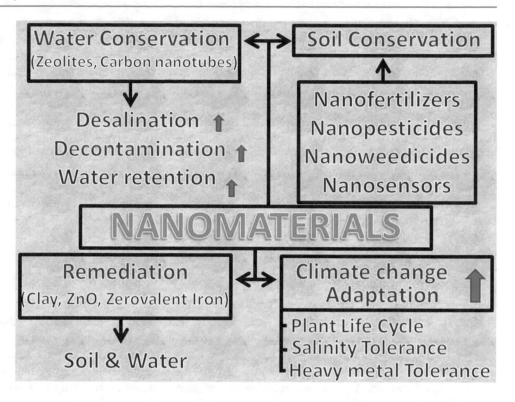

degree in severity of contamination. Technologies, like nanofiltration, have been used for decontamination of polluted water (Bora and Dutta 2014; Kumar et al. 2014). NMs, like CNTs and nano-alumina, have been utilized for nanofiltration intending to purify irrigation water. Nanofiltration could help in removing solids and harmful microorganisms. Interestingly, solar energy has been incorporated with nanofiltration technologies for desalination and facilitation of irrigation water. Further, plants grown using desalinated water, needed less water and agrochemicals (Kumar et al. 2014; Singh Sekhon 2014). NPs could be applied to reduce heavy metals, inclusive of uranium and lead. Fe_3O_4 NPs possess magnetic properties and have been reported to interact with arsenic ions. They could be helpful in the purification of arsenic-rich water (Yavuz et al. 2006). NMs, like graphene, have been used for removing sodium chloride from saline water, with efficiency greater than conventional desalination technologies, like reverse osmosis (Cohen-Tanugi and Grossman 2012; O'Hern et al. 2015).

7 Soil Conservation and Management

Apart from water, soil is another important prerequisite essential for growing crops. Soil health has to be immaculate to support plant growth (Pramanik and Pramanik 2016). A fertile soil makes nutrients available to crops, provides environment for soil biota to exist and assists infiltration of air, water and establishment of roots. Further, the sustenance of animals and human beings is dependent on consumption of food crops (Amundson et al. 2015). The animals and human beings could get exposed to hazardous elements through these crops. Therefore, it becomes important to manage and conserve the soil systems (Medina et al. 2015). As explained in previous sections, NMs and NPs could be used for improving the soil health by its application in form of NFs (Dwivedi et al. 2016; Shang et al. 2019), and removing the pests and weeds from soil systems by applying NPCs and NHs/NWs, as mentioned in Fig. 4 (Rai and Ingle 2012; Bhattacharyya et al. 2016).

8 Contaminant Remediation

Nanomaterials could be used to remove the pollutants from soil and water, as shown in Fig. 4. Clay NMs have been used for removing arsenic from water. Arsenic-rich water is passed through hydrotalcite for its removal. This technique could be incorporated with passing polluted water across filter candles or pervious pots, which is used widely in the developing world (Gillman 2006). ZnO NPs could be used for arsenic removal in the water filtration devices. NMs, like zerovalent iron, are used on a large scale for pollution remediation from the soil and water. Examples of NMs reported to remediate soil and water are zeolite NPs, metallic oxide NPs (like TiO_2 NPs), CNTs and bimetallic noble element NPs. NP-rich filters could be installed for removal of the organic pollutants and agrochemicals, like DDT, from water (Karn et al. 2009).

Interestingly, use of sorbents, like biochar, has increased in the recent past to remove inorganic and organic pollutants from the soil and water. Nanotechnology could be synchronized with biochar application to enhance the efficacy of both (Wang et al. 2017). NMs, like CuO, ZnO and SiO_2, have been doped onto the biochar in various studies. Biochar-based nanocomposites were reported to minimize the pollutants inclusive of heavy metals, inorganic contaminants and organic contaminants, from contaminated soil and water (Wang et al. 2015a; Tan et al. 2016). Removal of contaminants could be achieved by incorporating mechanisms like physisorption, chemisorption, ion exchange, diffusion and others. Additionally, it could also be used for enhancing the nutritional status of the soil by providing macro- and micro-nutrients to the plants. Biochars are rich in organic matter which could help in boosting plant growth. Biochar could also help in enhancement of water retention capacity of the soils in which they are applied (Liu et al. 2019). Therefore, NP-doped biochar would aid in the remediation of contaminants and the enhancement of crop production, simultaneously.

9 Harvesting Nanoparticles

Plants could be used as an agency for production of NPs in a green eco-friendly manner, as shown in Fig. 5. Such NPs could be used in a variety of industries (Mokerov et al. 2001). The plants are grown in the soils impinged with specific NPs. These NPs are sucked up by the plants, and later these NPs are extracted. A few of the NPs synthesized include Au, Ag, Cu, Zn and Pt (Quintanar-Guerrero et al. 1998; Xu et al. 2015). Plants like *Sesbania* and *Medicago sativa* have been used to generate gold NPs. Similarly, *Helianthus annuus* and *Brassica juncea* were utilized for producing an extensive range of NPs including Ag, Ni, Co, Zn and Cu (Iravani 2011). However, the biosynthesis routes are still in their initial stages with prevalent issues of NP stability, size distribution and extraction (Perlatti et al. 2013; Giongo et al. 2016; Wani and Kothari 2018).

Nanomaterials, like nanocellulose, could be produced cheaply from wastes like wheat straw, soy hulls, potato pulp or sugar beet pulp, as shown in Fig. 5 (Sankar et al. 2016). Nanocellulose are bio-based NMs with large surface area, great strength and exceptional optical properties (Anastas and Eghbali 2010; Wanyika et al. 2012). Nanocellulose could be used for a wide variety of uses as nanocomposites (Shen 2017). Rice husks are rich in silica and could be used for production of silica NPs (Muramatsu et al. 2014). Graphene oxide NPs could be produced from agricultural waste

materials (Somanathan et al. 2015). In a nut shell, various agro-related waste materials could be used for producing the NMs (Bruce et al. 2005; Baker et al. 2017).

Interestingly, microorganisms could be used for synthesis of the NMs, as shown in Fig. 5. The microorganisms are inclusive of *Pseudomonas stutzeri*, *Klebsiella aerogenes* and *Clostridium thermoaceticum*. They have been used for synthesizing NMs like Au, Si, ZnS and CdS (Park et al. 2016). Fungi, like *Verticillium*, *Aspergillus* and *Fusarium oxysporum*, have been reported to be efficient producers of NPs (Kitching et al. 2015). The synthesized NPs could include metal or metal sulphide NPs such as Au and ZnS NPs.

10 Adapting to Climate Change

The weather patterns have been changing across the globe. There has been a rise in the prevailing atmospheric temperatures; drastic alterations in rainfall patterns; and frequent occurrences of extreme weather events and hazards. These cumulatively affect agriculture in a negative manner. For example, crops may face drought periods, excessive rainfall phases and extremes of temperature, which detrimentally affects crop production (Anwar et al. 2007). Hazards have the potential to destroy the standing crops and damage the soil health in the longer run. Problems of water shortage could arise. Therefore, it becomes very critical to look for technologies that enhance the adaptability to the existential threats of climate change (Vermeulen et al. 2012). Increasing the adaptation potential in crops requires managing genetic expression during stress, hormonal and enzymatic alterations and decreasing the plant life without compromising with the yield. Technologies have been developed to reduce the adverse impact of climate change on the crop production (Pretty 2008). Nanotechnology could play a humongous role in helping the plants tackle climate change, as given in Fig. 4.

Nanomaterials could increase crop production in adverse environmental conditions. Nano-SiO_2 was stated to boost seed germination and plant growth in *Lycopersicon esculentum* and *Cucurbita maxima*, when grown under salt stress (Haghighi et al. 2012; Siddiqui and Al-Whaibi 2014). Activated carbon-based TiO_2 was employed in appropriate concentrations and was reported to decrease germination time and boost germination of seed in tomato and mung bean (Singh et al. 2016). $FeSO_4$ NPs were demonstrated to uplift the salinity tolerance in *Helianthus annuus* apart from increasing the quality and quantity of plants by enhancing the photosynthetic efficiency, leaf area and CO_2 assimilation (Torabian et al. 2017). Silicon NPs have been reported to reduce the stress caused by UV-B in *Triticum aestivum* (Tripathi et al. 2017). Zeolite NMs boost nutrient facilitation

Fig. 5 Harvesting nanomaterials and nanoparticles

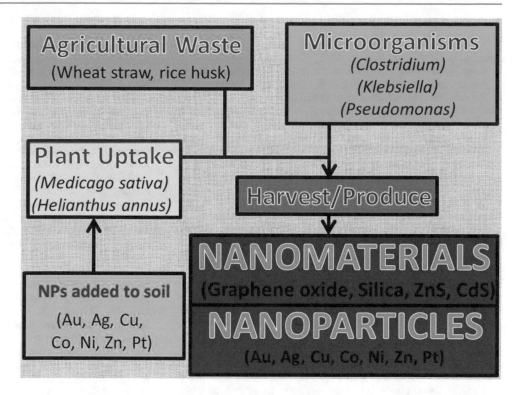

in plants, thereby assisting plant growth (Manjaiah et al. 2018). Application of NFs in *Triticum aestivum* decreased the life cycle in plants from 170 to 130 days. This decrease could be utilized to harvest the crops quicker in scenario of climate change (Abdel-Aziz et al. 2016). NMs could be applied to soils to remove the pollutants. Application of Si NPs to *Oryza sativa* enhances its tolerance to cadmium stress (Wang et al. 2015b). Si NPs were stated to improve tolerance to cadmium, lead, copper and zinc stress in *Oryza sativa* (Wang et al. 2016).

Pest attack could increase during the outbreak of an environmental hazard. NMs have been reported to tackle pests efficiently. Silver NPs have strong antibacterial properties helping the plants in facing pathogenic attack (Vanti et al. 2019). NMs, like ZnO, CuO and MgO, and NPs, like Cu, Ag and Zn, have been stated to control spread and occurrence of multiple diseases caused by microorganisms in the plants (Giannousi et al. 2013; Malandrakis et al. 2019; Vanti et al. 2019). Additionally, nanocomposites, like silver laden chitosan NMs, enriched with fungicides could enhance antifungal properties in the plants and soil (Le et al. 2019).

The various NMs could increase enzymatic activity, enabling the plants to tolerate stress. NPs, like ZnO and SiO$_2$, boost activity of stress relieving enzymes, like peroxidase and superoxide dismutase, thereby aiding in stress tolerance escalation (Shalaby et al. 2016). NMs have been stated to regulate expression of the genes under stress, helping in plant protection (Onaga and Wydra 2016). Silver

NPs were shown to aid regulation of genetic expression in *Arabidopsis*, which could aid in protecting the plants from stress (Banerjee and Kole 2016).

11 Conclusion

Nanotechnology could be used in various ways to enhance the crop production across the world. Nanofertilizers, nanopesticides, nanoherbicides and nanosensors application to soils could be preferred over conventional use of agrochemicals, to tackle food security and boost the production of crops. They would achieve this by enhancing seed germination, boosting nutrient levels, detecting pest attack, identifying disease prevalence, enhancing photosynthesis, remediating polluted lands, filtering polluted water and facing changes in climatic conditions. Nano-robots could be utilized for identifying the deficiency of nutrients and moisture content in soil. The NPs harvested from the plants could be used in various industries for a wide variety of applications. Although NMs and NPs have been shown to be harmless to the environment and living organisms, nanotechnology is an unfolding and expanding field, and there could be potential damages waiting to unfurl because of the ambiguous chemical properties of NMs. It is very critical to proceed with the nanoscale technologies keeping in the mind the hidden issues involved and the plethora of advantages it possesses.

References

Abdel-Aziz HMM, Hasaneen MNA, Omer AM (2016) Nano chitosan-NPK fertilizer enhances the growth and productivity of wheat plants grown in sandy soil. Spanish J Agric Res 14:e0902. https://doi.org/10.5424/sjar/2016141-8205

Acosta C, Barat JM, Martínez-Máñez R et al (2018) Toxicological assessment of mesoporous silica particles in the nematode Caenorhabditis elegans. Environ Res 166:61–70. https://doi.org/10.1016/j.envres.2018.05.018

Afsharinejad A, Davy A, Jennings B, Brennan C (2016) Performance analysis of plant monitoring nanosensor networks at THz frequencies. IEEE Internet Things J 3:59–69. https://doi.org/10.1109/JIOT.2015.2463685

Ahmed F, Arshi N, Kumar S et al (2013) Nanobiotechnology: Scope and potential for crop improvement. Crop Improv Under Advers Cond 245–269

Al-Askar AA, Hafez EE, Kabeil SA, Meghad A (2013) Bioproduction of silver-nano particles by Fusarium oxysporum and their antimicrobial activity against some plant pathogenic bacteria and fungi. Life Sci J 10:2470–2475

Alejandro PDL, Rubiales D (2009) Nanotechnology for parasitic plant control. Pest Manag Sci 65:540–545. https://doi.org/10.1002/ps.1732

Alfadul SM, Altahir OS, Khan M (2017) Application of nanotechnology in the field of food production. Acad J Sci Res 5:143–154

Amthor JS (2001) Effects of atmospheric CO2 concentration on wheat yield: review of results from experiments using various approaches to control CO2 concentration. F Crop Res 73:1–34. https://doi.org/10.1016/s0378-4290(01)00179-4

Amundson R, Berhe AA, Hopmans JW et al (2015) Soil and human security in the 21st century. Science (80) 348:1261071. https://doi.org/10.1126/science.1261071

Anastas P, Eghbali N (2010) Green chemistry: Principles and practice. Chem Soc Rev 39:301–312. https://doi.org/10.1039/b918763b

Anwar MR, O'Leary G, McNeil D et al (2007) Climate change impact on rainfed wheat in south-eastern Australia. F Crop Res 104:139–147. https://doi.org/10.1016/j.fcr.2007.03.020

Aouada FA, De Moura MR (2015) Nanotechnology applied in agriculture: Controlled release of agrochemicals. In: Nanotechnologies in food and agriculture, pp 103–118

Ardakani AS (2013) Toxicity of silver, titanium and silicon nanoparticles on the root-knot nematode, Meloidogyne incognita, and growth parameters of tomato. Nematology 15:671–677. https://doi.org/10.1163/15685411-00002710

Aruoja V, Dubourguier HC, Kasemets K, Kahru A (2009) Toxicity of nanoparticles of CuO, ZnO and TiO2 to microalgae Pseudokirchneriella subcapitata. Sci Total Environ 407:1461–1468. https://doi.org/10.1016/j.scitotenv.2008.10.053

Atha DH, Wang H, Petersen EJ et al (2012) Copper oxide nanoparticle mediated DNA damage in terrestrial plant models. Environ Sci Technol 46:1819–1827. https://doi.org/10.1021/es202660k

Aziz N, Faraz M, Pandey R et al (2015) Facile Algae-derived route to biogenic silver nanoparticles: synthesis, antibacterial, and photocatalytic properties. Langmuir 31:11605–11612. https://doi.org/10.1021/acs.langmuir.5b03081

Baker S, Volova T, Prudnikova SV et al (2017) Nanoagroparticles emerging trends and future prospect in modern agriculture system. Environ Toxicol Pharmacol 53:10–17. https://doi.org/10.1016/j.etap.2017.04.012

Banerjee J, Kole C (2016) Plant nanotechnology: an overview on concepts, strategies, and tools. Plant Nanotechnol Princ Pract 1–14

Barik TK, Sahu B, Swain V (2008) Nanosilica—from medicine to pest control. Parasitol Res 103:253–258. https://doi.org/10.1007/s00436-008-0975-7

Barker AV, Pilbeam DJ (2015) Handbook of plant nutrition, 2nd edn. CRC Press

Batsmanova LM, Gonchar LM, Taran NY, Okanenko AA (2013) Using a colloidal solution of metal nanoparticles as micronutrient fertiliser for cereals. In: Proceedings of the 2nd International Conference—nanomaterials: applications and properties, pp 2–3

Berahmand AA, Panahi AG, Sahabi H et al (2012) Effects silver nanoparticles and magnetic field on growth of fodder maize (Zea mays L.). Biol Trace Elem Res 149:419–424. https://doi.org/10.1007/s12011-012-9434-5

Bhattacharyya A, Bhaumik A, Rani PU et al (2010) Nano-particles—a recent approach to insect pest control. Afr J Biotechnol 9:3489–3493. https://doi.org/10.5897/AJB2010.000-3206

Bhattacharyya A, Duraisamy P, Govindarajan M, et al (2016) Nano-biofungicides: emerging trend in insect pest control. Adv Appl Through Fungal Nanobiotechnol 307–319

Bheemidi VS (2011) novel applications of nanotechnology in life sciences. J Bioanal Biomed 03: https://doi.org/10.4172/1948-593x.s11-001

Bora T, Dutta J (2014) Applications of nanotechnology in wastewater treatment—a review. J Nanosci Nanotechnol 14:613–626. https://doi.org/10.1166/jnn.2014.8898

Bruce DM, Hobson RN, Farrent JW, Hepworth DG (2005) High-performance composites from low-cost plant primary cell walls. Compos Part A Appl Sci Manuf 36:1486–1493. https://doi.org/10.1016/j.compositesa.2005.03.008

Brunel F, El Gueddari NE, Moerschbacher BM (2013) Complexation of copper(II) with chitosan nanogels: toward control of microbial growth. Carbohydr Polym 92:1348–1356. https://doi.org/10.1016/j.carbpol.2012.10.025

Cavalcanti A, Wood WW, Kretly LC, Shirinzadeh B (2003) Computational nanorobotics: agricultural and environmental perspectives. Nanomedicine 2:82–87

Chakravarthy A (2012) DNA-tagged nano gold: a new tool for the control of the armyworm, Spodoptera litura Fab. (Lepidoptera: Noctuidae). Afr J Biotechnol 11. https://doi.org/10.5897/ajb11.883

Changmei L, Chaoying Z, Junqiang W et al (2002) Research of the effect of nanometer materials on germination and growth enhancement of glycine max and its mechanism. Soybean Sci 21:168–171

Chen H, Yada RY (2011) International conference on food and agriculture applications of nanotechnologies. NanoAgri 2010, São Pedro, SP, Brazil, June 20 to 25, 2010. Trends Food Sci Technol 22:583–584. https://doi.org/10.1016/j.tifs.2011.10.007

Chen YW, Lee HV, Juan JC, Phang SM (2016) Production of new cellulose nanomaterial from red algae marine biomass Gelidium elegans. Carbohydr Polym 151:1210–1219. https://doi.org/10.1016/j.carbpol.2016.06.083

Cohen-Tanugi D, Grossman JC (2012) Water desalination across nanoporous graphene. Nano Lett 12:3602–3608. https://doi.org/10.1021/nl3012853

Collins J (2006) Taking RFID to new depths. RFID J

Conway GR, Barbie EB (1988) After the Green Revolution. Sustainable and equitable agricultural development. Futures 20:651–670. https://doi.org/10.1016/0016-3287(88)90006-7

Corradini E, de Moura MR, Mattoso LHC (2010) A preliminary study of the incorparation of NPK fertilizer into chitosan nanoparticles. Express Polym Lett 4:509–515. https://doi.org/10.3144/expresspolymlett.2010.64

Cropper M, Griffiths C (1994) The interaction of population growth and environmental quality. Am Econ Rev 84:250–254. https://doi.org/10.2307/2117838

Czarnobai De Jorge B, Bisotto-de-Oliveira R, Pereira CN, Sant'Ana J (2017) Novel nanoscale pheromone dispenser for more accurate evaluation of Grapholita molesta (Lepidoptera: Tortricidae)

attract-and-kill strategies in the laboratory. Pest Manag Sci 73:1921–1926. https://doi.org/https://doi.org/10.1002/ps.4558

Das CK, Srivastava G, Dubey A et al (2016) Nano-iron pyrite seed dressing: a sustainable intervention to reduce fertilizer consumption in vegetable (beetroot, carrot), spice (fenugreek), fodder (alfalfa), and oilseed (mustard, sesamum) crops. Nanotechnol Environ Eng 1. https://doi.org/10.1007/s41204-016-0002-7

Dasgupta N, Ranjan S, Mundekkad D et al (2015) Nanotechnology in agro-food: from field to plate. Food Res Int 69:381–400. https://doi.org/10.1016/j.foodres.2015.01.005

Davarpanah S, Tehranifar A, Davarynejad G et al (2016) Effects of foliar applications of zinc and boron nano-fertilizers on pomegranate (*Punica granatum* cv. Ardestani) fruit yield and quality. Sci Hortic (Amsterdam) 210:57–64. https://doi.org/10.1016/j.scienta.2016.07.003

De La Torre-Roche R, Hawthorne J, Deng Y et al (2013) Multiwalled carbon nanotubes and C60 fullerenes differentially impact the accumulation of weathered pesticides in four agricultural plants. Environ Sci Technol 47:12539–12547. https://doi.org/10.1021/es4034809

Debnath N, Das S, Seth D et al (2011) Entomotoxic effect of silica nanoparticles against Sitophilus oryzae (L.). J Pest Sci (2004) 84:99–105. https://doi.org/10.1007/s10340-010-0332-3

Derosa MC, Monreal C, Schnitzer M et al (2010) Nanotechnology in fertilizers. Nat Nanotechnol 5:91. https://doi.org/10.1038/nnano.2010.2

Ding WK, Shah NP (2009) Effect of various encapsulating materials on the stability of probiotic bacteria. J Food Sci 74:M100–M107. https://doi.org/10.1111/j.1750-3841.2009.01067.x

Ditta A (2012) How helpful is nanotechnology in agriculture? Adv Nat Sci Nanosci Nanotechnol 3:33002. https://doi.org/10.1088/2043-6262/3/3/033002

Dufresne A, Dupeyre D, Vignon MR (2000) Cellulose microfibrils from potato tuber cells: processing and characterization of starch-cellulose microfibril composites. J Appl Polym Sci 76:2080–2092. https://doi.org/10.1002/(SICI)1097-4628(20000628)76:14%3c2080::AID-APP12%3e3.0.CO;2-U

Dwivedi S, Saquib Q, Al-Khedhairy AA, Musarrat J (2016) Understanding the role of nanomaterials in agriculture. In: Microbial inoculants in sustainable agricultural productivity. Functional Applications, vol 2, pp 271–288

Eichert T, Goldbach HE (2008) Equivalent pore radii of hydrophilic foliar uptake routes in stomatous and astomatous leaf surfaces—further evidence for a stomatal pathway. Physiol Plant 132:491–502. https://doi.org/10.1111/j.1399-3054.2007.01023.x

El Beyrouthya M (2014) Nanotechnologies: novel solutions for sustainable agriculture. Adv Crop Sci Technol 02: https://doi.org/10.4172/2329-8863.1000e118

Elfeky SA, Mohammed MA, Khater MS, Osman YAH (2013) Effect of magnetite nano-fertilizer on growth and yield of *Ocimum basilicum* L. Int J Indig Med Plants 64:1286–1293

Elmer WH, White JC (2016) The use of metallic oxide nanoparticles to enhance growth of tomatoes and eggplants in disease infested soil or soilless medium. Environ Sci Nano 3:1072–1079. https://doi.org/10.1039/c6en00146g

Espinosa E, Tarrés Q, Delgado-Aguilar M et al (2016) Suitability of wheat straw semichemical pulp for the fabrication of lignocellulosic nanofibres and their application to papermaking slurries. Cellulose 23:837–852. https://doi.org/10.1007/s10570-015-0807-8

Evans JR (2013) Improving photosynthesis. Plant Physiol 162:1780–1793. https://doi.org/10.1104/pp.113.219006

Ganeshkumar R, Sopiha KV, Wu P et al (2016) Ferroelectric KNbO$_3$ nanofibers: Synthesis, characterization and their application as a humidity nanosensor. Nanotechnology 27:395607. https://doi.org/10.1088/0957-4484/27/39/395607

Gao F, Hong F, Liu C et al (2006) Mechanism of nano-anatase TiO$_2$ on promoting photosynthetic carbon reaction of spinach: inducing complex of Rubisco-Rubisco activase. Biol Trace Elem Res 111:239–253. https://doi.org/10.1385/BTER:111:1:239

Gao F, Liu C, Qu C et al (2008) Was improvement of spinach growth by nano-TiO$_2$ treatment related to the changes of Rubisco activase? Biometals 21:211–217. https://doi.org/10.1007/s10534-007-9110-y

Gao J, Wang Y, Folta KM et al (2011) Polyhydroxy fullerenes (fullerols or fullerenols): Beneficial effects on growth and lifespan in diverse biological models. PLoS ONE 6:e19976. https://doi.org/10.1371/journal.pone.0019976

Ghormade V, Deshpande MV, Paknikar KM (2011) Perspectives for nano-biotechnology enabled protection and nutrition of plants. Biotechnol Adv 29:792–803. https://doi.org/10.1016/j.biotechadv.2011.06.007

Giannousi K, Avramidis I, Dendrinou-Samara C (2013) Synthesis, characterization and evaluation of copper based nanoparticles as agrochemicals against Phytophthora infestans. RSC Adv 3:21743–21752. https://doi.org/10.1039/c3ra42118j

Gillman GP (2006) A simple technology for arsenic removal from drinking water using hydrotalcite. Sci Total Environ 366:926–931. https://doi.org/10.1016/j.scitotenv.2006.01.036

Giongo AMM, Vendramim JD, Forim MR (2016) Evaluation of neem-based nanoformulations as alternative to control fall armyworm. Cienc E Agrotecnologia 40:26–36. https://doi.org/10.1590/S1413-70542016000100002

Giraldo JP, Landry MP, Faltermeier SM et al (2014) Plant nanobionics approach to augment photosynthesis and biochemical sensing. Nat Mater 13:400–408. https://doi.org/10.1038/nmat3890

Gleick PH (1993) Water and Conflict: Fresh Water Resources and International Security. Int Secur 18:79. https://doi.org/10.2307/2539033

Gogos A, Knauer K, Bucheli TD (2012) Nanomaterials in plant protection and fertilization: Current state, foreseen applications, and research priorities. J Agric Food Chem 60:9781–9792. https://doi.org/10.1021/jf302154y

González-Fernández R, Prats E, Jorrín-Novo JV (2010) Proteomics of plant pathogenic fungi. J Biomed Biotechnol 2010:1–36. https://doi.org/10.1155/2010/932527

Gottschalk F, Lassen C, Kjoelholt J et al (2015) Modeling flows and concentrations of nine engineered nanomaterials in the Danish environment. Int J Environ Res Public Health 12:5581–5602. https://doi.org/10.3390/ijerph120505581

Gubbins EJ, Batty LC, Lead JR (2011) Phytotoxicity of silver nanoparticles to *Lemna minor* L. Environ Pollut 159:1551–1559. https://doi.org/10.1016/j.envpol.2011.03.002

Habibi Y, Vignon MR (2008) Optimization of cellouronic acid synthesis by TEMPO-mediated oxidation of cellulose III from sugar beet pulp. Cellulose 15:177–185. https://doi.org/10.1007/s10570-007-9179-z

Haghighi M, Afifipour Z, Mozafarian M (2012) The alleviation effect of silicon on seed germination and seedling growth of tomato under salinity stress. Veg Crop Res Bull 76:119–126. https://doi.org/10.2478/v10032-012-0008-z

He X, Deng H, Hwang H, min, (2019) The current application of nanotechnology in food and agriculture. J Food Drug Anal 27:1–21. https://doi.org/10.1016/j.jfda.2018.12.002

Hibberd JM, Whitbread R, Farrar JF (1996) Effect of elevated concentrations of CO$_2$ on infection of barley by *Erysiphe graminis*. Physiol Mol Plant Pathol 48:37–53. https://doi.org/10.1006/pmpp.1996.0004

Hong F, Yang F, Liu C et al (2005) Influences of nano-TiO2 on the chloroplast aging of spinach under light. Biol Trace Elem Res 104:249–260. https://doi.org/10.1385/BTER:104:3:249

Hong F, Zhou J, Liu C et al (2005) Effect of Nano-TiO2 on photochemical reaction of chloroplasts of spinach. Biol Trace Elem Res 105:269–279. https://doi.org/10.1385/BTER:105:1-3:269

Hou R, Zhang Z, Pang S et al (2016) Alteration of the nonsystemic behavior of the pesticide ferbam on tea leaves by engineered gold nanoparticles. Environ Sci Technol 50:6216–6223. https://doi.org/10.1021/acs.est.6b01336

Iavicoli I, Leso V, Ricciardi W et al (2014) Opportunities and challenges of nanotechnology in the green economy. Environ Heal A Glob Access Sci Source 13. https://doi.org/10.1186/1476-069X-13-78

Imada K, Sakai S, Kajihara H et al (2016) Magnesium oxide nanoparticles induce systemic resistance in tomato against bacterial wilt disease. Plant Pathol 65:551–560. https://doi.org/10.1111/ppa.12443

Iravani S (2011) Green synthesis of metal nanoparticles using plants. Green Chem 13:2638–2650. https://doi.org/10.1039/c1gc15386b

Jaberzadeh A, Moaveni P, Tohidi Moghadam HR, Zahedi H (2013) Influence of bulk and nanoparticles titanium foliar application on some agronomic traits, seed gluten and starch contents of wheat subjected to water deficit stress. Not Bot Horti Agrobot Cluj-Napoca 41:201–207. https://doi.org/10.15835/nbha4119093

Jo YK, Kim BH, Jung G (2009) Antifungal activity of silver ions and nanoparticles on phytopathogenic fungi. Plant Dis 93:1037–1043. https://doi.org/10.1094/PDIS-93-10-1037

Joshi A, Kaur S, Dharamvir K et al (2018) Multi-walled carbon nanotubes applied through seed-priming influence early germination, root hair, growth and yield of bread wheat (Triticum aestivum L.). J Sci Food Agric 98:3148–3160. https://doi.org/10.1002/jsfa.8818

Kah M, Hofmann T (2014) Nanopesticide research: current trends and future priorities. Environ Int 63:224–235. https://doi.org/10.1016/j.envint.2013.11.015

Kah M, Beulke S, Tiede K, Hofmann T (2013) Nanopesticides: state of knowledge, environmental fate, and exposure modeling. Crit Rev Environ Sci Technol 43:1823–1867. https://doi.org/10.1080/10643389.2012.671750

Kale AP, Gawade SN (2016) Studies on nanoparticle induced nutrient use eficiency of fertilizer and crop productivity. Green Chem Technol Lett 2:88. https://doi.org/10.18510/gctl.2016.226

Karn B, Kuiken T, Otto M (2009) Nanotechnology and in situ remediation: a review of the benefits and potential risks. Environ Health Perspect 117:1823–1831. https://doi.org/10.1289/ehp.0900793

Khiari R (2017) Valorization of agricultural residues for cellulose nanofibrils production and their use in nanocomposite manufacturing. Int J Polym Sci 2017:1–10. https://doi.org/10.1155/2017/6361245

Khodakovskaya M, Dervishi E, Mahmood M, et al (2012a) Erratum: Carbon nanotubes are able to penetrate plant seed coat and dramatically affect seed germination and plant growth (ACS Nano (2009) 3:3221–3227. https://doi.org/10.1021/nn900887m). ACS Nano 6:7541. https://doi.org/10.1021/nn302965w

Khodakovskaya MV, De Silva K, Biris AS et al (2012) Carbon nanotubes induce growth enhancement of tobacco cells. ACS Nano 6:2128–2135. https://doi.org/10.1021/nn204643g

Kim SW, Jung JH, Lamsal K et al (2012) Antifungal effects of silver nanoparticles (AgNPs) against various plant pathogenic fungi. Mycobiology 40:53–58. https://doi.org/10.5941/MYCO.2012.40.1.053

Kitching M, Ramani M, Marsili E (2015) Fungal biosynthesis of gold nanoparticles: mechanism and scale up. Microb Biotechnol 8:904–917. https://doi.org/10.1111/1751-7915.12151

Kole C, Kole P, Randunu KM et al (2013) Nanobiotechnology can boost crop production and quality: First evidence from increased plant biomass, fruit yield and phytomedicine content in bitter melon (Momordica charantia). BMC Biotechnol 13. https://doi.org/10.1186/1472-6750-13-37

Kumar S, Ahlawat W, Bhanjana G et al (2014) Nanotechnology-based water treatment strategies. J Nanosci Nanotechnol 14:1838–1858. https://doi.org/10.1166/jnn.2014.9050

Kumar S, Bhanjana G, Sharma A et al (2017) Development of nanoformulation approaches for the control of weeds. Sci Total Environ 586:1272–1278. https://doi.org/10.1016/j.scitotenv.2017.02.138

Lahiani MH, Dervishi E, Chen J et al (2013) Impact of carbon nanotube exposure to seeds of valuable crops. ACS Appl Mater Interfaces 5:7965–7973. https://doi.org/10.1021/am402052x

Lateef A, Nazir R, Jamil N et al (2016) Synthesis and characterization of zeolite based nano-composite: An environment friendly slow release fertilizer. Microporous Mesoporous Mater 232:174–183. https://doi.org/10.1016/j.micromeso.2016.06.020

Le VT, Bach LG, Pham TT et al (2019) Synthesis and antifungal activity of chitosan-silver nanocomposite synergize fungicide against Phytophthora capsici. J Macromol Sci Part A Pure Appl Chem 56:522–528. https://doi.org/10.1080/10601325.2019.1586439

Lei Z, Mingyu S, Chao L et al (2007) Effects of nanoanatase TiO$_2$ on photosynthesis of spinach chloroplasts under different light illumination. Biol Trace Elem Res 119:68–76. https://doi.org/10.1007/s12011-007-0047-3

Lélé SM (1991) Sustainable development: a critical review. World Dev 19:607–621

Li ZZ, Chen JF, Liu F et al (2007) Study of UV-shielding properties of novel porous hollow silica nanoparticle carriers for avermectin. Pest Manag Sci 63:241–246. https://doi.org/10.1002/ps.1301

Lindblade KA, Walker ED, Onapa AW et al (1999) Highland malaria in Uganda: prospective analysis of an epidemic associated with El Niño. Trans R Soc Trop Med Hyg 93. https://doi.org/10.1016/S0035-9203(99)90344-9

Linglan M, Chao L, Chunxiang Q et al (2008) Rubisco activase mRNA expression in spinach: modulation by nanoanatase treatment. Biol Trace Elem Res 122:168–178. https://doi.org/10.1007/s12011-007-8069-4

Liu R, Lal R (2014) Synthetic apatite nanoparticles as a phosphorus fertilizer for soybean (Glycine max). Sci Rep 4. https://doi.org/10.1038/srep05686

Liu R, Lal R (2015) Potentials of engineered nanoparticles as fertilizers for increasing agronomic productions. Sci Total Environ 514:131–139. https://doi.org/10.1016/j.scitotenv.2015.01.104

Liu F, Wen LX, Li ZZ et al (2006) Porous hollow silica nanoparticles as controlled delivery system for water-soluble pesticide. Mater Res Bull 41:2268–2275. https://doi.org/10.1016/j.materresbull.2006.04.014

Liu R, Kang Y, Pei L et al (2016) Use of a new controlled-loss-fertilizer to reduce nitrogen losses during winter wheat cultivation in the Danjiangkou reservoir area of China. Commun Soil Sci Plant Anal 47:1137–1147. https://doi.org/10.1080/00103624.2016.1166245

Liu X, Liao J, Song H et al (2019) A biochar-based route for environmentally friendly controlled release of nitrogen: urea-loaded biochar and bentonite composite. Sci Rep 9:9548. https://doi.org/10.1038/s41598-019-46065-3

Madusanka N, Sandaruwan C, Kottegoda N et al (2017) Urea–hydroxyapatite-montmorillonite nanohybrid composites as slow release nitrogen compositions. Appl Clay Sci 150:303–308. https://doi.org/10.1016/j.clay.2017.09.039

Malandrakis AA, Kavroulakis N, Chrysikopoulos CV (2019) Use of copper, silver and zinc nanoparticles against foliar and soil-borne plant pathogens. Sci Total Environ 670:292–299. https://doi.org/10.1016/j.scitotenv.2019.03.210

Manjaiah KM, Mukhopadhyay R, Paul R et al (2018) Clay minerals and zeolites for environmentally sustainable agriculture. Modif Clay Zeolite Nanocompos Mater Environ Pharm Appl 309–329

Marchiol L (2018) Nanotechnology in agriculture: new opportunities and perspectives. New Visions Plant Sci. https://doi.org/10.5772/intechopen.74425

Mariano M, El Kissi N, Dufresne A (2014) Cellulose nanocrystals and related nanocomposites: Review of some properties and challenges. J Polym Sci Part B Polym Phys 52:791–806. https://doi.org/10.1002/polb.23490

McLamore ES, Diggs A, Calvo Marzal P et al (2010) Non-invasive quantification of endogenous root auxin transport using an integrated flux microsensor technique. Plant J 63:1004–1016. https://doi.org/10.1111/j.1365-313X.2010.04300.x

Medina J, Monreal C, Barea JM et al (2015) Crop residue stabilization and application to agricultural and degraded soils: a review. Waste Manag 42:41–54. https://doi.org/10.1016/j.wasman.2015.04.002

Millán G, Agosto F, Vázquez M et al (2008) Use of clinoptilolite as a carrier for nitrogen fertilizers in soils of the Pampean regions of Argentina. Cienc E Investig Agrar 35:245–254. https://doi.org/10.4067/S0718-16202008000300007

Mokerov VG, Fedorov YV, Velikovski LE, Scherbakova MY (2001) New quantum dot transistor. Nanotechnology 12:552–555. https://doi.org/10.1088/0957-4484/12/4/336

Mukhopadhyay SS (2005) Weathering of soil minerals and distribution of elements: pedochemical aspects. Clay Res 24:183–199

Muramatsu H, Kim YA, Yang KS et al (2014) Rice husk-derived graphene with nano-sized domains and clean edges. Small 10:2766–2770. https://doi.org/10.1002/smll.201400017

Naderi MR, Abedi A (2012) Application of nanotechnology in agriculture and refinement of environmental pollutants. J Nanotechnol 11:18–26

Naderi M, Danesh-Shahraki A (2011) The application of nanotechnology in the formulation optimization of chemical fertilizers. J Nano 106:20–22

Najafi Disfani M, Mikhak A, Kassaee MZ, Maghari A (2017) Effects of nano Fe/SiO₂ fertilizers on germination and growth of barley and maize. Arch Agron Soil Sci 63:817–826. https://doi.org/10.1080/03650340.2016.1239016

Namasivayam SKR, Aruna A, Gokila, (2014) Evaluation of silver nanoparticles-chitosan encapsulated synthetic herbicide paraquate (AgNp-CS-PQ) preparation for the controlled release and improved herbicidal activity against Eichhornia crassipes. Res J Biotechnol 9:19–27

Navrotsky A (2000) Nanomaterials in the environment, agriculture, and technology (NEAT). J Nanoparticle Res 2:321–323. https://doi.org/10.1023/A:1010007023813

Nin-Pratt A (2016) Agricultural intensification and fertilizer use. International Food Policy Research Institute

Nuruzzaman M, Rahman MM, Liu Y, Naidu R (2016) Nanoencapsulation, nano-guard for pesticides: a new window for safe application. J Agric Food Chem 64:1447–1483. https://doi.org/10.1021/acs.jafc.5b05214

O'Hern SC, Jang D, Bose S et al (2015) Nanofiltration across defect-sealed nanoporous monolayer graphene. Nano Lett 15:3254–3260. https://doi.org/10.1021/acs.nanolett.5b00456

Ohlsson I (1996) Site-specific management for agricultural systems. F Crop Res 48:91–92. https://doi.org/10.1016/0378-4290(96)82398-7

Onaga G, Wydra K (2016) Advances in plant tolerance to biotic stresses. Plant Genomics

Österholm P, Åström M (2004) Quantification of current and future leaching of sulfur and metals from Boreal acid sulfate soils, western Finland. Aust J Soil Res 42:547–551. https://doi.org/10.1071/sr03088

Pal S, Tak YK, Song JM (2007) Does the antibacterial activity of silver nanoparticles depend on the shape of the nanoparticle? A study of the gram-negative bacterium Escherichia coli. Appl Environ Microbiol 73:1712–1720. https://doi.org/10.1128/AEM.02218-06

Pandey G (2018) Challenges and future prospects of agri-nanotechnology for sustainable agriculture in India. Environ Technol Innov 11:299–307. https://doi.org/10.1016/j.eti.2018.06.012

Panpatte DG, Jhala YG, Shelat HN, Vyas RV (2016) Nanoparticles: the next generation technology for sustainable agriculture. In: Microbial inoculants in sustainable agricultural productivity. Functional Applications 289–300

Parisi C, Vigani M, Rodríguez-Cerezo E (2015) Agricultural nanotechnologies: what are the current possibilities? Nano Today 10:124–127. https://doi.org/10.1016/j.nantod.2014.09.009

Park TJ, Lee KG, Lee SY (2016) Advances in microbial biosynthesis of metal nanoparticles. Appl Microbiol Biotechnol 100:521–534. https://doi.org/10.1007/s00253-015-6904-7

Patil CD, Borase HP, Suryawanshi RK, Patil SV (2016) Trypsin inactivation by latex fabricated gold nanoparticles: a new strategy towards insect control. Enzyme Microb Technol 92:18–25. https://doi.org/10.1016/j.enzmictec.2016.06.005

Perlatti B, de Souza Bergo PL, Fernandes da Silva MF das G et al (2013) Polymeric Nanoparticle-Based Insecticides: A Controlled Release Purpose for Agrochemicals. Insectic Dev Safer More Eff Technol

Petosa AR, Rajput F, Selvam O et al (2017) Assessing the transport potential of polymeric nanocapsules developed for crop protection. Water Res 111:10–17. https://doi.org/10.1016/j.watres.2016.12.030

Philip D (2011) Mangifera Indica leaf-assisted biosynthesis of well-dispersed silver nanoparticles. Spectrochim Acta Part a Mol Biomol Spectrosc 78:327–331. https://doi.org/10.1016/j.saa.2010.10.015

Pokropivny V, Hussainova I, Vlassov S (2007) Introduction to nanomaterials. Tartu University, Tartu 3330

Postel SL, Daily GC, Ehrlich PR (1996) Human appropriation of renewable fresh water. Science (80) 271:785–788. https://doi.org/10.1126/science.271.5250.785

Pramanik S, Pramanik G (2016) Nanotechnology for sustainable agriculture in India. In: Ranjan S et al (eds) Nanoscience in food and agriculture. Sustainable agriculture reviews, vol 3, pp 243–280

Prasad R (2014) Synthesis of silver nanoparticles in photosynthetic plants. J Nanoparticles 2014:1–8. https://doi.org/10.1155/2014/963961

Prasad R, Swamy VS (2013) Antibacterial activity of silver nanoparticles synthesized by bark extract of Syzygium cumini. J Nanoparticles 2013:1–6. https://doi.org/10.1155/2013/431218

Prasad TNVKV, Sudhakar P, Sreenivasulu Y et al (2012) Effect of nanoscale zinc oxide particles on the germination, growth and yield of peanut. J Plant Nutr 35:905–927. https://doi.org/10.1080/01904167.2012.663443

Prasad R, Bhattacharyya A, Nguyen QD (2017) Nanotechnology in sustainable agriculture: recent developments, challenges, and perspectives. Front Microbiol 8:1014. https://doi.org/10.3389/fmicb.2017.01014

Presley DR, Ransom MD, Kluitenberg GJ, Finnell PR (2004) Effects of thirty years of irrigation on the genesis and morphology of two semiarid soils in Kansas. Soil Sci Soc Am J 68:1916–1926. https://doi.org/10.2136/sssaj2004.1916

Pretty J (2008) Agricultural sustainability: concepts, principles and evidence. Philos Trans R Soc B Biol Sci 363:447–465. https://doi.org/10.1098/rstb.2007.2163

Qi M, Liu Y, Li T (2013) Nano-TiO₂ improve the photosynthesis of tomato leaves under mild heat stress. Biol Trace Elem Res 156:323–328. https://doi.org/10.1007/s12011-013-9833-2

Qu X, Brame J, Li Q, Alvarez PJJ (2013) Nanotechnology for a safe and sustainable water supply: Enabling integrated water treatment and reuse. Acc Chem Res 46:834–843. https://doi.org/10.1021/ar300029v

Quintanar-Guerrero D, Allémann E, Fessi H, Doelker E (1998) Preparation techniques and mechanisms of formation of biodegradable nanoparticles from preformed polymers. Drug Dev Ind Pharm 24:1113–1128. https://doi.org/10.3109/03639049809108571

Rai M, Ingle A (2012) Role of nanotechnology in agriculture with special reference to management of insect pests. Appl Microbiol Biotechnol 94:287–293. https://doi.org/10.1007/s00253-012-3969-4

Rai V, Acharya S, Dey N (2012) Implications of nanobiosensors in agriculture. J Biomater Nanobiotechnol 03:315–324. https://doi.org/10.4236/jbnb.2012.322039

Ram P, Vivek K, Kumar SP (2014) Nanotechnology in sustainable agriculture: present concerns and future aspects. Afr J Biotechnol 13:705–713. https://doi.org/10.5897/ajbx2013.13554

Rico CM, Majumdar S, Duarte-Gardea M et al (2011) Interaction of nanoparticles with edible plants and their possible implications in the food chain. J Agric Food Chem 59:3485–3498. https://doi.org/10.1021/jf104517j

Rodell M, Velicogna I, Famiglietti JS (2009) Satellite-based estimates of groundwater depletion in India. Nature 460:999–1002. https://doi.org/10.1038/nature08238

Sabbour MM (2012) Entomotoxicity assay of two Nanoparticle Materials 1—(Al_2O_3 and TiO_2) against Sitophilus oryzae under laboratory and store conditions in Egypt. J Nov Appl Sci 103–108

Sabir A, Yazar K, Sabir F et al (2014) Vine growth, yield, berry quality attributes and leaf nutrient content of grapevines as influenced by seaweed extract (Ascophyllum nodosum) and nanosize fertilizer pulverizations. Sci Hortic (Amsterdam) 175:1–8. https://doi.org/10.1016/j.scienta.2014.05.021

Saharan V, Sharma G, Yadav M et al (2015) Synthesis and in vitro antifungal efficacy of Cu-chitosan nanoparticles against pathogenic fungi of tomato. Int J Biol Macromol 75:346–353. https://doi.org/10.1016/j.ijbiomac.2015.01.027

Sankar S, Sharma SK, Kaur N et al (2016) Biogenerated silica nanoparticles synthesized from sticky, red, and brown rice husk ashes by a chemical method. Ceram Int 42:4875–4885. https://doi.org/10.1016/j.ceramint.2015.11.172

Sasson Y, Levy-Ruso G, Toledano O, Ishaaya I (2007) Nanosuspensions: emerging novel agrochemical formulations. In: Ishaaya I, Nauen R, Horowitz AR (eds) Insecticides design using advanced technologies. Springer, Berlin, pp 1–39

Satapanajaru T, Anurakpongsatorn P, Pengthamkeerati P, Boparai H (2008) Remediation of atrazine-contaminated soil and water by nano zerovalent iron. Water Air Soil Pollut 192:349–359. https://doi.org/10.1007/s11270-008-9661-8

Scrinis G, Lyons K (2007) The emerging nano-corporate paradigm: nanotechnology and the transformation of nature, food and agri-food systems. Int J Sociol Food Agric 15:22–44

Servin A, Elmer W, Mukherjee A et al (2015) A review of the use of engineered nanomaterials to suppress plant disease and enhance crop yield. J Nanoparticle Res 17:1–21. https://doi.org/10.1007/s11051-015-2907-7

Shalaby TA, Bayoumi Y, Abdalla N et al (2016) Nanoparticles, soils, plants and sustainable agriculture. In: Shivendu R, Nandita D, Eric L (eds) Nanoscience in food and agriculture, vol 1, pp 283–312

Shang Y, Kamrul Hasan M, Ahammed GJ et al (2019) Applications of nanotechnology in plant growth and crop protection: A review. Molecules 24. https://doi.org/10.3390/molecules24142558

Sharifi-Rad J, Sharifi-Rad M, Teixeira da Silva JA (2016) Morphological, physiological and biochemical responses of crops (Zea mays L., Phaseolus vulgaris L.), medicinal plants (Hyssopus officinalis L., Nigella sativa L.), and weeds (Amaranthus retroflexus L., Taraxacum officinale F. H. Wigg) exposed to SiO_2 nanoparticles. J Agric Sci Technol 18:1027–1040

Sharma VK, Yngard RA, Lin Y (2009) Silver nanoparticles: Green synthesis and their antimicrobial activities. Adv Colloid Interface Sci 145:83–96. https://doi.org/10.1016/j.cis.2008.09.002

Sharma K, Sharma R, Shit S, Gupta S (2012) Nanotechnological application on diagnosis of a plant disease. In: International conference on advances in biological and medical sciences, pp 149–150

Sharma H, Dhirta B, Shirkot P (2017) Evaluation of biogenic iron nano formulations to control Meloidogyne incognita in okra. Int J Chem Stud 5:1278–1284

Shen Y (2017) Rice husk silica derived nanomaterials for sustainable applications. Renew Sustain Energy Rev 80:453–466. https://doi.org/10.1016/j.rser.2017.05.115

Shenashen M, Derbalah A, Hamza A et al (2017) Antifungal activity of fabricated mesoporous alumina nanoparticles against root rot disease of tomato caused by Fusarium oxysporium. Pest Manag Sci 73:1121–1126. https://doi.org/10.1002/ps.4420

Sheykhbaglou R, Sedghi M, Shishevan MT, Sharifi RS (2010) Effects of nano-iron oxide particles on agronomic traits of soybean. Not Sci Biol 2:112–113. https://doi.org/10.15835/nsb224667

Shojaei TR, Salleh MAM, Tabatabaei M, et al (2018) Applications of nanotechnology and carbon nanoparticles in agriculture. Synth Technol Appl Carbon Nanomater 247–277

Shrivastava S, Dash D (2009) Agrifood nanotechnology: a tiny revolution in food and agriculture. J Nano Res 6:1–14. https://doi.org/10.4028/www.scientific.net/JNanoR.6.1

Siddiqui MH, Al-Whaibi MH (2014) Role of nano-SiO_2 in germination of tomato (Lycopersicum esculentum seeds Mill.). Saudi J Biol Sci 21:13–17. https://doi.org/10.1016/j.sjbs.2013.04.005

Singh R, Singh R, Singh D et al (2010) Effect of weather parameters on karnal bunt disease in wheat in karnal region of Haryana. J Agrometeorol 12:99–101

Singh S, Singh BK, Yadav SM, Gupta AK (2015a) Applications of nanotechnology in agricultural and their role in disease management. Res J Nanosci Nanotechnol 5:1–5. https://doi.org/10.3923/rjnn.2015.1.5

Singh A, Singh NB, Hussain I et al (2015b) Plant-nanoparticle interaction : an approach to improve agricultural practices and plant productivity. Int J Pharm Sci Invent 4:25–40

Singh P, Singh R, Borthakur A et al (2016) Effect of nanoscale TiO_2—activated carbon composite on Solanum lycopersicum (L.) and Vigna radiata (L.) seeds germination. Energy Ecol Environ 1:131–140. https://doi.org/10.1007/s40974-016-0009-8

Singh Sekhon B (2014) Nanotechnology in agri-food production: an overview. Nanotechnol Sci Appl 7:31–53. https://doi.org/10.2147/NSA.S39406

Somanathan T, Prasad K, Ostrikov KK et al (2015) Graphene oxide synthesis from agro waste. Nanomaterials 5:826–834. https://doi.org/10.3390/nano5020826

Sousa GFM, Gomes DG, Campos EVR et al (2018) Post-emergence herbicidal activity of nanoatrazine against susceptible weeds. Front Environ Sci 6. https://doi.org/10.3389/fenvs.2018.00012

Srivastava G, Das CK, Das A et al (2014) Seed treatment with iron pyrite (FeS_2) nanoparticles increases the production of spinach. RSC Adv 4:58495–58504. https://doi.org/10.1039/c4ra06861k

Stadler T, Buteler M, Weaver DK (2010) Novel use of nanostructured alumina as an insecticide. Pest Manag Sci 66:577–579. https://doi.org/10.1002/ps.1915

Stamp P, Visser R (2012) The twenty-first century, the century of plant breeding. Euphytica 186:585–591. https://doi.org/10.1007/s10681-012-0743-8

Subramanian KS, Manikandan A, Thirunavukkarasu M, Rahale CS (2015) Nano-fertilizers for balanced crop nutrition. Nanotechnol Food Agric 69–80

Sun CQ (2007) Size dependence of nanostructures: Impact of bond order deficiency. Prog Solid State Chem 35:1–159. https://doi.org/10.1016/j.progsolidstchem.2006.03.001

Suresh AK, Pelletier DA, Doktycz MJ (2013) Relating nanomaterial properties and microbial toxicity. Nanoscale 5:463–474. https://doi.org/10.1039/c2nr32447d

Szargut J, Zibik A, Stanek W (2002) Depletion of the non-renewable natural exergy resources as a measure of the ecological cost. Energy Convers Manag 43:1149–1163. https://doi.org/10.1016/S0196-8904(02)00005-5

Tan X, Liu Y, Gu Y et al (2016) Biochar-based nano-composites for the decontamination of wastewater: a review. Bioresour Technol 212:318–333. https://doi.org/10.1016/j.biortech.2016.04.093

Tapan A, Biswas AK, Kundu S (2010) Nano-fertiliser-a new dimension in agriculture. Indian J Fertil 6:22–24

Tarafdar JC, Raliya R, Mahawar H, Rathore I (2014) Development of zinc nanofertilizer to enhance crop production in pearl millet (*Pennisetum americanum*). Agric Res 3:257–262. https://doi.org/10.1007/s40003-014-0113-y

Torabian S, Zahedi M, Khoshgoftar AH (2017) Effects of foliar spray of nano-particles of $FeSO_4$ on the growth and ion content of sunflower under saline condition. J Plant Nutr 40:615–623. https://doi.org/10.1080/01904167.2016.1240187

Torney F (2009) Nanoparticle mediated plant transformation. In: Emerging technologies in plant science research. Interdepartmental plant physiology major fall seminar series, p 696

Torney F, Trewyn BG, Lin VSY, Wang K (2007) Mesoporous silica nanoparticles deliver DNA and chemicals into plants. Nat Nanotechnol 2:295–300. https://doi.org/10.1038/nnano.2007.108

Tripathi DK, Singh S, Singh VP et al (2017) Silicon nanoparticles more effectively alleviated UV-B stress than silicon in wheat (Triticum aestivum) seedlings. Plant Physiol Biochem 110:70–81. https://doi.org/10.1016/j.plaphy.2016.06.026

Tripathi M, Kumar S, Kumar A (2018) Agro-nanotechnology: a future technology for sustainable agriculture. Int J Curr Microbiol Appl Sci 196–200

Ul Haq I, Ijaz S (2019) Use of metallic nanoparticles and nanoformulations as nanofungicides for sustainable disease management in plants. In: Prasad R, Kumar V, Kumar M, Choudhary D (eds) Nanobiotechnology in bioformulations. Springer, Cham, pp 289–316

Ulrichs C, Mewis I, Goswami A (2006) Crop diversification aiming nutritional security in West Bengal—biotechnology of stinging capsules in nature's water-blooms. . Ann Tech Issue State Agri Technol Serv Assoc Govt West Bengal India 10:1–18

Van Bavel J (2013) The world population explosion: causes, backgrounds and -projections for the future. Facts, Views Vis ObGyn 5:281–291

Dijk van M, Meijerink G (2014) A review of food security scenario studies: gaps and ways forward. In: Achterbosch TJ, Dorp M, van Driel WF, van Groot JJ, Lee J, van der Verhagen A, Bezlepkina I (eds) The food puzzle: pathways to securing food for all, pp 30–32

Vanathi P, Rajiv P, Sivaraj R (2016) Synthesis and characterization of Eichhornia-mediated copper oxide nanoparticles and assessing their antifungal activity against plant pathogens. Bull Mater Sci 39:1165–1170. https://doi.org/10.1007/s12034-016-1276-x

Vanti GL, Nargund VB, Basavesha KN et al (2019) Synthesis of Gossypium hirsutum-derived silver nanoparticles and their antibacterial efficacy against plant pathogens. Appl Organomet Chem 33: e4630. https://doi.org/10.1002/aoc.4630

Vermeulen SJ, Aggarwal PK, Ainslie A et al (2012) Options for support to agriculture and food security under climate change. Environ Sci Policy 15:136–144. https://doi.org/10.1016/j.envsci.2011.09.003

Vidhyalakshmi R, Bhakyaraj R, Subhasree RS (2009) Encapsulation "the future of probiotics"—a review. Adv Biol Res (Rennes) 3:96–103

Wang L, Li X, Zhang G et al (2007) Oil-in-water nanoemulsions for pesticide formulations. J Colloid Interface Sci 314:230–235. https://doi.org/10.1016/j.jcis.2007.04.079

Wang H, Wick RL, Xing B (2009) Toxicity of nanoparticulate and bulk ZnO, Al_2O_3 and TiO_2 to the nematode Caenorhabditis elegans. Environ Pollut 157:1171–1177. https://doi.org/10.1016/j.envpol.2008.11.004

Wang S, Wang F, Gao S (2015) Foliar application with nano-silicon alleviates Cd toxicity in rice seedlings. Environ Sci Pollut Res 22:2837–2845. https://doi.org/10.1007/s11356-014-3525-0

Wang S, Gao B, Zimmerman AR et al (2015) Removal of arsenic by magnetic biochar prepared from pinewood and natural hematite. Bioresour Technol 175:391–395. https://doi.org/10.1016/j.biortech.2014.10.104

Wang S, Wang F, Gao S, Wang X (2016) Heavy metal accumulation in different rice cultivars as influenced by foliar application of nano-silicon. Water Air Soil Pollut 227. https://doi.org/10.1007/s11270-016-2928-6

Wang P, Tang L, Wei X et al (2017) Synthesis and application of iron and zinc doped biochar for removal of p-nitrophenol in wastewater and assessment of the influence of co-existed Pb(II). Appl Surf Sci 392:391–401. https://doi.org/10.1016/j.apsusc.2016.09.052

Wani KA, Kothari R (2018) Agricultural nanotechnology: applications and challenges. Ann Plant Sci 7:2146. https://doi.org/10.21746/aps.2018.7.3.9

Wanyika H, Gatebe E, Kioni P et al (2012) Mesoporous silica nanoparticles carrier for urea: potential applications in agrochemical delivery systems. J Nanosci Nanotechnol 12:2221–2228. https://doi.org/10.1166/jnn.2012.5801

Xu C, Peng C, Sun L et al (2015) Distinctive effects of TiO_2 and CuO nanoparticles on soil microbes and their community structures in flooded paddy soil. Soil Biol Biochem 86:24–33. https://doi.org/10.1016/j.soilbio.2015.03.011

Yamanaka M, Hara K, Kudo J (2005) Bactericidal actions of a silver ion solution on Escherichia coli, studied by energy-filtering transmission electron microscopy and proteomic analysis. Appl Environ Microbiol 71:7589–7593. https://doi.org/10.1128/AEM.71.11.7589-7593.2005

Yang FL, Li XG, Zhu F, Lei CL (2009) Structural characterization of nanoparticles loaded with garlic essential oil and their insecticidal activity against *Tribolium castaneum* (Herbst) (Coleoptera: Tenebrionidae). J Agric Food Chem 57:10156–10162. https://doi.org/10.1021/jf9023118

Yavuz CT, Mayo JT, Yu WW et al (2006) Low-field magnetic separation of monodisperse Fe_3O_4 nanocrystals. Science (80) 314:964–967. https://doi.org/10.1126/science.1131475

Zheng L, Hong F, Lu S, Liu C (2005) Effect of nano-TiO2 on strength of naturally aged seeds and growth of spinach. Biol Trace Elem Res 104:83–91. https://doi.org/10.1385/BTER:104:1:083

Interaction of Titanium Dioxide Nanoparticles with Plants in Agro-ecosystems

Ranjana Singh, Kajal Patel, and Indu Tripathi

Abstract

The remarkable progress in nanotechnology has significantly augmented the utilization of nanoscaled (≤ 100 nm) nanomaterials (NMs) in wide range of products. Titanium dioxide NPs (TiO$_2$-NPs) are the most commonly used NMs among all that are unabatedly used in a variety of industrial and consumer products. This demand-based excessive production and application of TiO$_2$-NPs ultimately lead way to their release in the environment which causes potential risks to environmental components and their functioning. Plants, being the primary component of any ecosystem, are the initial point for the NPs' interaction that is requisite feature for risk assessment. Researchers over the globe have observed that these NPs pose both detrimental and beneficial effects on plants; however, these impacts are determined by their mode of interaction depending on experimental conditions. Besides, due to their photocatalytic and growth-promoting activities, researchers are paying attention on developing some TiO$_2$-NP-based formulations so that they can be used in agriculture to improve crop yield and quality and also to protect plants from various pests and pathogens. It is reported that TiO$_2$-NPs also improve the plant performance under various abiotic stresses. Under the light of current knowledge, this chapter provides an overview of TiO$_2$-NPs, their interactions, i.e., uptake, translocation and accumulation with plants, feedback of plants at different levels (viz. morphological, physiological, biochemical and molecular) with an overview on intrinsic mechanism of TiO$_2$-NPs detoxification within plants. Eventually, it gives a brief knowledge on application of TiO$_2$-NPs in agriculture as growth boosters and protecting agents against various biotic and abiotic stresses.

Keywords

Detoxification mechanism • Ecosystem • Environmental risks • Nanomaterials • Nanotechnology • Plant interaction • Phytotoxicity

1 Introduction

Nanotechnology is a revolutionary science that includes synthesis and manipulation of the matter at an atomic or molecular scale (i.e. 10^{-9} m) with an objective to produce materials with novel functionalities and improved characters stemming from their surface area, shape, electronic and optical behavior. Among the various nanomaterials (NMs), nanoparticles (NPs) play a special role in a broad array of applications in variety of fields like production and synthesis of materials, remediation techniques for environment, cosmetics, agriculture and medicines. Due to escalating demands, the value of these NMs is expected to be increased about 30 billion USD by the end of 2020 in the global market (Wang et al. 2013).

Titanium dioxide (TiO$_2$), an oxide of titanium (Ti), occurs on Earth mainly in three crystalline forms, viz. anatase (tetragonal), rutile (tetragonal) and brookite (orthorhombic) with brookite having no commercial value (Fig. 1) (Fries and Simkó 2012; Waghmode et al. 2019). According to thermodynamics calculations, both anatase and brookite get transformed into rutile which is found to be chemically very active and highly stable at all temperature and pressure

R. Singh (✉)
Department of Botany, Government Model Degree College, Arniya, Bulandshahr, U.P. 203131, India
e-mail: vermaranjana09@gmail.com

K. Patel
Department of Botany, University of Delhi, New Delhi, 110007, India
e-mail: kajal78672@gmail.com

I. Tripathi
Department of Environmental Studies, University of Delhi, New Delhi, 110007, India
e-mail: indutripathi9@gmail.com

P. Singh et al. (eds.), *Plant-Microbes-Engineered Nano-particles (PM-ENPs) Nexus in Agro-Ecosystems*,
Advances in Science, Technology & Innovation,
https://doi.org/10.1007/978-3-030-66956-0_4

Crystal	Structure	Lattice structure	Density	Molecules	Specific properties and applications

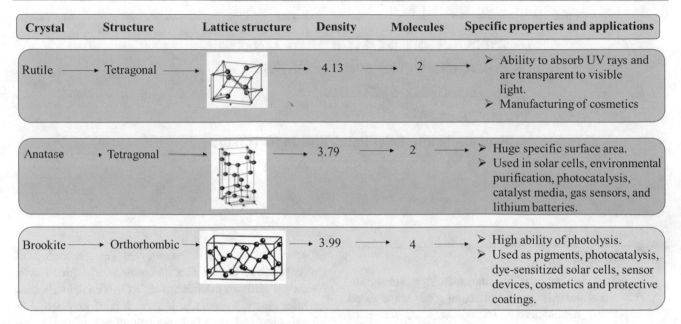

Fig. 1 Summarization of types of crystal form of TiO_2 nanoparticles, their properties and applications

(Nyamukamba et al. 2018). Normally, a mixture of two crystal forms, viz. anatase and rutile, is one of the most commonly used and manufactured NMs. Commercial production of TiO_2-NPs has been increased from 5000 metric tons per year (MT yr^{-1}) to more than 10,000 MT yr^{-1} during 2006–2014 that is expected to increased uninterruptedly up to approximately 2.5 million MT by 2025 (Menard et al. 2011; Ziental et al. 2020).

Each form of TiO_2-NPs is associated with some specialized properties based on which they are utilized for production of wide range of consumer goods (Fig. 1). A mixture of crystalline forms (anatase and rutile) of TiO_2-NPs is extensively used as coloring agent in coating, plastics and glass. It is also observed that this mixture exhibits higher efficiency for conversing solar energy to electrical energy hence, used in nanocrystalline solar cells (Riu et al. 2006). By virtue of some unique physical and chemical properties like brightness with high refractive index ($n = 2.4$) high stability, anticorrosive, UV attenuating (includes both UV light absorbing and scattering) and photocatalytic activity, TiO_2-NPs are broadly used in myriad of consumer and industrial product, including sunscreens and toothpaste, paints, lacquers and paper, plastics, pharmaceuticals, textiles gas sensor and in photocatalytic processes such as water treatment to eliminate hazardous industrial by-products (Fig. 2) (Riu et al. 2006; Keller et al. 2013; Wang et al. 2014; Waghmode et al. 2019). There are some emerging future applications that include their use for self-cleaning and anti-fogging purposes (Montazer and Seifollahzadeh 2011), as potential photosensitizers in photodynamic therapy (PDT) for cancer treatment, etc. (Shi et al. 2013). Besides

this, TiO_2-NPs also show antibacterial and antiviral disinfectants properties under UV light irradiation (Montazer and Seifollahzadeh 2011). The Food and Drug Administration (FDA), USA, has given approval to use TiO_2 as a food pigment additives and preservatives; therefore, it is used in food products like candies, oils, beverages, sweeteners and other processed foods (Weir et al. 2012).

Due to excessive production and improper handling, TiO_2-NPs find their way to different sections of environment (water, soil and air) and hence, considered as an emerging environmental contaminant. It is also estimated that 9–37% of engineered NPs are emitted directly into the atmosphere, whereas the remaining 63–91% eventually ends up in landfills (Keller et al. 2013). TiO_2-NPs deposited in soils and in landfills with the dominant fraction of 80.6% (Nowack et al. 2015). Annual input of TiO_2-NPs into soil in Europe may reach up to 0.13 $\mu g\ kg^{-1}$ that may further increase as high as 1200 $\mu g\ kg^{-1}$ if fields are exposed to sewage sludge (Sun et al. 2014). Future application of agrochemicals formulations in agriculture may lead to an additional annual deposition (3 or more than 5000 $\mu g\ kg^{-1}$) of TiO_2-NPs into soils (Gogos et al. 2012; Moll et al. 2016). The excessive emission of TiO_2-NPs presents the most significant exposure avenues to the ecosystem where they are taken up by the aquatic and terrestrial plants and may cause damage (Zhu et al. 2010; Goswami et al. 2017; Shah et al. 2017). Therefore, concern over the potential risks of these NPs to environment has been raised (Ghosh et al. 2010; Khot et al. 2012; Shah et al. 2017). An improved knowledge of TiO_2-NPs toxicity in plants and other organisms will help in evaluating risks and also in their benign use in agriculture.

Fig. 2 Potential applications of TiO$_2$ nanoparticles

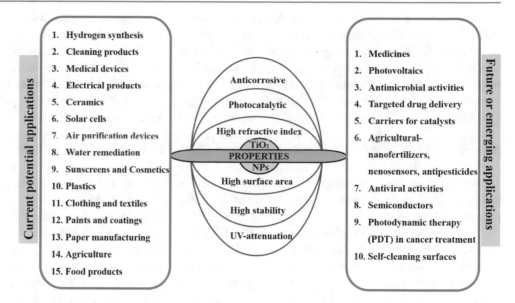

1.1 Nanoparticles in Agro-ecosystems

Soil is known as the sink of NPs and portrays first site of interactions between NPs and plants (Cornelis et al. 2014). Plants being a primary producer play a critical role for any ecosystem to function. As plants are first point of entry of NPs, thus opens a potential pathway for TiO$_2$-NPs in food chain through the uptake and transportation and can be accumulated in high trophic level consumers (Zhu et al. 2010; Rico et al. 2011). Air is another means of NPs contamination through which they primarily interact with leaf and other aerial parts of the plant.

Over the past several years, research has been focused on the NPs' interaction with plants and their impact on ecology, food chain and human health (Baan et al. 2006; Cox et al. 2016; Ali et al. 2017; Goswami et al. 2017; Ziental et al. 2020). Excessive accumulation of NPs inside the plant system adversely affects the various physiological and metabolic processes by regulating genes and cellular components, and consequently affects plant yield and productivity (Cornelis et al. 2014; Siddiqui et al. 2015; Tripathi et al. 2017; Tan et al. 2018). A large number of toxicological studies conducted on various plants reported contrasting effects (detrimental and beneficial) of TiO$_2$-NPs on plants growth and development (Tripathi et al. 2017; Chaudhary and Singh 2020). However, their impacts were dependent on size, chemical structure and concentration of NPs (Mukherjee et al. 2016) and show variation with species and stages of plant growth (Du et al. 2011; Servin et al. 2012).

Despite of the availability of rich source of information on the toxicity of TiO$_2$-NPs in various organisms, still there are little researches which have been performed on terrestrial plants. Further, mechanisms by which TiO$_2$-NPs exert contradictory effects on growth and development have not yet

been completely elucidated. Indeed, for monitoring environmental risks, it is very essential to know the absorption, uptake and accumulation of TiO$_2$-NPs in plants as well as their interaction with plant cells and biomolecules which is overviewed in this chapter. Besides, recently in the perspective of sustainable agriculture, application of TiO$_2$-NPs is considered as one of the challenging approaches to augment plant performance and productivity and to meet emerging demand for food. However, the application of NMs in the field of agriculture is in nascent stage, but over the past few years few studies have been conducted with an objective to promote commercial applications of TiO$_2$-NPs in agriculture. TiO$_2$-NPs are found to be useful to improve plants performance under different abiotic stresses (like low temperature, heat, drought, salinity, heavy metal) and biotic stresses (pests, pathogens infections) (Khan 2016; Singh and Lee 2016; Gohari et al. 2020). It has been suggested that TiO$_2$-NPs' treatment induces production of secondary metabolites and alleviates the oxidative damage caused by reactive oxygen species (ROS) by activation of antioxidant defense system which ultimately improve plant performance under stress condition. This chapter will provide an overview of the different roles and applications of TiO$_2$-NPs with their future perspectives in agriculture sector.

2 Uptake, Translocation and Accumulation of TiO$_2$ Nanoparticles in Plants

Uptake and translocations of TiO$_2$-NPs in plant are complex processes that are in novice stage. Nowadays, researchers have begun to elucidate the mechanism of their uptake and translocation in plants. NPs are introduced to different sections of the environment with an estimate of 13.8% into soil,

18.5% into water and 2.2% in the air (Keller et al. 2013). Soils that are final sink or main source of NPs' pollutant (Cornelis et al. 2014) play a significant role in transformation and modulation of TiO$_2$-NPs. NPs interact with soil as well as the other environmental components and affect their properties and their behavior (Fig. 3). Soil being a negatively charged due to presence of hydroxyl ions and natural organic material (NOM) may effectively attract the positive-charged NPs and may affect the solubility, mobility and plant availability of TiO$_2$-NPs. The factors that are known to affect these processes include chemical characteristics, pH, cation-exchange capacity (CEC), redox potential and NOM content. An increase in the hydrogen ion (H$^+$), i.e., lowering of pH in soil, increases the TiO$_2$-NPs availability since H$^+$ has higher affinity for negative charges on clay particles and soil colloids, thus competing with the TiO$_2$-NPs, and releasing the NPs. High organic content including fulvic and humic acids can lead to an improved stability and hence, better bioavailability of TiO$_2$-NPs.

In general, high clay and/or NOM content in soil along with high pH reduces the TiO$_2$-NPs mobility and availability to plants (Pachapur et al. 2016). The attractions between various functional groups like –COOH, –OH present on organic matters and the TiO$_2$-NPs decrease the zeta potential that increase the stability of TiO$_2$-NPs in soil. While low pH and high redox potential/ zeta potential facilitate the release of firmly attached TiO$_2$-NPs in rhizospheric region from where they can be easily taken up by plants root. Mudunkotuwa and Grassian (2010) observed an aggregation of TiO$_2$-NPs at different pH levels that may ultimately reduce their uptake through cell wall because now they are bigger than the pore size of cell wall/plasma membrane. An increase in salt concentration in soil might induce aggregation and precipitation of TiO$_2$-NPs, which may produce differing effects (Navarro et al. 2008).

The existence of microorganisms such as bacteria and fungi in soil also manipulate the NPs uptake, primarily if these organisms are symbiotically associated with plants just like mycorrhizal fungi (Feng et al. 2013; Wang et al. 2016). Moreover, plants show different mechanisms for low and high uptake of TiO$_2$-NPs. Plants reduce the NPs' uptake strategically by evolving a method to increase the pH in the rhizosphere, which in turn reduce TiO$_2$-NPs mobility in soil. Mucilaginous secretion and exudates excreted either from the plants or microorganisms acidify the rhizosphere which promotes dissolution of NPs (Ma et al. 2010; Zhang et al. 2012). On contrary, to increase NPs uptake, plants decrease the pH around the root zone by releasing more H$^+$

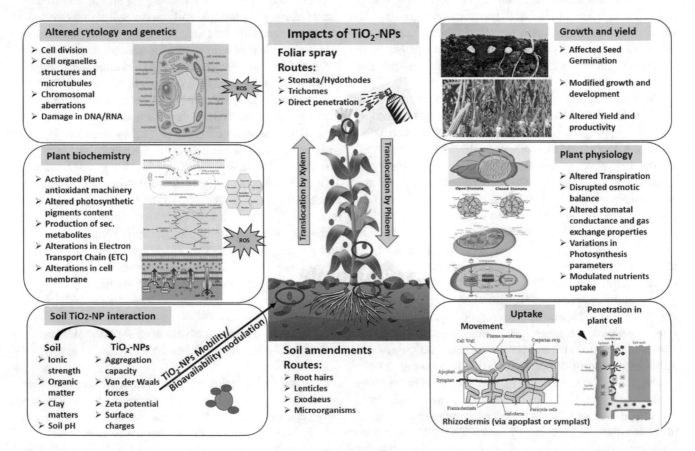

Fig. 3 TiO$_2$ nanoparticle uptake, translocation, accumulation and impacts on plant

(Monshausen et al. 2007; Kurepa et al. 2010) which increases the bioavailable TiO_2-NPs and thereby the uptake. The uptake of NPs in plants is determined by various factors such as plant species and age, and morphological and chemical properties of NPs (Nair et al. 2010; Rico et al. 2011). In addition, their uptake also depends on exposure pathways.

2.1 Uptake and Translocation of TiO₂ Nanoparticles Through Roots

Nanoparticles after released into soil biotransformed by the interaction of humic acid and root exudates, and then uptaken by the surface pores of root cells (Rico et al. 2011). TiO_2-NPs first move into root through apoplast. Then, some of the total amount of TiO_2-NPs is transported further into the cell, and some become bound to the cell wall substances. '*The mechanism of NPs uptake is generally considered as an active-transport mechanism that includes several other cellular processes such as signaling, recycling and the regulation of plasma membrane*' (Tripathi et al. 2017).

The TiO_2-NPs' movement from outside medium into the root cell wall is a non-metabolic and passive route that is determined by diffusion or mass flow. After getting entry through cell wall and plasma membrane in root epidermis, TiO_2-NPs reach to stelar vascular tissues (xylem) through apoplastic or symplastic movements or by both (Larue et al. 2012b; Kurepa et al. 2010) (Fig. 3). Then, from root tissue they are transported to other regions in plants through unidirectional movement using xylem tissue. Raliya et al. (2015a, b) reported TiO_2-NPs accumulation in roots, shoots and leaves and suggested that once TiO_2-NPs are uptaken by tomato plants (either through leaf epidermis or root cell), they are translocated throughout the plant using xylem and phloem tissues of plant. TiO_2-NPs uptaken by leaf cells follow bidirectional pathways where NPs transported by phloem tissue.

Wang et al. (2014) suggested that upon foliar application, due to small size NPs enter in to the plant cells either by direct penetration or by the mechanism of gaseous uptake. The rate of NPs translocation and their accumulation in various plant tissues is dissimilar for both foliar and soil application and depends on the shape and size of NPs and the size of pores present on cell wall (Carpita et al. 1979; Asli and Neumann 2009; Judy et al. 2012) as the root cell wall is the main site of NPs entry into the plant cells (Kurepa et al. 2010). Size seems to be one of the main factors that limit the movement and accumulation of NPs. NPs having dimension up to 40–50 nm only can move and accumulate within the cell (Gonzalez-Melendi et al. 2008; Taylor et al. 2014) while large-sized TiO_2-NPs cannot enter into plant cells and thus sieved out (Larue et al. 2011).

However, there are some studies showing the accumulation/internalization of bigger-sized TiO_2-NPs of about 450 nm dimension in plant cells (Santos Filho et al. 2019). It seems that they might have followed another path to get entry into the cells.

Like some other NPs, TiO_2-NPs might have induced the formation of new and large-sized pores on cell wall (Navarro et al. 2008; Wang et al. 2016; Yan and Chen 2019) and may directly reach to cytosol without forming endosomes or encapsulating in any organelle (Serag et al. 2011). Besides, in order to get successful entry and internalization within the cell TiO_2-NPs may bind to some surrounding proteins that could behave as carrier proteins (Nel et al. 2009). In this regard, aquaporins have been identified as a potential transporters for NPs within the plant cell (Rico et al. 2011), but because of very small size of aquaporins (2.8–3.4 A°) (Wu et al. 2017), make them dubious path for NPs entry (Schwab et al. 2015). Besides, they might be integrated into the cell through invagination of the plasma membrane forming a vesicle that can move to various cell compartments as endosomes (Etxeberria et al. 2006; Kurepa et al. 2010). Kurepa et al. (2010) demonstrate the presence of TiO_2-50% Alizarin red S (ARS) nanocomposites in form of endosomes (globular bodies) in cotyledons cells and epidermis of petioles and hypocotyls of *Arabidopsis thaliana*. They proposed that TiO_2-NPs internalized both by clathrin-dependent and independent endocytic pathways as observed by Onelli et al. (2008) in *Nicotiana tabaccum*, for gold NPs. Additionally, the type of NPs and their morphology and physico-chemical properties have also been observed to play a determining role in NPs uptake (Ma et al. 2010; Rico et al. 2011; Raliya et al. 2016). TiO_2-NPs within the plant cell are found to be transported through the plasmodesmata show symplastic movement (Kurepa et al. 2010; Tripathi et al. 2017; Yan and Chen 2019).

In *Arabidopsis*, TiO_2-NPs are found to aggregate in plasmodesmata and in the cell wall (Kurepa et al. 2010) suggesting that there may be obstruction of intercellular communication, due to accumulation of TiO_2-NPs. Kurepa et al. (2010) also proposed that the release of H^+ by the plant root cells resulted into adsorption of TiO_2-50% ARS nanocomposites on the surface of root that promotes micronutrients uptake from the surrounding environment consequently by lowering the pH of the root zone (Monshausen et al. 2007). Mattiello and Marchiol (2017) reported TiO_2-NPs uptake by root tissue and a subsequent translocation and accumulation in barley seedling tissues and in stroma of chloroplast. Similarly, Kurepa et al. (2010) observed their accumulation in the shoots of *Arabidopsis thaliana*. TiO_2-NPs was uptaken and accumulated increasingly in tip of root passing when passing through the various root tissues like root cap, epidermis, columella and initials of root meristem, sequentially. A transmission electron microscopy (TEM) study confirmed that NPs maintained their morphology and size during accumulation in various plant parts (Raliya et al. 2016). Following a well-known apoplastic pathway, TiO_2-NPs may diffuse in the spaces present between the cell wall and plasma membrane (Lin et al. 2009). Gonzalez-Melendi et al. (2008) reported some NPs in extracellular space and within some cells in *Cucurbita* plants. Aggregates of TiO_2-NPs tend to accumulate in endodermal cells due to presence of Casparian strips which act as barrier for apoplastic movement of NPs (Larue et al. 2012a; Patrick et al. 2015). For efficient translocation, NPs that are following apoplastic pathway must enter into the symplast of the cell to reach to vascular tissues (xylem and phloem). Ultra-small TiO_2-NPs disturbed the structural integrity of microtubular networks of plasmodesmata indicating its symplastic movement in *Arabidopsis* (Wang et al. 2011). Further, binding of TiO_2-NPs with array of carrier proteins such as aquaporins helps in accomplishment of their internalization into cell (Rico et al. 2011; Patrick et al. 2015).

2.2 Uptake and Translocation of TiO_2 Nanoparticles Through Leaves

In addition to root pathway, TiO_2-NPs can also enter the plants by means of leaves through foliar spray. The entry of TiO_2-NPs through aerial parts may involve stomata, trichomes, hydathodes, lenticels and cuticle wounds or they may directly penetrate through the foliar cells as found in tomato leaves; then, translocated to other plant tissues along with the sugar and nutrients through phloem (Nair et al. 2010; Raliya et al. 2015a). The current studies regarding the explained mechanisms are scanty, but many researchers are working toward it. Wang et al. (2013) reported that upon application of 100 mg l^{-1} of TiO_2-NPs on leaves, the recovery rate of TiO_2-NPs translocated into leaf, stem and roots was observed 61.25%, 33.30% and 5.45%, respectively. This led them to assume that uptake of TiO_2-NPs was mediated via stomata. Similar studies were also reported where TiO_2-NPs got distributed from leaves to other parts in lettuce (Larue et al. 2011). Studies showed that foliar applications of TiO_2-NPs at reproductive stage increased pigment content and photosynthesis in maize (Morteza et al. 2013). However, antagonistic effects of TiO_2-NPs were also reported, for instance, application of TiO_2-NPs decreased net photosynthetic rate in long raceme elm (Gao et al. 2013). A size-dependent uptake of TiO_2-NPs in *Triticum* and rapeseed leaves was also reported by many researchers (Kurepa et al. 2010; Larue et al.2012b; Chichiricò and Poma 2015).

Overall, it is concluded that some morphological and chemical dissimilarities among plant species, such as difference in hydraulic conductivity and pore size of cell wall may manipulate translocation and accumulation of NPs (Judy et al. 2012). However, more research is needed to understand the mechanisms of TiO_2-NPs uptake and their intracellular accumulation and distribution.

3 Impacts of TiO_2 Nanoparticles

3.1 In Ecosystem

Large-scale use of TiO_2-NPs in consumer products contribute their exposure to both biotic (flora and fauna) as well as abiotic factors (soil, air, water) of environment. Recent estimate shows 13.8% to soil, 18.5% to water and 2.2% release of TiO_2-NPs in soil, water and air, respectively (Tan et al. 2018). Boxall et al. (2007) anticipated their concentration to be 24.5 mg l^{-1} for water and 1030 mg kg^{-1} for soil. It is clear that TiO_2-NPs present in a significant amount in each compartment of ecosystem which acts as the sink for them. Plants being a primary producer are an important component of any food chain and occupy first trophic level. Moreover, they provide first point of entry of TiO_2-NPs through which they may further transferred to different trophic levels of food chain occupied by consumers like invertebrates and vertebrates and ultimately affect its functioning (Federici et al 2007; Blaise et al. 2008; Binh et al. 2015; Cox et al. 2016; Tripathi et al. 2017; Tan et al. 2018). Some of the impacts of TiO_2-NPs on different components of ecosystem are listed in Table 1.

The excessive accumulation of TiO_2-NPs in soil badly affects its composition, quality and fertility by inhibiting soil microbial enzyme activities and diversity (Du et al. 2011; Simonin et al. 2016; Tan et al. 2018). Contrary to this, Menard et al. (2011) reported alteration and improvement in water properties due to aggregation, partition and increased suspended particulate matter on TiO_2-NPs exposure in water bodies. High concentrations of TiO_2-NPs in air may further amalgamate with other environmental pollutants (Shah et al. 2017). In addition, they are also reported to affect various biotic communities by altering their biological processes and growth (Sharma 2009; Roh et al. 2010; Lapied et al. 2011; Menard et al. 2011; Shi et al. 2013; Hou et al. 2019). Ranjan and Ramalingam (2016) reported inhibition in bacterial growth on exposure to TiO_2-NPs due to production of reactive oxygen species (ROS) and alternation of membrane integrity. TiO_2-NPs are observed to inhibit growth and colonization in fungi (Markowska-Szczupak et al. 2011). Mosses or bryophytes also respond to TiO_2-NPs and are considered as good TiO_2-NPs accumulator, thus can be utilized for monitoring pollution (Motyka et al. 2019).

Inhibition in fertility, sustainability, enhancement in mortality and apoptotic frequency are some of prominent effects of ecotoxicity observed in fishes and invertebrates on

Table 1 Impacts of TiO$_2$ nanoparticles exposure on various components of the ecosystem

S. No.	Component	Impacts	References
1	**Soil**	1. Activated cascading negative effects on denitrification enzyme activity	Simonin et al. (2016)
		2. Intricate alterations of the bacterial community structure	Du et al. (2011)
		3. Soil enzyme activities were inhibited	Tan et al. (2018)
		4. Soil quality and health were affected	
		5. Reduced microbial diversity of soil	
2	**Water**	1. Aggregation and partition to sediment	Menard et al. (2011)
		2. Increased suspended particulate matter in water	
		3. Alteration and improvement in water properties	
3	**Air**	1. Oxidize organic and inorganic compounds	Binas et al. (2017)
		2. Alter and improve air quality	
4	**Flora**		
i	Bacteria	1. Increased reactive oxygen generation	Ranjan and Ramalingam (2016)
		2. Alteration of membrane integrity and permeability	
		3. Growth inhibition and died due to inner wall rupture	
ii	Algae	1. Reduced the availability of light to entrapped algal cells	Menard et al. (2011)
		2. Inhibiting their growth	Sharma (2009)
		3. Lipid peroxidation was induced	
iii	Fungi	1. Prevents the fungal colonization	Markowska-Szczupak et al. (2011)
		2. Inhibition of growth under UV irradiance	
iv	Bryophytes	1. Accumulation in the various compartments of the moss shoots	Motyka et al. (2019)
		2. It can be utilized for biomonitoring of the TiO$_2$-NP pollution	
v	Higher plants	1. Both positive and negative impacts observed	Menard et al. (2011)
		2. Affected physiology, biochemistry and genetic constitution	Raliya et al. (2015a, b), Santos Filho et al. (2019)
		3. Affected plant growth and productivity	
5	**Fauna**		
i	Invertebrates	1. Provoke ecotoxicity on *Caenorhabditis elegans* fertility and survival	Roh et al. (2010)
		2. No mortality, but an enhanced apoptotic frequency for *Lumbricus terrestris*	Lapied et al. (2011)
ii	Vertebrates	1. Increased oxidative stress, damages lipids, carbohydrates, proteins and DNA	Ramsden et al. (2013)
		2. Neuronal dysfunction and neurodegenerative diseases	Vignardi et al. (2015)
		3. Vacuolar degeneration, necrosis and apoptosis of liver cells	Hou et al. (2019)
		4. Induce genotoxicity and carcinogenesis	

TiO$_2$-NPs exposure (Roh et al. 2010; Lapied et al. 2011). TiO$_2$-NPs reported to have induced immunotoxicity, cytotoxicity and genotoxicity, and thus affect physiology and reproductive processes of fishes and their larvae (Jovanović and Palić 2012; Ramsden et al. 2013; Vignardi et al. 2015). Potential impacts of TiO$_2$-NPs on vertebrates or high-level organisms were also investigated where they led to cell injury, apoptosis, oxidation and DNA damage and aging (Baan et al. 2006; Trouiller et al. 2009; Shah et al. 2017; Hou et al. 2019). In conclusion, the different impacts of NPs on various components of ecosystem are very complex and diverse, and depend on various physico-chemical characters of NPs. However, their behavior and fate in natural systems must be evaluated well to understand the environmental hazards associated with them.

3.2 Phytotoxicity of TiO$_2$-NPs

As plants play a crucial role in food chain of any ecosystem, thus any damage to them would be noxious for each factor and ultimately leads to imbalance in ecosystem (Ma et al.

2010; Tan et al. 2018). Considering this, a large number of studies have been done to assess TiO_2-NPs induced toxicity on seed germination, growth and development of plants and revealed that these NPs caused contrasting (both positive and negative) impacts on growth and development of plants by affecting their metabolic processes. Impacts of TiO_2-NPs on various plants are documented in Table 2. The literature related to impacts of TiO_2-NPs on plants is still emerging. It is evident from various toxicological studies of TiO_2-NPs, it can be concluded that the interaction between plants and TiO_2-NPs is very complicated and influenced by properties of both TiO_2-NPs and plants as well (Mukherjee et al. 2016). Here we will classify our studies in following four main sections:

Table 2 Summary of various studies for the phytotoxicity of titanium dioxide nanoparticles in plants

S. No.	Plants	Application of TiO_2-NPs	Concentrations	Impacts	References
I	**Growth and development**				
1	Spinacia oleracea	5 nm, anatase, seeds soaked and foliar application	0.25%	Improved spinach growth and leaf area	Yang et al. (2007)
2	Allium cepa	21 nm, anatase, suspended on distilled water	10, 100, and 1000 mg l^{-1}	Reduced seed germination rate and root development	Santos Filho et al. (2019)
3	Zea mays	30 nm, suspended in medium	30 and 1000 mg l^{-1}	Growth inhibition, inhibited leaf and roots growth	Asli and Neuman (2009)
4	Solanum lycopersicum	22–28.5 nm, anatase, soil and foliar application	0–1000 mg kg^{-1} or mg l^{-1}	No effect on seed germination but promoted growth	Raliya et al. (2015b)
5	Agropyron desertorum	21 nm, mixture of anatase and rutile. Seeds soaked in np treated water	0, 5, 20, 40, 60 and 80 mg l^{-1}	Improved seed germination and seedling growth	Azimiet al. (2013)
6	Linum usitatissimum	10–25 nm, anatase, foliar application	0, 10, 100, and 500 mg l^{-1}	Increased plant height, number of subsidiary branches per plant	Aghdam et al. (2016)
7	Oryza sativa	20 nm, anatase, suspended in Hoagland nutrient solution	0, 100, 250, and 500 mg l^{-1}	reduction in biomass	Wu et al. (2017)
8	Arachis hypogaea	5 nm, anatase, NP powders blended with soil mixture	50 and 500 mg·kg^{-1}	Increased root and shoot biomass	Rui et al. (2018)
9	Triticum aestivum	< 20 nm, anatase, mixed in soil	0, 20, 40, 60, 80, 100 mg kg^{-1}	Promoted growth, increased biomass	Rafique et al. (2018)
10	Hordeum vulgare	< 25 nm, anatase, nTiO$_2$ powder suspensions	0, 500, 1000, and 2000 mg l^{-1}	reduction in the development and germination	Mattiello et al. (2015)
11	Lemna minor	5 -10 nm, anatase, suspended in medium	0, 10, 50, 100, 200, 1,000, and 2,000 mg l^{-1}	Inhibited plant growth	Song et al. (2012)
12	Avena sativa	22- 25 nm, suspended in distilled water	0, 250, 500 and 1000 mg l^{-1}	Promotion in seed germination and seedlings growth	Andersen et al. (2016)
13	Cucumis sativus	22- 25 nm, suspended in distilled water	0, 250, 500 and 1000 mg l^{-1}	Promoted seed germination rate and seedling growth	Andersen et al. (2016)
14	Spinacia oleracea	Rutile, seeds soaked in TiO$_2$.NPs solution	0, 0.25, 0.5, 1.0, 1.5, 2.0, 2.5, 4.0, and 6.0%	Increased seed vigor index, germination rate and biomass	Zheng et al. (2005)
15	Brassica campestris	27 nm, anatase + rutile, seeds treatment	0, 100, 500, 1,000, 2,500, and 5,000 mg l^{-1}	Seed germination and seedling growth enhanced but not significantly	Song et al. (2013)
16	Arabidopsis thaliana	21 and 33 nm, anatase, seeds treatment	0 and 500 mg l^{-1}	Increased seed germination and developmental process	Tumburu et al. (2015)
17	Hordeum vulgare	foliar application of nTiO$_2$ using spray	0 and 2000 ppm	Increased cell growth and enhanced fresh and dry wt	Janmohammadi et al. (2016)
18	Foeniculum vulgare	21 nm, anatase, seeds treatment	0, 5, 20, 40, 60,80 mg l^{-1}	Germination rate and biomass increased	Feizi et al. (2013)

(continued)

Table 2 (continued)

S. No.	Plants	Application of TiO$_2$-NPs	Concentrations	Impacts	References
II	**Biochemical and Physiological**				
1	*Spinacia oleracea*	5 nm, anatase, seeds treatment	0.25%	Improved N$_2$ cycle, oxygen evolution, chlorophyll synthesis and photosynthesis	Yang et al. (2007)
2	*Zea mays*	30 nm, suspended in medium	30, 1000 mg l^{-1}	Reduction of cell wall pore size, reduced transpiration	Asli and Neumann (2009)
3	*Solanum lycopersicum*	22–28.5 nm, anatase, soil and foliar application	0–1000 mg kg^{-1}	Relative chlorophyll in leaves and lycopene content of fruits increased	Raliya et al. (2015b)
4	*Linum usitatissimum*	10–25, anatase, foliar application	0, 10, 100, and 500 mg l^{-1}	Enhanced chlorophyll, carotenoids contents, reduced MDA, inhibited H$_2$O$_2$ accumulation	Aghdam et al. (2016)
5	*Arachis hypogaea*	5 nm, anatase, NP powders blended with soil mixture	50 and 500 mg kg^{-1}	Increased photosynthetic efficiency, altered biochemical profile	Rui et al. (2018)
6	*Triticum aestivum*	< 20 nm, anatase, mixed in soil	0, 20, 40, 60, 80, 100 mg kg^{-1}	Increased chlorophyll overproduction of H$_2$O$_2$	Rafique et al. (2018)
7	*Nicotiana tabaccum*	90–110 nm, spherical, suspended in distilled water	0, 2, 4, 6, 8 and 10 mM	Increased MDA content and lipid peroxidation and decrease in concentration dependent manner	Ghosh et al. (2010)
8	*Ulmus elongate*	6.22 nm, anatase, TiO$_2$-NPs powder suspended in water	0.1 g, 0.2 g, or 0.4 g/ 100 ml	Reduced photosynthetic rate, chlorophyll fluorescence and transpiration	Gao et al. (2013)
9	*Spinacia oleracea*	Rutile, seeds soaked in TiO$_2$-NPs solution	0, 0.25, 0.5, 1.0, 1.5, 2.0, 2.5, 4.0, and 6.0‰	Enhanced rubisco activity, chlorophyll content, photosynthesis	Zheng et al. (2005)
10	*Phaseolus vulgaris*	21 nm, anatase, application by spraying	0.01%, 0.02%, 0.03% and 0.05%	Increased antioxidant enzymes, inhibited ROS accumulation, reduced chlorophyll degradation	Ebrahimi et al. (2016)
11	*Hordeum vulgare*	foliar application of nTiO$_2$ using spray	0 and 2000 ppm	Improving chlorophyll content, hormones synthesis and photosynthetic complexes	Janmohammadi et al. (2016)
12	*Cicer arietinum*	7–40 nm, anatase, foliar application on potted plants	0, 2, 5, and 10 ppm	Reduced electrolyte leakage, membrane damage and increase cold stress tolerance	Mohammadi et al. (2014)
13	*Solanum lycopersicum*	16.04 nm, anatase, seeds treated with TiO$_2$- NPs solution	0.05, 0.1, 0.2 g l^{-1}	Increased net photosynthesis (PSII activity), transpiration and conductance	Qi et al. (2013)
14	*Scenedesmus sp. and Chlorella sp.*	<25 nm, anatase, suspended in culture media	3, 6, 12, 24, 48, 96 and 192 mg l^{-1}	Chlorophyll content decrease in concentration-dependent manner	Sadiq et al. 2011
15	*Picochlorum sp.*	21 nm, anatase, suspended in culture media	10 mg l^{-1}	High chlorophyll a concentration	Hazeem et al. (2016)
III	**Genetic or molecular**				
1	*Allium cepa*	21 nm, anatase, suspended on distilled water	10, 100, and 1000 mg l^{-1}	Increased lytic vacuoles, oil bodies, nucleolar alterations and damaged DNA	Santos Filho et al. (2019)
2	*Triticum aestivum*	< 20 nm, mixed in soil	0, 20, 40, 60, 80, 100 mg kg^{-1}	Higher micronuclei (MN) formation	Rafique et al. (2018)
3	*Hordeum vulgare*	< 25 nm, anatase, nTiO2 powder suspensions	0, 500, 1000, and 2000 mg l^{-1}	Higher percentage of mitotic index but reduction of cell divisions	Mattiello et al. (2015)
4	*Nicotiana tabaccum*	90–110 nm, spherical, suspended in distilled water	0, 2, 4, 6, 8 and 10 mM	DNA damage, increase micronuclei formation, chromosomal aberrations	Ghosh et al. (2010)
5	*Allium cepa*	90–110 nm, spherical, suspended in distilled water	0, 2, 4, 6, 8 and 10 mM	DNA damage, increase micronuclei formation, chromosomal aberrations	Ghosh et al. (2010)

(continued)

Table 2 (continued)

S. No.	Plants	Application of TiO$_2$-NPs	Concentrations	Impacts	References
6	*Zea mays*	<100 nm, mixture (rutile and anatase), seeds soaked	0.2, 1.0, 2.0, and 4.0‰	Damaged structure of DNA, reduction in mitotic index	Castiglione et al. (2011)
7	*Vician arbonensis*	<100 nm, mixture (rutile and anatase), seeds soaked	0.2, 1.0, 2.0, and 4.0‰	Chromosomal structure fragmentations and damage	Castiglione et al. (2011)
8	*Cucurbita pepo*	23–31 nm, mixture (rutile and anatase), suspended in media	50 mg l^{-1}	Damaged genomic DNA	Moreno-Olivas et al. (2014)
IV	**Yields and productivity**				
1	*Solanum lycopersicum*	22–28.5 nm, anatase, soil and foliar application	0–1000 mg kg^{-1}	Enhanced fruit yields, biomass and productivity	Raliya et al. (2015b)
2	*Linum usitatissimum*	10–25 nm, anatase, foliar application	0, 10, 100 and 500 mg l^{-1}	Enhanced seed oil, yield and protein contents	Aghdam et al. (2016)
3	*Arachis hypogaea*	5 nm, anatase, NP powders blended with soil mixture	50 and 500 mg kg^{-1}	Promoted crop yield and its nutritional quality	Rui et al. (2018)
4	*Hordeum vulgare*	foliar application of nTiO$_2$ using spray	0 and 2000 ppm	Increased grain yield and biomass	Janmohammadi et al. (2016)
5	*Triticum aestivum*	foliar application of nTiO$_2$ using spray	0.01%, 0.02%, 0.03%	Enhanced plant growth, yield and quality (gluten and starch content)	Jaberzadeh et al. (2013)
6	*Spinacia oleracea*	5 nm, anatase, seeds soaked and foliar application	0.25%	Enhanced biomass, nutritional quality and improved yield	Yang et al. (2007)
7	*Spinacia oleracea*	Rutile, seeds soaked in TiO$_2$-NPs solution	0, 0.25, 0.5, 1.0, 1.5, 2.0, 2.5, 4.0, 6.0%	Improved biomass and yield	Zheng et al. (2005)
8	*Oryza sativa*	20 nm, anatase, suspended in Hoagland nutrient solution	0, 100, 250, and 500 mg l^{-1}	Reduction in biomass	Wu et al. (2017)

3.2.1 Phytotoxicity at Morphological Level

Upon interaction with plants, TiO$_2$-NPs significantly influence the morphological characteristics of plants such as seed germination, growth potential and biomass (Zheng et al. 2005; Raliya et al. 2015a, b; Santos Filho et al. 2019). An extensive research revealed that TiO$_2$-NPs show both beneficial and detrimental effects on seed germination and growth (Andersen et al. 2016). However, there are some studies reported no significant impacts on seed germination or growth parameters in plant like *Lactuca sativa*, *Brassica campestris*, *Phaseolus vulgaris* (Song et al. 2013), *Solanum lycopersicum* (Raliya et al. 2015b), *Zea mays* (Asli and Neumann 2009) and *Hordeum vulgare* (Mattiello et al. 2015). Andersen et al. (2016) also reported seed germination, root elongation and growth in plant species varying from one species to other. Results showed that TiO$_2$-NPs can penetrate the seed coat and accumulate within tissues, but it did not show toxicity. Studies reported that TiO$_2$-NPs at its optimal concentration showed positive impacts on root and shoot growth of the *Triticum aestivum*. The biomass as well as dry weight of the root was also extremely affected by TiO$_2$-NPs (Feizi et al. 2012; Mahmoodzadeh and Aghili 2014). In a study, Zheng et al. (2005) observed positive impacts of TiO$_2$-NPs on seed germination rate, germination index, seed vigor index and seedling growth of spinach.

Similar results for these parameters for TiO$_2$-NPs were also reported in several plant species, including *Spinacia oleracea* (Zheng et al. 2005; Yang et al. 2007), cucumber (Servin et al. 2012), *Agropyron desertorum* (Azimi et al. 2013), *Foeniculum vulgare* (Feizi et al. 2013), *Brassica napus* (Mahmoodzadeh et al. 2013), *Solanum lycopersicum* (Raliya et al. 2015b), *Arabiodpis thaliana* (Tumburu et al. 2015), *Linum usitatissimum* (Aghdam et al. 2016), *Vigna radiata* (Singh et al. 2016), *Hordeum vulgare* (Janmohammadi et al. 2016) and *Arachis hypogaea* (Rui et al. 2018). According to them, TiO$_2$-NPs' treatment improves seed performance and various morphological parameters such as root elongation, shoot growth, seedling growth, number of lateral roots, leaf area and plant biomass. TiO$_2$-NPs mediate photo-generation of superoxide and hydroxide anions and can reactivate the aged seeds. Besides they may enhance penetrability of the seed capsule and induce oxidation–reduction reactions that would improve the water and oxygen imbibition in seeds, and thus accelerate the metabolism and promote seed germination (Zheng et al. 2005; Azimi et al. 2013).

Raliya et al. (2015b) found that differences in TiO$_2$-NPs treatments (through foliar and soil amendment) showed contrasting effects on root length. They reported that foliar treatments of TiO$_2$-NPs significantly reduce the root length in tomato plants in all concentrations except at 1000 mg kg^{-1}. TiO_2-NPs treatment (up to 250 mg kg^{-1}) of soil increases the length of root at low concentration but the higher concentrations do not show any significant variations in the root growth. Asli and Neumann (2009) reported that TiO_2-NPs treatments (30 and 1000 mg l^{-1}) promote the leaves growth in stunted maize plant.

On the contrary, the study of Da Costa and Sharma (2015) pointed toward phytotoxicity of TiO_2-NPs on *Oryza sativa* and observed a decrease in seedling growth on exposure to TiO_2-NPs at 1000 ppm. Song et al. (2013) observed the inhibitory effects of TiO_2-NPs on germination and seedling growth of *Brassica napus*, *Lactuca sativa* L. and *Phaseolus vulgaris*. Additionally, TiO_2-NPs application under different experimental conditions negatively affects the growth and development of some plants like *Allium cepa, Lemna minor, Oryzae sativa, Hordeum vulgare* and *Zea mays* by altering physiological and biochemical processes (Table 2).

3.2.2 Phytotoxicity at Physiological Level

The impacts of TiO_2-NPs application on plants at the physiological level could be inspected by observing the chlorophyll content, nutrients uptake, transpiration rate, photosynthetic rate, stomatal conductivity and alteration of hormones. TiO_2-NPs regulate the activities of all enzymes related to nitrogen metabolism, including nitrate reductase, glutamine synthase, glutamate dehydrogenase and glutamine–pyruvic transaminase, etc. which facilitate the absorption of active nitrogen in plants in form of nitrate and help in the conversion process of inorganic nitrogen into organic nitrogen in form of protein and chlorophyll molecules, and improve the rate of photosynthesis that could ultimately reflect as improved biomass and dry mass of treated plant (Yang et al. 2007; Mishra et al., 2014). TiO_2-NPs treatment at low concentration facilitate the absorbance of minerals which in turn promotes the chlorophyll formation and activation of key enzymes for carbon fixation, but at high dose TiO_2-NPs produce ROS under light which would disrupt the membrane structure, therefore, reduce photosynthesis leading to reduced biomass in spinach (Zheng et al. 2005).

Raliya et al. (2015b) reported that foliar spray of 10 mg l^{-1} concentration of TiO_2-NPs on 14-day-old *Vigna radiata* plants significantly improve chlorophyll content and total soluble protein content in leaves by 446.4 and 94%, respectively. TiO_2-NPs treatment also enhanced chlorophyll content in *Solanum lycopersicum* (Raliya et al. 2015b), *Linum usitatissimum* (Aghdam et al. 2016), *Triticum aestivum* (Rafique et al. 2018), *Phaseolus vulgaris* (Ebrahimi et al. 2016) and *Hordeum vulgare* (Janmohammadi et al. 2016) too. It is reported that appropriate doses of TiO_2-NPs protect chlorophyll from degradation and increase its

synthesis by reducing H_2O_2 content in the cells and also enhance seed oil and protein contents (Aghdam et al. 2016; Ebrahimi et al. 2016). Due to photocatalytic properties, TiO_2-NPs can stimulate photosynthetic efficiency by enhancing light absorption capacity and they can also alter profiling of biochemicals such as amino acid and fatty acids (Rui et al. 2018). Zheng et al. (2005) observed that the rutile TiO_2-NPs showed photo-oxidation–reduction reactions which could accelerate the electron transport and the transformation from electric energy to active chemical energy like ATP, promoting the activity of the rubisco activase, and therefore, enhancing the photosynthetic activity. In *Hordeum vulgare*, foliar application of TiO_2-NPs reported to improved defense mechanism, and increase biosynthesis of phytohormones, photosynthetic pigments and photosynthetic efficiency (Janmohammadi et al. 2016).

Seeds of *Solanum lycopersicum* soaked in TiO_2-NPs solution also showed an enhancement of net photosynthesis (PSII activity), transpiration and conductance (Qi et al. 2013). Some studies reported the positive impacts of TiO_2-NPs on chlorophyll content in algal system, including *Picochlorum* sp. (Hazeem et al. 2016), *Scenedesmus* sp. and *Chlorella* sp. (Sadiq et al. 2011). Gao et al. (2013) demonstrated that the treatment of anatase-TiO_2-NPs reduced photosynthetic rate and transpiration rate by reducing the photosynthetic efficiency of mesophyll cells of leaves not by regulating stomatal activity. In *Zea mays*, TiO_2-NPs exposure remarkably decreased the transpiration rate by reducing cell wall pore size and root hydraulic conductivities in a concentration-dependent manner (Asli and Neumann 2009). In addition, TiO_2-NPs can affect the membrane integrity which consequently modifies the uptake mechanism of water as well as of nutrients. Apart from this, TiO_2-NPs are reported to be augmenting the performance in plants grown under various stresses, which will be discussed later.

The risks of TiO_2-NPs to plants due to its high sensitivity toward cytotoxic and genotoxic effects, clogging of pores and barriers in apoplast stream leads to interruption in nutrients uptake, and thus, causes toxicity (Mattiello et al. 2015; Santos Filho et al. 2019). Despite, TiO_2-NPs' treatment in seedling augment growth and developmental processes by increasing light absorption, chlorophyll content and photosynthesis in treated plants (Yang et al. 2007; Raliya et al. 2015b). Further, Tumburu et al. (2015) provided a genetic basis and observed that TiO_2-NPs treatment increased the expression of various transcripts responsible for root development and cell differentiation.

3.2.3 Phytotoxicity at Biochemical Level

The mechanism of plant responses to TiO_2-NPs exposure could be better assessed by observing ROS generation, oxidative damages H_2O_2 content, malondialdehyde (MDA) level, electrolyte leakage and enzymatic and non-enzymatic antioxidants. It is now well-known fact that plants under severe stress accelerate the production of ROS that leads to oxidative damage in cell by breaking the equilibrium between ROS and their scavenger antioxidants. Increased level of MDA and H_2O_2 in plant cells are the indices of oxidative damages caused by free radicals/ROS produced during stress (Singh et al. 2012). MDA that is a product of lipid peroxidation resulted in low activities of antioxidant enzymes. Ghosh et al. (2010) observed an increase in MDA level at 4 mM of treatment in *Allium cepa* suggesting that the lipid peroxidation caused the DNA damage. Wang et al. (2011) pointed out an increase in ROS on TiO_2-NPs exposure caused DNA damage in *Arabidopsis thaliana*. Plants have antioxidant systems both enzymatic (such as superoxide dismutase (SOD), catalase (CAT), ascorbate peroxidase (APX), guaiacol peroxidase (GPX), dehydroascorbate reductase (DHAR) and glutathione reductase (GR) and non-enzymatic (ascorbic acid, proline, cysteine, non-protein thiols, etc.) to protect themselves from damaging effects of ROS produced under the adverse situation. Jacob et al. (2013) reported that TiO_2-NPs modify the activities of enzymatic antioxidants in *Phaseolus vulgaris* at 10 and 30 ppm. Moreover, in spinach seedlings treated with colloidal solution of 0.25% TiO_2-NPs cause oxidative stress in chloroplasts (Lei et al. 2008).

TiO_2-NPs have found to affect antioxidant activities of SOD, POD and CAT in duckweed (Song et al. 2012), ascorbate peroxidase (APX) in faba bean (Foltête et al. 2011), MDA in onion and tobacco (Ghosh et al. 2010) and thiols (GSH) in bean (Castiglione et al. 2014). Song et al. (2012) observed that in duckweed, TiO_2-NPs increased the enzyme activity at concentration lower than 200 mg l^{-1}, while high concentration (500 mg l^{-1}) caused serious damage to plant cells. Moreover, TiO_2-NPs exposure also ameliorates plant tolerance to various stresses by alleviating the toxicity induced by ROS (Lei et al. 2008; Gohari et al. 2020).

3.2.4 Phytotoxicity at Gene Level

In plants, NPs cause cytotoxicity and genotoxicity in terms of alteration of cell structure, DNA structure, cell division, micronuclei formation, chromosomal aberrations and DNA damage (Yan and Chen 2019; Santos Filho et al. 2019). During the process of entry into a plant cell, TiO_2-NPs may use several transporters present on cell wall (Nel et al. 2009; Rico et al. 2011). The induction of cytotoxic and genotoxic responses in cell lines of plants and animals on TiO_2-NPs exposure evident recently. However, till now the knowledge in context to molecular mechanisms of TiO_2-NP-mediated toxicity in plants is less but still emerging. However, the progressive technology especially transcriptomics can help to understand the link between regulation of genes and their impacts in response to TiO_2-NPs' exposure. Ghosh et al.

(2010) evaluated the genotoxic response of *Allium cepa* and *Nicotiana tabaccum* to TiO_2-NPs using two classical techniques, i.e., comet assay and the DNA laddering. It is concluded that the genotoxic potential of TiO_2-NPs on onion and tobacco increased with increasing numbers of micronuclei formation, chromosomal aberrations and DNA damage. TiO_2-NPs exposure to plants or seeds at particular concentration, resulted in the genotoxicity, mutagenicity and cytotoxicity.

Castiglione et al. (2011) used TiO_2-NPs soaked seeds to know the potential hazards of these NPs on monocots and dicots, i.e., *Zea mays* and *Vician arbonensis*. Results showed reduction of mitotic index and concentration-dependent increase in the DNA damage, chromosomal aberrations and fragmentations. The application of TiO_2-NPs (seeds soaking, soil amendments or foliar application) reported to have genotoxic and cytotoxic impacts in *Hordeum vulgare* (Matiellio et al. 2015), *Triticum aestivum* (Rafique et al. 2018) and *Allium cepa* (Santos Filho et al. 2019). Zhao et al. (2016) observed the plasma membrane damage, the presence of oil bodies and changes in the number of vacuoles in response to TiO_2-NPs. Contrary to this, Frazier et al. (2014) confirmed the activation of the expression profiles of microRNAs (miRNAs) and gene regulators in response to TiO_2-NPs that are known to improve plant development by increasing plant tolerance to abiotic stresses. Besides, selected area electron diffraction (SAED) with TEM analysis done in roots of onion treated with 1000 mg l^{-1} of TiO_2-NPs (anatase phase; 25 nm) revealed internalization of TiO_2-NPs in the vacuole but in the brookite phase that is less reactive and with comparatively bigger in size (orthorhombic format with 450 nm) and predicted this as a protective mechanism (Santos Filho et al. 2019).

3.2.5 Impacts on Overall Plant Productivity and Yield

The overall plant productivity and yield are the resultant of various parameters of plant, i.e., physiological, biochemical and molecular. From the above section, it can be observed that TiO_2-NPs have a crucial role in regulation of these processes. Authors who conducted long-term studies on life cycle of plants are able to conclude results in context to productivity hence, very few studies are reported. Like, Raliya et al. (2015b) demonstrated constructive role of TiO_2-NPs in plant processes and fecundity of plants. It showed enhanced fruit yields, biomass and productivity in *Solanum lycopersicum*. Another full life cycle experiments on *Linum usitatissimum* also reported enhanced seed oil, yield and protein contents on TiO_2-NPs exposure (Aghdam et al. 2016). Similarly, favorable results obtained in case of *Arachis hypogaea* (peanut) planted in soil amended with anatase TiO_2-NPs by Rui et al. (2018). Janmohammadi et al. (2016) observed increased grain yield and biomass for *Hordeum vulgare.* Enhanced plant growth, yield and quality (in terms of gluten and starch content) under foliar application of TiO$_2$-NPs solution in *Triticum aestivum* under water-deficit stress conditions were observed by Jaberzadeh et al. (2013). Results from spinach with TiO$_2$-NPs treatment also showed increased biomass, nutritional quality and improved yield (Zheng et al. 2005; Yang et al. 2007).

From the above discussion, it can be concluded that the impacts posed by TiO$_2$-NPs exhibit a dual characteristic, as it could be either toxic or beneficial which is determined by various factors such as experimental conditions, features of targeted plants and also the nature of NPs used (Zheng et al. 2005; Andersen et al. 2016; Cox et al. 2016; Chaudhary and Singh 2020). However, the results showed promising contribution to support TiO$_2$-NPs applications, but still subtle methodology and conditions are accountable factors.

3.3 Detoxification Mechanism vs TiO$_2$ Nanoparticles Phytotoxicity

It is clear from above discussion that the plants have a defense system that protects them when they are exposed to adverse conditions. At toxic concentration, NPs interact with the cellular moiety and accelerate the production of ROS (Rico et al. 2015; Tripathi et al. 2017). Yin et al. (2012) demonstrated that the excess ROS led to cytotoxicity in a cell in response to TiO$_2$-NPs. These ROS in low concentration act as signaling molecules and thus activate plant antioxidant defense system to alleviate the toxicity caused by different free radicals. Antioxidant defense system of plants has both enzymatic (SOD, CAT, POD, GPX, APX) and non-enzymatic antioxidants including such as ascorbic acid, glutathione, non-protein thiols, polyphenols, carotenoids, etc. (Singh et al. 2016; Rico et al. 2015). SOD detoxifies the superoxide anions by dismutating it into H$_2$O$_2$, while CAT and GPX significantly scavenge the both ROS and peroxy radicals (Rico et al. 2015). In addition to this, enzymes like APX, DHAR and GR that are chief components of ascorbate–glutathione cycle (AA-GSH cycle) help in maintaining ROS concentration and redox status of plant cell (Rico et al. 2015). A large number of research information revealed the direct influence of NPs on the activities of various antioxidant enzymes (Mohammadi et al. 2014; Rico et al. 2015), but there is still lack of information that could associate the activation of antioxidant with the chemical properties of NPs. Indeed, the information is unreliable and irregular that show variations. Lei et al. (2008) and Song et al. (2012) observed that TiO$_2$-NPs enhanced the enzymatic activities of several enzymes such as SOD, CAT, APX and GPX in *Spinacia oleracea* and *Lemna minor,* respectively. On contrary to this, Foltête et al. (2011) reported that TiO$_2$-NPs treatment decreased GR and APX enzymes activities in *Vicia faba.* Besides enzymatic defense system, plants have some cellular defense mechanisms also in which they protect themselves from the toxic effects of TiO$_2$-NPs by increasing the number of oil bodies and lytic vacuoles (Santos Filho et al. 2019). Furthermore, transformation of TiO$_2$-NPs from anatase form (25 nm) to another brookite form (orthorhomboid; 450 nm) may also be a protective mechanism to mitigate the toxic effects of TiO$_2$-NPs in *Allium cepa* (Santose Filho et al. 2019).

4 Scope of TiO$_2$ Nanoparticles in Agriculture

Various anthropogenic activities have led to increased pollution of soil all over the world which ultimately affects the crop yield and productivity. Further, a rapid growth in world's population may eventually end with increasing demand for food supply globally. Thus, there is an urgent need to adapt some innovative technologies in the field of agriculture to ensure food security. Nanotechnology has enormous potential in the field of agriculture. Because of their markedly different physiochemical properties than their bulk counterparts, NPs are of special interest to combat real life agricultural issues and provide better foods globally.

Based on nature, application of TiO$_2$-NPs in agriculture (Fig. 4) can be outlined as:

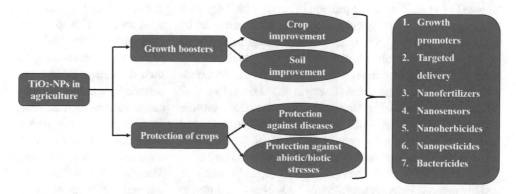

Fig. 4 Applications of TiO$_2$ nanoparticle in agriculture in the diverse forms of nanoformulations to promote growth and to protect plants from various biotic and abiotic stresses

- *Growth boosters*: TiO$_2$-NPs interact with plants and influence their metabolic activities, plant growth and development. Besides this, they can also alter soil properties like physico-chemical and biological which can improve soil health and its quality. So, they can be used as nanofertilizers and nanosensors.

- *Protecting agents*: TiO$_2$-NPs are found to be effectively used in plant protection against various stresses like cold stress, drought, salinity stress, heavy metal toxicity and others. Also, due to their unique properties and chemical activities they can be utilized for crop protection from pests and pathogens. So, they can be utilized as bactericides, nanoherbicides and nanopesticides.

For the achievement of above-mentioned benefits, TiO$_2$-NPs are used as carrier agents for smart targeted delivery of various nutrients and biological molecules which is supposed to get indulged within soil and plant. The detailed discussion and studies to support the utilization of TiO$_2$-NPs in order to promote agriculture production are given in the following sections (Table 3).

4.1 Nanoparticles as Growth Promoter

With an objective to promote TiO$_2$-NPs use for agricultural applications, their effects in seed germination as well as in seedling growth have been investigated in several crop plants (Tables 2 and 3A) and found contradictory (i.e., both positive and negative results). There is no agreement on the responses of plants to TiO$_2$-NPs exposure because different plant species, their growth stages and treatment conditions seem to display diverse responses (Cox et al. 2016; Mukherjee et al. 2016; Mattiello et al. 2015; Tan et al. 2018; Chaudhary and Singh 2020).

TiO$_2$–NPs have been reported to promote crop yield by improving seed germination, growth and development of wide varieties of plants, viz. *Cicer* sp., cabbage, *Brassica napus*, cucumber, spinach, onion, soybean, maize, tomato, wheat and many more. Entry of NPs into the seeds plays a critical significant role in increasing the rate of seed germination (Fan et al. 2016). The priming of seeds with TiO$_2$–NPs could promote seed germination and plant growth by increasing chlorophyll contents, photosynthesis performance by improving light absorbance, activate the photochemical reaction, induction of rubisco activase and nitrate reductase activity and water conduction (Zheng et al. 2005; Raliya et al. 2015a; Singh et al. 2016). Studies have reported that exposure to definite doses of TiO$_2$-NPs accelerated germination in aged seeds of spinach and the germination time in *Triticum aestivum* seeds (Zheng et al. 2005; Feizi et al. 2012; Jiang et al. 2017). Size of NPs plays a very important role in its behavior, toxicity and reactivity. A treatment of TiO$_2$-NPs ($\sim$20 nm) to

Canola seeds, improved seed germination and seedling vigor at 2000 mg l^{-1} concentration (Mahmoodzadeh et al. 2013). Lu et al. (2006) reported an improvement in seed germination and growth at low concentrations of mixture of TiO$_2$-NPs and nSiO$_2$.

By virtue of their photocatalytic properties, TiO$_2$-NPs induce oxidation–reduction reactions and increase growth and development of plant (Hong et al. 2005; Lei et al. 2008; Li et al. 2015). Besides, photo-attenuation activity of TiO$_2$-NPs protects the chloroplasts from photo-oxidation and aging and hence, extends the photosynthetic activity of chloroplasts (Yang et al. 2008). Overall, seed priming and application of TiO$_2$-NPs via roots or leaves (soil amendment or foliar spray) at low concentrations could be a better alternative to improve the quality, biomass and yields of many cultivars (Moaveni et al. 2011; Morteza et al. 2013; Raliya et al. 2015a, b; Rafique et al. 2018). Du et al. (2011) had demonstrated that the concentrations of TiO$_2$-NPs significantly affect the enzymatic activities of soil enzymes which in turn affect the soil properties and health. Therefore, TiO$_2$-NP can be utilized as nanofertilizers to improve agriculture output.

Besides, they have a potential to be used as nanosensors to detect and measure crop's nutrient content, pest and pathogens, weeds, moisture level and fertility of soil and others to increase crop yield. The large surface-to-volume ratio and high reactivity of TiO$_2$-NPs considered suitable as delivery agents for different molecules, proteins, nucleotides and other chemicals in plants that may help in crop improvement. They can be used as a carrier of agrochemicals and facilitate site targeted, controlled and slow release of nutrients for better growth and high yield.

4.2 TiO$_2$ Nanoparticles and Crop Protection

Owing to their photo-protective and photocatalytic characteristics, TiO$_2$-NPs are known to have potential application in different formulations in plant protection as they amend the life span of chemically active ingredients present in plants (Gogos et al. 2012; Khot et al. 2012). Under biotic stress, various pests and pathogens involved which infect a wide species of crop plants and cause a huge crop loss with reduced quantity and quality of plant products. Many chemicals as pesticides, bactericides, fungicides are being used to manage these phyto-pathogens in agriculture to get better crop yield and efficiency. But they are highly toxic and cause various human diseases and environmental threats. Thus, there is an urgent need to find some green alternatives that can effectively control pests and pathogens resistant to chemicals/pesticides, in an environmentally friendly manner. Therefore, NP-mediated green pesticides are now become of unusual significance in crop protection (Chowdappa and

Table 3 Summary of studies related to the TiO$_2$-nanoparticles applications in agriculture

Plant species	Particles size and Treatment	Impacts	References
3A. TiO$_2$ nanoparticles as plant growth promoters			
Vigna radiata L. (Mung bean)	10–15 nm; 10 ppm foliar spray	Enhanced germination rate, reduced germination time, increased growth of root and shoot, number of root nodules and chlorophyll content; improved population and enzymatic activity of rhizospheric microbes.	Raliya et al. (2015a)
Glycine max L. (Soybean)	0.03–0.05%	0.05% treatment enhanced height as well as dry weight	Rezaei et al. (2015)
Triticum aestivum L. (Wheat)	<100 nm; 0, 20,40, 60, 80 100 mg kg^{-1}, 60 days, soil	Increased chlorophyll content and root-shoot biomass up to 80 mg kg^{-1}	Rafique et al. (2018)
Nicotiana tabaccum L. (Tobacco)	<25nm, 0, 0.1, 1, 2.5, 5%	Reduction in biomass, inhibition of seed germination and root growth, up-regulation of gene for alcohol dehydrogenase and APX	Frazier et al. (2014)
Spinacia oleracea L. (Spinach)	0.25–6 %, rutile	Increased rate of seed germination and vigor index in aged seeds; enhanced chlorophyll content, Rubisco enzyme activity, rate of photosynthetic and plant growth.	Zheng et al. (2005)
Cicer arietinum L. (Chickpea)	7–40 nm, 2–10 mg kg^{-1}	Improved plant growth and development by reducing electrolytic leakage ; increased MDA content	Mohammadi et al. (2014)
Triticum aestivum L. (Wheat)	21 nm; 0,1,2,10,100, 500 mg l^{-1}	Lower concentrations increased growth of shoot and root in seedlings	Feizi et al. (2012)
Cucumis sativus L. (Cucumber)	27 ± 4 nm, 0, 250, 500, 750 mg kg^{-1}	Increased CAT activity in leaves; increased P and K content in fruit	Servin et al. (2013)
Brassica napus, (Canola)	20 nm; 2000 mg l^{-1}	Improved seed germination and seed vigor; increased chlorophyll content	Mahmoodzadeh et al. (2013)
Solanum lycopersicum (Tomato)	16 nm, 0.05–0.2 mg l^{-1}	Enhanced photosynthesis, transpiration rate and stomatal conductance	Qi et al. (2013)
Spinacea oleracea L. (Spinach)	4–6 nm, 0.25%	Enhanced plant growth and activity of enzymatic activity of glutamate dehydrogenase, glutamine synthetase and glutamic pyruvic transaminase	Yang et al. (2006)
Allium cepa L. (Onion), *Lycopersicum esulentum*L. (Tomato), *Raphanus sativus* L. (Radish)	0, 100, 200, and 400 mg l^{-1}	Showed positive effect on seed germination and seedling growth with 100% germination at 100 mg kg^{-1} in tomato while at 400 mgkg^{-1} in radish.	Haghighi and Silva (2014)
*Solanum lycopersicum*L. (Tomato)	25 nm, concentrations 0–1000 mg kg^{-1}; Aerosol and soil-mediated application	Promoted plant's growth and development at low concentrations; increased seed germination; root length plant height of soil-treated plants significantly up to 250 mg kg^{-1} while foliar application decreased root length; chlorophyll increased up to 500 mg kg^{-1}	Raliya et al. (2015b)
Zea mays L. (Maize)	titanium dioxide bulk and 0.01% and 0.03%., spray	Increase chlorophyll content and carotenoids, anthocyanin; facilitate an increase in crop yield as number of male and female flowers increased	Morteza et al. (2013)
*Brassica napus*L.	anatase/rutile −80:20; 27 nm; Foliar spray, 0, 500, 2500, 4000 mg l^{-1}	Enhanced root length, short length, fresh weight; increased photosynthesis, chlorophyll content, nitrate reductase activity, enzymatic and non-enzymatic antioxidants.	Li et al. (2015)
Solanum lycopersicum L. and *Vigna radiata* L.	activated carbon-based TiO$_2$, 30-50 nm, concentrations 0–500 mg l^{-1}	Promote seed germination and reduce germination time	Singh et al. (2016)
Forms of TiO$_2$-NPs	**Pathogen/disease**	**Impacts**	**References**
3B. TiO$_2$ nanoparticles as protecting agents against pathogens and diseases			
Light activated nanoscale formulation of TiO$_2$with Ag and Zn	n-TiO$_2$formulation with Ag and Zn on *Xanthomonas perforans* to control bacterial spot disease in tomato	TiO$_2$/Ag and TiO$_2$/Zn showed higher photocatalytic activity in comparison to control against *X. perforans;* mixture effectively controlled the disease without impacting tomato yield.	Paret et al. (2012)

(continued)

Table 3 (continued)

Plant species	Particles size and Treatment	Impacts	References
Light-activated TiO$_2$-NPs formulation with Zn	Nanocomposite to control bacterial leaf spot disease in Rosa Noare	Field applications of TiO$_2$/Zn on Rosa Noare significantly reduced bacterial spot.	Paret et al. (2013a, b)
TiO$_2$ -NPs (1). anatase- hydrophilic (2). Anatase—(3) hydrophobic rutile	Application of solid TiO$_2$-NPs against rice weevil *Sitophilus oryzae*	Application of solid TiO$_2$-NPs (anatase-hydrophilic, anatase-hydrophobic, rutile) separately, at 0.5–2.0 g Kg^{-1} showed up to 90% mortality against rice weevil *Sitophilus oryzae* after 7 day of treatment.	Goswami et al. (2010)
TiO$_2$-NPs	Use of TiO$_2$-NPs treated leaves for insecticidal activity against *Spodoptera littoralis* (Egyptian cotton leaf worm)	TiO$_2$-NPs was more effective against the 2nd instar larvae (LC$_{50}$62.5 ppm); than the 4th instar (LC$_{50}$-125 ppm); malformations in larvae, pupae and adult stages.	Shaker et al. (2017)
Ag-doped hollow and solid TiO$_2$-NPs	Against phytopathogens *Fusarium solani* and *Venturia inaequalis*	Hollow Ag doped TiO$_2$-NPs were found to be more efficient than solid. Visible light exposure further increased its antifungal activities; inhibited naphthoquinone pigment production: a pigment responsible for pathogenicity of *F. solani*.	Boxi et al., (2016)
TiO$_2$-NPs	Against—*Pectobacterium betavasculorum, Xanthomonas campestris* pv. *beticola* (Pammel), and *Pseudomonas syringae* pv. *Aptata*	Application TiO$_2$-NPs (0.25 and 0.50 mLL^{-1}) increased plant growth, chlorophyll, carotenoid, antioxidative enzymes, proline and H$_2$O$_2$ contents, but decreased MDA content in presence or absence of bacteria; also reduced the disease indices of beetroot (*Beta vulgaris* L) pathogens-Soft rot and bacterial pocket, leaf spot caused by pathogens.	Siddiqui et al. (2019)

Plant Species	**Stress**	**NPs size; Treatments-concentration**	**Impacts**	**References**
3C. TiO$_2$ nanoparticles as protecting agents under different stresses				
Linum usitatissimum (Flax)	Drought	10–25 nm; 0, 10, 100, and 500 mg l^{-1}, foliar treatment	TiO$_2$—NPs at low concentration enhanced photosynthetic pigment contents; reduced H$_2$O$_2$ and MDA levels in stressed plant at 10 mgL^{-1}; improved oil and protein values at 100mg L^{-1}	Aghdam et al. (2016)
Dracocephalum moldavica L.	Salinity (0, 50 and 100 mM NaCl)	20–30 nm; 0, 50, 100 and 200 mg l^{-1}); under hydroponic condition	Improvement of various agronomic traits and enhanced antioxidant enzymes activities, increased essential oil content (1.19%), 100 mg L^{-1} treatment significantly ameliorated salinity effects	Gohari et al. 2020
Glycin max (Soybean)	Heavy Metal- Cd- 50-150 mg kg^{-1})	<100 nm; 100–300 mg kg^{-1} to the soil	Inhibited Cd toxicity due to increased photosynthetic rate and growth parameters of plants	Singh and Lee (2016)
Lycopersicum esculentum L. (Tomato)	Heat	16.04 nm; Seed treatment with 0.05 0.1 and 0.2 g l^{-1}	TiO$_2$-NPs exposure enhanced photosynthesis, transpiration and stomatal conductance under heat stress while decreased chlorophyll fluorescence and electron transport in leaves	Qi et al. (2013)
Triticum aestivum L. (Wheat)	Drought PEG induced	10–25 nm; seeds 0, 500, 1000, and 2000 mg L^{-1}	TiO$_2$-NPs exposure increased seed germination and early growth of wheat by alleviating PEG-stimulated drought stress toxicity	Faraji and Sepehri (2019)
Triticum aestivum L. (Wheat)	Drought	Foliar spray at stem elongation and flowering stages, 0.01% and 0.03%	Various agronomic parameters such as plant height, number of seeds and weight, ear number of ears and weight, yield and biomass, gluten and starch content increased by 0.02%	Jaberzadeh et al. (2013)

(continued)

Table 3 (continued)

Plant species	Particles size and Treatment	Impacts		References
Oryza sativa (Rice)	Elevated CO_2 (570 µmol mol^{-1})	100 nm; Soil treatment 0, 50, and 200 mg kg^{-1}	Elevated CO_2 concentration increased negative impacts of TiO_2-NPs on growth and yield of rice, improved nutritional quality and increased accumulation of nutrients like Ca, Mg, Mn, P, Zn under elevated CO_2 levels in combination of TiO_2-NPs at 200 mg kg^{-1}; altered soil microbial composition.	Du et al. (2017)
Dracocephalum moldavica L. (Dragonhead)	Water deficit	Foliar spray 0, 10 and 40 ppm	Low concentration of TiO_2-NPs increased plant shoot dry mass and essential oils content; reduced MDA under stress, oxidative damage and membrane damage.	Mohammadi et al. (2016)
Lycopersicon esculentum Mill.	Salinity 200 mM	Foliar spray—0, 5, 10, 20, 40 mg l^{-1}	TiO_2-NPs treatments up to 20 mgL^{-1} improved plant growth and yield, as well as fruit quality in terms of enhanced lycopene content under salt stress	Khan (2016)
Cicer arietinum L. (Chickpea)	Cold	7- 40 nm, TiO_2 NPs suspension sprayed on seedlings; 5 mg l^{-1})	An increase in transcript—derived fragments in TiO_2 NPs treated plants, reduce electrolyte leakage index, transcriptional regulation of different genes involved in metabolism pathways, cell protection, signaling and chromosomal structure	Amini et al. (2017)
Oryza sativa (Rice)	Cd (0, 10 and 20 mg L^{-1})	0, 10, 100 and 1000 mg l^{-1})	Decreased Cd uptake and distribution in rice roots and leaves; increased chlorophyll content, photosynthetic rates in Cd stressed plant suggesting the positive impacts of TiO_2-NPs.	Ji et al. (2017)
Spinacia oleracea (Spinach)	UV–B radiation	5 nm; 0.25% nano-anatase in seeds and spray on leaves	TiO_2-NPs exposure decreased ROS, H_2O_2 and MDA content while increased SOD, CAT, APX, GPX enzymes activities and also elevate rate of oxygen evolution in chloroplasts under UV-B radiations	Lei et al. (2008)

Gowda 2013; Shaker et al. 2017). They can be used in two ways: either direct application in the field, killing insects and larvae, or can be used as nanocarriers that released commercial pesticides to enhance their efficiency. In the present decade, nanotechnology has involved in nanopesticides, fungicides, bactericides and so on formulations which are found to increase the solubility of less soluble active components and control their slow release and help in developing disease-free agricultural crops (Table 3B) (Debnath et al. 2011; Gogos et al. 2012).

The application of TiO_2 was found to be effective against *Curvularia, Cercospora, Pseudomonas* sp., *Xanthomonas* sp., and thus, mitigate the adverse effects of leaf spot in maize, bacterial leaf blight and blast disease in rice, spray molds in tomato, leaf spot and brown blotch disease in cowpea, cucumber powdery mildew and litchi downy blight (Chao and Choi 2005; Lu et al. 2006; Owolade and Ogunleti 2008; Choi et al. 2015). Kamran et al. (2011) reported that the nanosilver and TiO_2-NPs have a potential to be apply for eradication of the bacterial pathogens from the tobacco plant. NPs remain bound to the cell wall of pathogen and restrain the growth and development of conidia and conidiophores of fungal which eventually may cause death of fungal pathogen. TiO_2-NPs significantly inhibit the incidence of rice blast and tomato spray mold which are reported to increase grain weight by 20% because of the growth stimulatory effect of TiO_2-NPs (Mahmoodzadeh et al. 2000).

Spraying of TiO_2-NPs sol (average size of 30.6 nm) on cucumber leaves controlled bacterial angular spot and downy mildew diseases caused by *Pseudomonas syringe* pv *lachrymans* and *Xanthomonas vesicatoria* because it form an adhesive and thin transparent film like covering on the surface of leaf (Zhang et al. 2007). Siddiqui et al. (2019) observed that application of 0.25 and 0.50 ml l^{-1} of TiO_2-NPs to plants with, or without, bacterial strain had not only improved the growth, photosynthetic pigment contents (chlorophylls and carotenoids), antioxidative enzymes, proline and H_2O_2 contents, but also increased resistance to diseases like soft rot, bacterial pocket and leaf spot caused by *Pectobacterium betavasculorum, Xanthomonas campestris* pv. *Beticola* (Pammel) and *P. syringae* pv. *aptata* in beetroot. In the past few years, antimicrobial photocatalyst technology has been developed. Following this, many photoactivated NP formulations of TiO_2 with Zn and Ag have been developed and applied to manage bacterial leaf spot on Rosa 'Noare' and tomato caused by *Xanthomonas* sp. (Paret et al. 2012, 2013a, b).

Field applications of TiO_2-NPs formulations significantly reduced the survival of mentioned pathogens devoid of any adverse impacts on plants growth. Similarly, Boxi et al. (2016) reported Ag doped (hollow and solid) TiO_2-NPs to be effective against two strong plant pathogens, *Fusarium solani* (wilt disease) and *Venturia inaequalis* (causes apple scab disease) in presence of visible light that attributed to oxidative damage to the cell membrane caused by $^{\bullet}OH$ radicals that are generated during photocatalysis, and by reacting with sulfide and disulfide of cellular proteins (Lin et al. 2011; Gupta et al. 2013). TiO_2-NPs take part in catalytic oxidation reaction with oxygen and thiol (-SH) groups that ultimately lead to cell death by creating blockage of bacterial respiration. NPs also arrest the production of toxic pigment that is called naphthoquinone for *F. solani* which has a role in fungal pathogenicity (Boxi et al. 2016). Moreover, a combination of TiO_2, Al and SiO_2 was found to be useful in managing downy and powdery mildew in grapes (Bowen et al. 1992), probably due to their direct effects on the fungal hyphae, intervention with detection of plant surface and activation of plant antioxidant defense system. Besides, visible light-activated TiO_2-NPs co-doped with nitrogen and fluorine was observed to be effective against *F. oxysporum* and could be used as antifungal agents (Mukherjee et al. 2020).

Shaker et al. (2017) found TiO_2-NPs as an efficient larvicidal agent against the larvae of cotton leaf worm (*Spodoptera littoralis*). TiO_2-NPs also affected some biological parameters (larval period, pupation, adult emergence, productiveness, hatching of eggs, adult longevity and sex ratio) of this insect where it caused irregularities in larvae, pupae and adult stages. TiO_2-NPs' application may reduce the problems caused by *S. littoralis* of the host crops and improve yield. Goswami et al. (2010) reported the effects of Al_2O_3, ZnO, TiO_2 and Ag NPs against pest and pathogens. Al-Bartya and Hamzab (2015) that biosynthesized-TiO_2-NPs also were lethal to the larvae of red palm weevil *Rhynchophorus ferrugineus*. As they are biosynthesized and novel with respect to their surface coating and reactivity, they may be an effective alternative to control pesticide-resistant pests.

4.3 TiO_2 Nanoparticles and Plants Tolerance Under Various Stresses

Environmental pollution and climate change directly or indirectly affects the growth and development of plants which in turn reduce crop productivity by imposing various stresses on crop plants. Thus, it is essential to pave ways to ameliorate the negative effects of different stresses to obtain optimum yields. In recent years, TiO_2-NPs are emerging as potential source for plant improving plant performance under various abiotic stresses (Frazier et al. 2014). Some studies have been documented in Table 3C.

Studies observed that application of TiO_2-NPs improves plants growth and development by alleviating stress-induced toxicity by enhancing antioxidants (Song et al. 2012; Mohammadi et al. 2014). TiO_2-NPs application alleviate toxicity and increased tolerance against cold in chick pea (Mohammadi et al. 2014), heat in tomato (Qi et al. 2013), salinity in tomato and barley (Khan 2016; Karami and Sepehri 2018; Gohari et al. 2020), drought in *Triticum aestivum, Linum usitatissimum* (Jaberzadeh et al. 2013; Aghdam et al. 2016; Mohammadi et al. 2016) and cadmium toxicity in soybean and rice (Ji et al. 2017; Singh and Lee 2016). Exogenous application of anatase-TiO_2-NPs not only ameliorates the damage to flax seed plant under drought stress but also improves its drought stress tolerance by improving morphological and physiological traits (Aghdam et al. 2016; Mohammadi et al. 2016). An enhanced chlorophyll and carotenoids contents, and low levels of H_2O_2 and MDA were noticed in plants treated with low concentration (10 mg l^{-1}) of TiO_2-NPs, while an improved value of seed oil and protein contents were reported in plants treated with higher concentration of TiO_2-NPs (100 mg l^{-1}). Likewise, in another study, Faraji and Sepehri (2019) observed that the exposure of TiO_2-NPs to wheat plant increased its seed germination and early growth of seedling under polyethylene glycol (PEG)-stimulated drought stress via counteracting the adverse effects of drought on seed germination, seed vigor, root and shoot length and fresh weight of seedling. In addition, TiO_2-NPs application in soil significantly ameliorate tolerance against Cd stress in soybean and rice by increasing chlorophyll, photosynthetic rate and growth parameters; and regulating Cd accumulation (Ji et al. 2017;

Singh and Lee 2016). Salinity or salt stress decreases the various growth and physiological parameters in *Triticum aestivum* which was reversed upon application of NPs (Shalata et al. 2001; Darko et al. 2017).

TiO$_2$-NPs application ultimately improve plant tolerance to extreme climate events by increasing proline and other amino acids, nutrients level and water uptake, and activities of antioxidant enzymes (Ghosh et al. 2010; Ebrahimi et al. 2016). Besides, they might also control expression of stress-related genes. In a study, cDNA-AFLP analysis performed on two genotypes of chickpea (Sel96Th11439, cold tolerant, and ILC533, cold susceptible), Amini et al. (2017) reported an increased level of transcript–derived fragments (TDF) which control cold tolerance along with defense and damage indices like electrolyte leakage level in TiO$_2$-NP-treated chickpea during cold stress (4 °C). Overexpression or upregulation of some recognized genes in TiO$_2$-NP-treated plants may be regarded as efficient markers in adaptation process in *Vigna radiata* against cold stress. Plant's tolerance to stress involves groups of various genes participated in several metabolism pathways, cellular defense system, cell signaling, protein synthesis and chromosomal structure. Further, TiO$_2$-NPs effectively alleviate the stress-induced toxicity through changing the levels of phytochemical production, activation of antioxidants defense system and stability of plastid pigments. Thus, application of TiO$_2$-NPs suggested to control the damage due to climate change in fields and to increase crop productivity.

5 Conclusions and Future Recommendations

Due to wide array of application in various fields, release and accumulation of TiO$_2$-NPs into environment become unavoidable. Hence, TiO$_2$-NPs pollution arises as a potential threat for ecosystem structure and functioning that resulted in declining food quality and yield, and health of human being. In this respect, a comprehensive understanding of TiO$_2$-NPs transfer through the ecosystem and its impacts on plants is very important.

In the last few years, a large number of studies have been conducted to understand the phytotoxicity of TiO$_2$-NPs and their interaction to plants; however, there are still some area which needs to be more explanation.

1. Toxicological studies revealed both the beneficial and harmful impacts of TiO$_2$-NPs on morphological, physiological, cellular and molecular aspects of plants, but further research is required to provide more information regarding their uptake, translocation and internalization in plant cells. These conflicting results point out the complexity of plants response to TiO$_2$-NPs that are not

only vary with TiO$_2$-NPs properties (i.e., size, concentration, shape, surface coating size, shape, surface coating, etc.) but are also dependent on the plant species and its various developmental stages and experimental conditions (type of medium, exposure method, exposure duration, etc.).

2. A clear understanding about TiO$_2$-NPs uptake and translocation mechanisms, i.e., about how the NPs with different shapes and size get enter into the cells, how they translocate from cortex to stelar region, and how they cross Casparian strips present in endodermal cells, etc., are still need more explanation.

3. A number of studies are available related to impacts of TiO$_2$-NPs on plants, but most of these studies are short period of time and experiments were conducted under controlled conditions in laboratory settings that are likely very differ from actual field conditions. Therefore, there is a need of long-term, well-designed, plant life cycle experiments to evaluate TiO$_2$-NPs impacts on plants under real field conditions so that their environmentally relevant implications can be advocated.

4. From the foregoing studies, it is noticed that different plant species may activate different detoxification mechanisms in response to TiO$_2$-NPs exposure to mitigate its toxicity, however, our knowledge about the role of ROS as a signaling molecule in plants under TiO$_2$-NPs stress is still in primary stage.

5. In addition, paucity of literature is available on role of different 'omics' methodologies, such as transcriptomics, proteomics and metabolomics that can give an authentic data to comprehensively evaluate TiO$_2$-NPs toxicity and tolerance mechanisms in plants.

6. Besides, due to several unique properties, TiO$_2$-NPs have attracted attention for its potential application as a growth promoter, nanofertilizer, nanopesticides and so on in agriculture. However, our knowledge regarding TiO$_2$-NPs uptake capacity and its permissible limit is still sketchy. However, the application of TiO$_2$-NPs in agriculture is very new and in its evolving stage and progress in research is still at bench-top scale. There is an urgent need to unravel the fate and behavior of TiO$_2$-NPs applications in agriculture to enhance plant growth and productivity and also to assess unforeseeable risks on environments.

7. The foreseen potential of TiO$_2$-NPs in near future includes their application for controlled and targeted release of chemicals or fertilizers, as encapsulated pesticides, as nanosensor and nanoherbicides as well to develop TiO$_2$-NP-based formulations to improve crop quality and yield by improving resistance/tolerance against various abiotic as well as biotic stresses under the present scenario of climate change and to fulfill the unforeseen demand of food supply.

Acknowledgements Authors are grateful to the University Grand Commission, New Delhi, for financial support to Ms Kajal Patel and Indu Tripathi. Authors are also thankful to Professor K.S. Rao, Head, Department of Botany, University of Delhi, New Delhi, for his kind support.

References

Aghdam MTB, Mohammadi H, Ghorbanpour M (2016) Effects of nanoparticulate anatase titanium dioxide on physiological and biochemical performance of *Linum usitatissimum*(Linaceae) under well-watered and drought stress conditions. Braz J Bot 39:139–146

Al-Bartya AM, Hamzab RZ (2015) Larvicidal, antioxidant activities and perturbation of Transminases activities of Titanium dioxide NPs synthesized using *Moringa oleifera* leaves extract against the red palm weevil (*Rhynchophorus ferrugineus*). Eur J Pharm Med Res 2:49–54

Ali T, Tripathi P, Azam A, Raza W, Ahmed AS, Ahmed A, Muneer M (2017) Photocatalytic performance of Fe-doped TiO_2 NPs under visible-light irradiation. Mater Res Express 4:015022

Amini S, Maali-Amiri R, Mohammadi R, Kazemi-Shahandashti SS (2017) cDNA-AFLP analysis of transcripts induced in chickpea plants by TiO_2 NPs during cold stress. Plant Physiol Biochem 111:39–49

Andersen CP, King G, Plocher M, Storm M, Pokhrel LR, Johnson MG, Rygiewicz PT (2016) Germination and early plant development of ten plant species exposed to titanium dioxide and cerium oxide NPs. Environ Toxicol Chem 35:2223–2229

Asli S, Neumann PM (2009) Colloidal suspensions of clay or titanium dioxide NPs can inhibit leaf growth and transpiration via physical effects on root water transport. Plant Cell Environ 32:577–584

Azimi R, Feizi H, Hosseini MK (2013) Can bulk and nanosized titanium dioxide particles improve seed germination features of wheatgrass (*Agropyron desertorum*). not Sci Biol 5:325–331

Baan R, Straif K, Grosse Y, Secretan B, El Ghissassi F, Cogliano V, WHO International Agency for Research on Cancer Monograph Working Group (2006) Carcinogenicity of carbon black, titanium dioxide, and talc. Policy Watch 7:295–296

Binas V, Venieri D, Kotzias D, Kiriakidis G (2017) Modified TiO_2 based photocatalysts for improved air and health quality. J Materiomics 3:3–16

Binh CTT, Peterson CG, Tong T, Gray KA, Gaillard J-F, Kelly JJ (2015) Comparing acute effects of a nano-TiO2 pigment on cosmopolitan freshwater phototrophic microbes using high-throughput screening. PLoS ONE 10:e0125613

Blaise C, Gagné F, Ferard JF, Eullaffroy P (2008) Ecotoxicity of selected nano-materials to aquatic organisms. Environ Toxicol 23:591–598

Bowen P, Menzies J, Ehret D, Samuels L, Glass AD (1992) Soluble silicon sprays inhibit powdery mildew development on grape leaves. J Am Soc Hortic Sci 117:906–912

Boxall A, Tiede K, Chaudhry Q (2007) Engineered nanomaterials in soils and water: how do they behave and could they pose a risk to human health. Nanomedicine 2:919–927

Boxi SS, Mukherjee K, Paria S (2016) Ag doped hollow TiO_2 NPs as an effective green fungicide against *Fusarium solani* and *Venturia inaequalis* phytopathogens. Nanotechnology 27:085103

Carpita N, Sabularse D, Montezinos D, Delmer DP (1979) Determination of the pore size of cell walls of living plant cells. Science 205:1144–1147

Castiglione MR, Giorgetti L, Geri C, Cremonini R (2011) The effects of nano-TiO_2 on seed germination, development and mitosis of root tip cells of *Vician arbonensis* L. and *Zea mays* L. J Nanopart Res 13:2443–2449

Castiglione MR, Giorgetti L, Cremonini R, Bottega S, Spanò C (2014) Impact of TiO_2 nanoparticles on *Vicia narbonensis* L.: potential toxicity effects. Protoplasma251(6):1471–1479. https://doi.org/10.1007/s00709-014-0649-5

Chao SHL, Choi HS (2005) Method for providing enhanced photosynthesis. Korea Research Institute of Chemical Technology. Bulletin, South Korea Press, 10

Chaudhary I, Singh V (2020) Titanium dioxide NPs and its impact on growth, biomass and yield of agricultural crops under environmental stress: a review. J Nanosci Nanotechnol 10:1–8

Chichiriccò G, Poma A (2015) Penetration and toxicity of nanomaterials in higher plants. Nanomaterials 5:851–873

Choi HG, Moon BY, Bekhzod K, Park KS, Kwon JK, Lee JH, Cho MW, Kang NJ (2015) Effects of foliar fertilization containing titanium dioxide on growth, yield and quality of strawberries during cultivation. Hortic Environ Biotechnol 56:575–581

Chowdappa P, Gowda S (2013) Nanotechnology in crop protection: status and scope. Pest Manage Horticult Ecosyst 19:131–151

Cornelis G, Hund-Rinke K, Kuhlbusch T, Van den Brink N, Nickel C (2014) Fate and bioavailability of engineered NPs in soils: a review. Crit Rev Environ Sci Technol 44:2720–2764

Cox A, Venkatachalam P, Sahi S, Sharma N (2016) Silver and titanium dioxide nanoparticle toxicity in plants: a review of current research. Plant Physiol Biochem 107:147–163

DaCosta MVJ, Sharma PK (2015) Influence of titanium dioxide NPs on the photosynthetic and biochemical processes in *Oryza sativa*. Int J Recent Sci Res 6:2445–2451

Darko E, Gierczik K, Hudak O, Forgo P, Pal M, Türkösi E, Kovacs V, Dulai S, Majlath I, Molnar I, Janda T(2017) Differing metabolic responses to salt stress in wheat-barley addition lines containing different 7H chromosomal fragments. PLOS one 12(3)

Debnath N, Das S, Seth D, Chandra R, Bhattacharya SC, Goswami A (2011) Entomotoxic effect of silica NPs against *Sitophilus oryzae* (L.). J Pest Sci 84:99–105

Du W, Sun Y, Ji R, Zhu J, Wu J, Guo H (2011) TiO_2 and ZnO NPs negatively affect wheat growth and soil enzyme activities in agricultural soil. J Environ Monitor 13:822–828

Du W, Gardea-Torresdey Jorge L, Xie Y, Yin Y, Zhu J, Zhang X, Ji R, Gu K, Peralta-Videa Jose R, Guo H (2017) Elevated CO_2 levels modify TiO_2 nanoparticle effects on rice and soil microbial communities. Sci Total Environment 578:408–416. https://doi.org/10.1016/j.scitotenv.2016.10.197

Ebrahimi A, Galavi M, Ramroudi M, Moaveni P (2016) Effect of TiO_2 NPs on antioxidant enzymes activity and biochemical biomarkers in pinto bean (*Phaseolus vulgaris* L.). J Mol Biol Res 6:58–66

Etxeberria E, Gonzalez P, Baroja-Fernandez E, Romero JP (2006) Fluid phase endocytic uptake of artificial nano-spheres and fluorescent quantum dots by sycamore cultured cells: evidence for the distribution of solutes to different intracellular compartments. Plant Signal Behav 1:196–200

Fan W, Peng R, Li X, Ren J, Liu T, Wang X (2016) Effect of titanium dioxide NPs on copper toxicity to *Daphnia magna* in water: role of organic matter. Water Res 105:129–137

Faraji J, Sepehri A (2019) Ameliorative effects of TiO_2 NPs and sodium nitroprusside on seed germination and seedling growth of wheat under PEG-stimulated drought stress. J Seed Sci 41(3):309–317

Federici G, Shaw BJ, Handy RD (2007) Toxicity of titanium dioxide NPs to rainbow trout (*Oncorhynchus mykiss*): gill injury, oxidative stress, and other physiological effects. Aquat Toxicol 84:415–430

Feizi H, Moghaddam PR, Shahtahmassebi N, Fotovat A (2012) Impact of bulk and nanosized titanium dioxide (TiO_2) on wheat seedgermination and seedling growth. Biol Trace Elem Res 146:101–106

Feizi H, Kamali M, Jafari L, Moghaddam PR (2013) Phytotoxicity and stimulatory impacts of nanosized and bulk titanium dioxide on fennel (*Foeniculum vulgare* Mill). Chemosphere 91:506–511

Feng Y, Cui X, He S, Dong G, Chen M, Wang J, Lin X (2013) The role of metal NPs in influencing arbuscular mycorrhizal fungi effects on plant growth. Environ Sci Technol 47:9496–9504

Foltête AS, Masfaraud JF, Bigorgne E, Nahmani J, Chaurand P, Botta C, Labille J, Rose J, Férard JF, Cotelle S (2011) Environmental impact of sunscreen nanomaterials: ecotoxicity and genotoxicity of altered TiO_2 nanocomposites on *Vicia faba*. Environ Pollut 159:2515–2522

Frazier TP, Burklew CE, Zhang B (2014) Titanium dioxide NPs affect the growth and microRNA expression of tobacco (*Nicotiana tabacum*). Funct Integr Genomics 14:75–83

Fries R, Simkó M, (2012) Nano-titanium dioxide (Part I): basics, production, applications. Institute of Technology Assessment of the Austrian Academy of Sciences. NanoTrust-Dossiers No. 033en–November 2012. https://doi.org/10.1016/j.envpol.2013.10.004

Gao J, Xu G, Qian H, Liu P, Zhao P, Hu Y (2013) Effects of nano-TiO_2 on photosynthetic characteristics of *Ulmus elongata* seedlings. Environ Pollut 176:63–70

Ghosh M, Bandyopadhyay M, Mukherjee A (2010) Genotoxicity of titanium dioxide (TiO_2) NPs at two trophic levels: plant and human lymphocytes. Chemosphere 81:1253–1262

Gogos A, Knauer K, Bucheli TD (2012) Nanomaterials in plant protection and fertilization: current state, foreseen applications, and research priorities. J Agric Food Chem 60:9781–9792

Gohari G, Mohammadi A, Akbari A, Panahirad S, Dadpour MR, Fotopoulos V, Kimura S (2020) Titanium dioxide NPs (TiO_2-NPs) promote growth and ameliorate salinity stress effects on essential oil profile and biochemical attributes of *Dracocephalum moldavica*. Sci Rep 10:1–14

González-Melendi P, Fernández-Pacheco R, Coronado MJ, Corredor E, Testillano PS, Marquina RMC, C, Ibarra MR, Rubiales D, Pérez-de-Luque, A, (2008) NPs as smart treatment-delivery systems in plants: assessment of different techniques of microscopy for their visualisation in plant tissues. Ann Bot 101:187–195

Goswami A, Roy I, Sengupta S, Debnath N (2010) Novel applications of solid and liquid formulations of NPs against insect pests and pathogens. Thin Solid Films 519:1252–1257

Goswami L, Kim KH, Deep A, Das P, Bhattacharya SS, Kumar S, Adelodun AA (2017) Engineered nano particles: nature, behavior, and effect on the environment. J Environ Manage 196:297–315

Gupta K, Singh RP, Pandey A, Pandey A (2013) Photocatalytic antibacterial performance of TiO_2 and Ag-doped TiO_2 against *S. aureus., P. aeruginosa* and *E. coli*. Beilstein J. Nanotechnol 4:345–351

Haghighi M, da Silva JAT (2014) The effect of N-TiO_2 on tomato, onion, and radish seed germination. J Crop SciBiotechnol17:221–227.

Hazeem LJ, Bououdina M, Rashdan S, Brunet L, Slomianny C, Boukherroub R (2016) Cumulative effect of zinc oxide and titanium oxide NPs on growth and chlorophyll a content of *Picochlorum*sp. Environ Sci Pollut Res 23:2821–2830

Hong F, Yang F, Liu C, Gao Q, Wan Z, Gu F, Wu C, Ma Z, Zhou J, Yang P (2005) Influences of nano-TiO_2 on the chloroplast aging of spinach under light. Biol Trace Elem Res 104:249–260

Hou J, Wang L, Wang C, Zhang S, Liu H, Li S, Wang X (2019) Toxicity and mechanisms of action of titanium dioxide NPs in living organisms. Int J Environ Sci 75:40–53

Jaberzadeh A, Moaveni P, Moghadam HRT, Zahedi H (2013) Influence of bulk and NPs titanium foliar application on some agronomic traits, seed gluten and starch contents of wheat subjected to water deficit stress. Not Bot Horti Agrobot Cluj Napoca 41:201–207

Jacob DL, Borchardt JD, Navaratnam L, Otte ML, Bezbaruah AN (2013) Uptake and translocation of Ti from NPs in crops and wetland plants. Int J Phytoremediat. 15:142–153

Janmohammadi M, Amanzadeh T, Sabaghnia N, Dashti S (2016) Impact of foliar application of nano micronutrient fertilizers and titanium dioxide NPs on the growth and yield components of barley under supplemental irrigation. Acta Agric Slov 107:265–276

Ji Y, Zhou Y, Ma C, Feng Y, Hao Y, Rui Y, Wu W, Gui X, Han Y, Wang Y, Xing, B (2017) Jointed toxicity of TiO2 NPs and Cd to rice seedlings: NPs alleviated Cd toxicity and Cd promoted NPs uptake. Plant Physiol Biochem 110:82–93

Jiang F, Shen Y, Ma C, Zhang X, Cao W, Rui Y (2017) Effects of TiO_2 NPs on wheat (*Triticum aestivum L.*) seedlings cultivated under super-elevated and normal CO_2 conditions. PLoS ONE 12(5): e0178088

Jiang G, Li X, Lan M, Shen T, Lv X, Dong F, Zhang S (2017) Monodisperse bismuth NPs decorated graphitic carbon nitride: enhanced visible-light-response photocatalytic NO removal and reaction pathway. Appl Catal B: Environ 205:532–540

Jovanović B, Palić D (2012) Immunotoxicology of non-functionalized engineered NPs in aquatic organisms with special emphasis on fish —Review of current knowledge, gap identification, and call for further research. Aquat Toxicol 118:141–151

Judy JD, Unrine JM, Rao W, Wirick S, Bertsch PM (2012) Bioavailability of gold nanomaterials to plants: importance of particle size and surface coating. Environ Sci Technol 46:8467–8474

Karami A, Sepehri A (2018) Nano titanium dioxide and nitric oxide alleviate salt induced changes in seedling growth, physiological and photosynthesis attributes of barley. Zemdirbyste-Agric 105:123–132

Keller AA, McFerran S, Lazareva A, Suh S (2013) Global life cycle releases of engineered nanomaterials. J Nanoparticle Res 15:1692

Khan MN (2016) Nano-titanium Dioxide (Nano-TiO_2) mitigates NaCl stress by enhancing antioxidative enzymes and accumulation of compatible solutes in tomato (*Lycopersicon esculentum* Mill.). J Plant Sci 11:1–11

Khodakovskaya MV, Lahiani MH (2014) NPs and plants: from toxicity to activation of growth. Hand Nanotoxicol Nanomed Stem Cell Use Toxicol 121–130

Khot LR, Sankaran S, Maja JM, Ehsani R, Schuster EW (2012) Applications of nanomaterials in agricultural production and crop protection: a review. Crop Prot 35:64–70

Kurepa J, Paunesku T, Vogt S, Arora H, Rabatic BM, Lu J, Wanzer MB, Woloschak GE, Smalle JA (2010) Uptake and distribution of ultrasmall anatase TiO_2 Alizarin red S nanoconjugates in *Arabidopsis thaliana*. Nano Lett 10:2296–2302

Lapied E, Nahmani JY, Moudilou E, Chaurand P, Labille J, Rose J, Exbrayat JM, Oughton DH, Joner EJ (2011) Ecotoxicological effects of an aged TiO_2 nanocomposite measured as apoptosis in the anecic earthworm *Lumbricus terrestris* after exposure through water, food and soil. Environ Int 37:1105–1110

Larue C, Khodja H, Herlin-Boime N, Brisset F, Flank AM, Fayard B, Chaillou S, Carrière M, (2011) Investigation of titanium dioxide NPs toxicity and uptake by plants. J Phys: Conf Ser 304(1):012057 (IOP Publishing)

Larue C, Laurette J, Herlin-Boime N, Khodja H, Fayard B, Flank AM, Brisset F, Carriere M (2012a) Accumulation, translocation and impact of TiO_2 NPs in wheat (*Triticum aestivum* spp.): influence of diameter and crystal phase. Sci Total Environ 431:197–208

Larue C, Veronesi G, Flank AM, Surble S, Herlin-Boime N, Carriere M (2012b) Comparative uptake and impact of TiO_2 nano particles in wheat and rapeseed. J Toxicol Environ Health A 75:722–734

Lei Z, Mingyu S, Xiao W, Chao L, Chunxiang Q, Liang C, Hao H, Xiaoqing L, Fashui H (2008) Antioxidant stress is promoted by nano-anatase in spinach chloroplasts under UV-B radiation. Biol Trace Elem Res 121:69–79

Li J, Naeem MS, Wang X, Liu L, Chen C, Ma N et al (2015) Nano-TiO$_2$ is not phytotoxic as revealed by the oilseed rape growth and photosynthetic apparatus ultra-structural response. PLoS ONE 10(12):e0143885

Lin S, Reppert J, Hu Q, Hudson JS, Reid ML, Ratnikova TA, Rao AM, Luo H, Ke PC (2009). Uptake, translocation, and transmission of carbon nanomaterials in rice plants. Small 5:1128–1132

Lin Y, Qiqiang W, Xiaoming Z, Zhouping W, Wenshui X, Yuming D (2011) Synthesis of Ag/TiO$_2$ core/shell NPs with antibacterial properties. B Korean Chem Soc 32:2607–2610

Lindberg HK, Falck GCM, Suhonen S, Vippola M, Vanhala E, Catalán J, Savolainen K, Norppa H (2009) Genotoxicity of nanomaterials: DNA damage and micronuclei induced by carbon nanotubes and graphite nanofibres in human bronchial epithelial cells in vitro. Toxicol Lett 186:166–173

Lu JW, Li FB, Guo T, Lin LW, Hou MF, Liu TX (2006) TiO$_2$ photocatalytic antifungal technique for crops diseases control. J. Environ. Sci 18:397–401

Ma X, Geiser-Lee J, Deng Y, Kolmakov A (2010) Interactions between engineered NPs (ENPs) and plants: phytotoxicity, uptake and accumulation. Sci Total Environ 408:3053–3061

Mahmoodzadeh H, Nabavi M, Kashefi H (2000) Effect of nanoscale titanium dioxide particles on the germination andgrowth of Canola (Brassica napus). J Ornamental Hortic Plants 3(1):25–32

Mahmoodzadeh H, Aghili R (2014) Effect on germination and early growth characteristics in wheat plants (Triticum aestivum L.) seeds exposed to TiO$_2$ NPs. J Chem Health Risks 4:29–36

Mahmoodzadeh H, Nabavi M, Kashefi H (2013) Effect of nanoscale titanium dioxide particles on the germination and growth of canola (Brassica napus). J Ornamen Horti Plants 3:25–32

Mandeh M, Omidi M, Rahaie M (2012) In vitro influences of TiO$_2$ NPs on barley (Hordeum vulgare L.) tissue culture. Biol Trace Elem Res 150:376–380

Markowska-Szczupak A, Ulfig K, Morawski AW (2011) The application of titanium dioxide for deactivation of bioparticulates: an overview. Catal Today 169:249–257

Maruyama CR, Guilger M, Pascoli M, Bileshy-José N, Abhilash PC, Fraceto Navarro E, Baun A, Behra R, Hartmann NB, Filser J, Miao AJ (2008) Environmental behavior and ecotoxicity of engineered NPs to algae, plants, and fungi. Ecotoxicol 17:372–386

Mattiello A, Filippi A, Pošćić F, Musetti R, Salvatici MC, Giordano C, Vischi M, Bertolini A, Marchiol L (2015) Evidence of phytotoxicity and genotoxicity in Hordeum vulgare L. exposed to CeO$_2$ and TiO$_2$ NPs. Front Plant Sci 6:1043

Mattiello A, Marchiol L (2017) Application of nanotechnology in agriculture: assessment of TiO$_2$ nanoparticle effects on Barley. In: Janus M (ed) Application of titanium dioxide. In Tech: London, UK, pp 23–39

Menard A, Drobne D, Jemec A (2011) Ecotoxicity of nanosized TiO$_2$. Review of in vivo data. Environ Pollut 159:677–684

Mishra V, Mishra RK, Dikshit A, Pandey AC (2014) Interactions of NPs with plants: an emerging prospective in the agriculture industry. In: Ahmad P, Rasool S (eds) Emerging technologies and management of crop stress tolerance: biological techniques, vol 1. Elsevier Academic Press, New York, pp 159–180

Moaveni P, Farahani HA, Maroufi K (2011) Effect of Ti [O. sub. 2] NPs spraying on wheat (Triticum aestivum L.) under field condition. Adv Environ Biol 2208–2211.

Mohammadi R, Maali-Amiri R, Mantri NL (2014) Effect of TiO$_2$ NPs on oxidative damage and antioxidant defense systems in chickpea seedlings during cold stress. Russ J Plant Physiology 61:768–775

Mohammadi H, Esmailpour M, Gheranpaye A (2016) Effects of TiO$_2$ nanoparticles and water-deficit stresson morpho-physiological

characteristics of dragonhead (Dracocephalum moldavica L.) plants. Acta Agriculturae Slovenica 107(2):385–396

Moll J, Gogos A, Bucheli TD, Widmer F, van der Heijden MG (2016) Effect of NPs on red clover and its symbiotic microorganisms. J Nanotechnol 14:36

Monshausen GB, Bibikova TN, Messerli MA, Shi C, Gilroy S (2007) Oscillations in extracellular pH and reactive oxygen species modulate tip growth of Arabidopsis root hairs. Proc Natl Acad Sci USA 20996–21001

Montazer M, Seifollahzadeh S (2011) Pretreatment of wool/polyester blended fabrics to enhance titanium dioxide nanoparticle adsorption and self-cleaning properties. Color Technol 127:322–327

Moreno-Olivas F, Gant VU, Johnson KL, Peralta-Videa JR, Gardea-Torresdey JL (2014) Random amplified polymorphic DNA reveals that TiO$_2$ NPs are genotoxic to Cucurbita pepo. J Zhejiang Univ Sci A 15:618–623

Morteza E, Moaveni P, Farahani HA, Kiyani M (2013) Study of photosynthetic pigments changes of maize (Zea mays L.) under nano TiO$_2$ spraying at various growth stages. Springer Plus 2:247

Motyka O, Chlebíková L, Kutláková KM, Seidlerová J (2019) Ti and Zn Content in Moss shoots after exposure to TiO$_2$ and ZnO NPs: biomonitoring possibilities. B Environ Contam Tox 102:218–223

Mudunkotuwa IA, Grassian VH (2010) Citric acid adsorption on TiO$_2$ NPs in aqueous suspensions at acidic and circumneutral pH: surface coverage, surface speciation, and its impact on nanoparticle—nanoparticle interactions. J Am Chem Soc 132:14986–14994

Mukherjee A, Sun Y, Morelius E, Tamez C, Bandyopadhyay S, Niu G, White JC, Peralta-Videa JR, Gardea-Torresdey JL (2016) Differential toxicity of bare and hybrid ZnO NPs in green pea (Pisum sativum L.): A life cycle study. Front Plant Sci 6:1242

Mukherjee K, Acharya K, Biswas A, Jana NR (2020) TiO$_2$ NPs Co-doped with nitrogen and fluorine as visible light-activated antifungal agents. ACS Appl Nano Mater 1–29

Nair R, Varghese SH, Nair BG, Maekawa T, Yoshida Y, Kumar DS (2010) Nanoparticulate material delivery to plants. Plant Sci 179:154–163

Navarro E, Baun A, Behra R, Hartmann NB, Filser J, Miao AJ, Quigg A, Santschi PH, Sigg L (2008) Environmental behavior and ecotoxicity of engineered NPs to algae, plants, and fungi. Ecotoxicol 17:372–386

Nel AE, Mädler L, Velegol D, Xia T, Hoek EM, SomasundaranP KF, Castranova V, Thompson M (2009) Understanding biophysico-chemical interactions at the nano-bio interface. Nat Mater 8:543–557

Nowack B, Baalousha M, Bornhöft N, Chaudhry Q, Cornelis G, Cotterill J, Gondikas A, Hassellöv M, Lead J, Mitrano DM, von der Kammer F (2015) Progress towards the validation of modeled environmental concentrations of engineered nanomaterials by analytical measurements. Environ Sci Nano 2:421–428

Nyamukamba P, Okoh O, Mungondori H, Taziwa R, Zinya S (2018) Synthetic methods for titanium dioxide NPs: a review. In: Yang D (eds) Titanium dioxide—Material for a sustainable environment. Intech Open, pp 151–175

Onelli E, Prescianotto-Baschong C, Caccianiga M, Moscatelli A (2008) Clathrin-dependent and independent endocytic pathways in tobacco protoplasts revealed by labelling with charged nanogold. J Exp Bot 59:3051–3068

Owolade OF, Ogunleti DO (2008) Effects of titanium dioxide on the diseases, development and yield of edible cowpea. J PltProt Res 48:329–335

Pachapur VL, Larios AD, Cledón M, Brar SK, Verma M, Surampalli RY (2016) Behavior and characterization of titanium dioxide and silver NPs in soils. Sci Total Environ 563:933–943

Paret M, Vallad G, Averett D, Jones J, Olson S (2012) Photocatalysis: Effect of light-activated nanoscale formulations of TiO$_2$ on

Xanthomonas perforans and control of bacterial spot of tomato. Phytopathology. 103: https://doi.org/10.1094/PHYTO-08-12-0183-R

Paret ML, Palmateer AJ, Knox GW (2013a) Evaluation of a light-activated nanoparticle formulation of titanium dioxide with zinc for management of bacterial leaf spot on rosa 'Noare.' Hort Sci 48:189–192

Paret ML, Vallad GE, Averett DR, Jones JB, Olson SM (2013b) Photocatalysis: effect of light-activated nanoscale formulations of TiO_2 on Xanthomonas perforans and control of bacterial spot of tomato. Phytopathology 103:228–236

Patrick JW, Tyerman SD, Bel AJE (2015) Long-distance transport. In: Buchanan BB, Gruissem W, Jones RL (eds) Biochemistry and molecular biology of plants, 2nd edn. Wiley, West Sussex, pp 658–710

Qi M, Liu Y, Li T (2013) Nano-TiO_2 improve the photosynthesis of tomato leaves under mild heat stress. Biol Trace Elem Res 156:323–328

Rafique R, Zahra Z, Virk N, Shahid M, Pinelli E, Park TJ, Kallerhoff J, Arshad M (2018) Dose-dependent physiological responses of Triticum aestivum L. to soil applied TiO_2 NPs: alterations in chlorophyll content, H_2O_2 production, and genotoxicity. Agric Ecosyst Environ 255:95–101

Raliya R, Biswas P, Tarafdar JC (2015) TiO_2 nanoparticle biosynthesis and its physiological effect on mung bean (Vigna radiata L.). Biotechno Rep 5:22–26

Raliya R, Nair R, Chavalmane S, Wang WN, Biswas P (2015) Mechanistic evaluation of translocation and physiological impact of titanium dioxide and zinc oxide NPs on the tomato (Solanum lycopersicum L.) plant. Metallomics 7:1584–1594

Raliya R, Franke C, Chavalmane S, Nair R, Reed N, Biswas P (2016) Quantitative understanding of nanoparticle uptake in watermelon plants. Front Plant Sci 7:1288

Ramsden CS, Henry TB, Handy RD (2013) Sub-lethal effects of titanium dioxide NPs on the physiology and reproduction of zebrafish. Aquat Toxicol 126:404–413

Ranjan S, Ramalingam C (2016) Titanium dioxide NPs induce bacterial membrane rupture by reactive oxygen species generation. Environ Chem Lett 14:487–494

Rezaei F, Moaveni P, Mozafari H, Morteza E (2015) Investigation of different concentrations and times of nano-TiO_2 foliar application on traits of soybean (Glycine max L.) at Shahr-e-Qods. Iran. Int J Biosci 6:109–114

Rico CM, Barrios AC, Tan W, Rubenecia R, Lee SC, Varela-Ramirez A, Peralta-Videa JR, Gardea-Torresdey JL (2015) Physiological and biochemical response of soil-grown barley (Hordeum vulgare L.) to cerium oxide NPs. Environ Sci Pollut 22:10551–10558

Rico CM, Majumdar S, Duarte-Gardea M, Peralta-Videa JR, Gardea-Torresdey JL (2011) Interaction of NPs with edible plants and their possible implications in the food chain. J Agric Food Chem 59:3485–3498

Riu J, Maroto A, Rius FX (2006) Nanosensors in environmental analysis. Talanta 69:288–301

Roh JY, Park YK, Park K, Choi J (2010) Ecotoxicological investigation of CeO_2 and TiO_2 NPs on the soil nematode Caenorhabditis elegans using gene expression, growth, fertility, and survival as endpoints. Environ Toxicol Pharmacol 29:167–172

Rui M, Ma C, White JC, Hao Y, Wang Y, Tang X, Yang J, Jiang F, Ali A, Rui Y, Cao W (2018) Metal oxide NPs alter peanut (Arachis hypogaea L.) physiological response and reduce nutritional quality: a life cycle study. Environ Sci Nano 5:2088–2102

Sadiq IM, Dalai S, Chandrasekaran N, Mukherjee A (2011) Ecotoxicity study of titania (TiO_2) NPs on two microalgae species: Scenedesmus sp. and Chlorella sp. Ecotoxicol Environ Saf 74:1180–1187

Santos Filho RD, Vicari T, Santos SA, Felisbino K, Mattoso N, Sant'Anna-Santos BF, Cestari MM, Leme DM (2019) Genotoxicity

of titanium dioxide NPs and triggering of defense mechanisms in Allium cepa. Genet Mol Biol 42:425–435

Schwab F, Zhai G, Kern M, Turner A, Schnoor JL, Wiesner MR (2015) Barriers, pathways and processes for uptake, translocation and accumulation of nanomaterials in plants-critical review. Nanotoxicology 10:257–278

Serag MF, Kaji N, Gaillard C, Okamoto Y, Terasaka K, JabasiniM TM, Mizukami H, Bianco A, Baba Y (2011) Trafficking and subcellular localization of multiwalled carbon nanotubes in plant cells. ACS Nano 5:493–499

Servin AD, Castillo-Michel H, Hernandez-Viezcas JA, Diaz BC, Peralta-Videa JR, Gardea-Torresdey JL (2012) Synchrotron micro-XRF and micro-XANES confirmation of the uptake and translocation of TiO_2 NPs in cucumber (Cucumis sativus) plants. Environ Sci Technol 46:7637–7643

Servin AD, Morales MI, Castillo-Michel H, Hernandez-Viezcas JA, Munoz B, Zhao LJ, Nunez JE, Peralta-Videa JR, Gardea-Torresdey JL (2013) Synchrotronverification of TiO accumulation in cucumber fruit: a possible pathway of TiO nanoparticle transfer from soil into the food chain. Environ Sci Technol 47:11592–11598

Shah SNA, Shah Z, Hussain M, Khan M (2017) Hazardous effects of titanium dioxide NPs in ecosystem. Bioinorg Chem Appl 2017:1–12

Shaker AM, Zaki AH, Abdel-Rahim EFM, Khedr MH (2017) TiO_2 NPs as an effective nanopesticide for cotton leaf worm. Agric Eng Int: CIGR J Special issue:61–68

Shalata A, Mittova V, Volokita M, Guy M, Tal M (2001) Response of the cultivated tomato and its wild salt-tolerant relative Lycopersicon pennellii to salt-dependent oxidative stress: The root antioxidative system. Physiol Plant 112(4):487–494

Sharma VK (2009) Aggregation and toxicity of titanium dioxide NPs in aquatic environment—a review. J Environ Sci Health A 44:1485–1495

Shi W, Yan Y, Yan X (2013) Microwave-assisted synthesis of nano-scale $BiVO_4$ photocatalysts and their excellent visible-light-driven photocatalytic activity for the degradation of ciprofloxacin. Chem Eng J. 215:740–746

Siddiqui MH, Al-Whaibi MH, Firoz M, Al-Khaishany MY (2015) Role of NPs in plants. In: SiddiquiMH, Al-Whaibi MH, Mohammad F (eds) Nanotechnology and plant sciences. Springer, Cham, pp 19–35

Siddiqui ZA, Khan MR, Abd Allah EF, Parveen A (2019) Titanium dioxide and zinc oxide NPs affect some bacterial diseases, and growth and physiological changes of beetroot. Int J Veg Sci 25:409–430

Simonin M, Richaume A, Guyonnet JP, Dubost A, Martins JM, Pommier T (2016) Titanium dioxide NPs strongly impact soil microbial function by affecting archaeal nitrifiers. Sci Rep 6:1–10

Singh R, Srivastava PK, Singh VP, Dubey G, Prasad SM (2012) Light intensity determines the extent of mercury toxicity in the cyanobacterium Nostoc muscorum. Acta Physiol Plant 34:1119–1131. https://doi.org/10.1007/s11738-011-0909-3

Singh J, Lee BK (2016) Influence of nano-TiO_2 particles on the bioaccumulation of Cd in soybean plants (Glycine max): A possible mechanism for the removal of Cd from the contaminated soil. J Environ Manage 170:88–96

Singh P, Singh R, Borthakur A, Srivastava P, Srivastava N, Tiwary D, Mishra PK (2016) Effect of nanoscale TiO_2-activated carbon composite on Solanum lycopersicum (L.) and Vigna radiata (L.) seeds germination. Energy Ecology Environ 1:131–140

Song G, Gao Y, Wu H, Hou W, Zhang C, Ma H (2012) Physiological effect of anatase TiO_2 NPs on Lemna minor. Environ Toxicol Chem 31:2147–2152

Song U, Shin M, Lee G, Roh J, Kim Y, Lee EJ (2013) Functional analysis of TiO_2 nanoparticle toxicity in three plant species. Biol Trace Elem Res 155:93–103

Sun TY, Gottschalk F, Hungerbühler K, Nowack B (2014) Comprehensive probabilistic modelling of environmental emissions of engineered nanomaterials. Environ Pollut 185:69–76

Tan W, Peralta-Videa JR, Gardea-Torresdey JL (2018) Interaction of titanium dioxide NPs with soil components and plants: current knowledge and future research needs–a critical review. Environ Sci Nano 5:257–278

Taylor AF, Rylott EL, Anderson CW, Bruce NC (2014) Investigating the toxicity, uptake, nanoparticle formation and genetic response of plants to gold. PLoS ONE 9:e93793

Tripathi DK, Singh S, Singh S, Pandey R, Singh VP, Sharma NC, Prasad SM, Dubey NK, Chauhan DK (2017) An overview on manufactured NPs in plants: uptake, translocation, accumulation and phytotoxicity. Plant Physiol Biochem 110:2–12

Trouiller B, Reliene R, Westbrook A, Solaimani P, Schiestl RH (2009) Titanium dioxide NPs induce DNA damage and genetic instability in vivo in mice. Cancer Res 69:8784–8789

Tumburu L, Andersen CP, Rygiewicz PT, Reichman JR (2015) Phenotypic and genomic responses to titanium dioxide and cerium oxide NPs in *Arabidopsis* germinants. Environ Toxicol Chem 34:70–83

Vignardi CP, Hasue FM, Sartório PV, Cardoso CM, Machado AS, Passos MJ, Santos TC, Nucci JM, Hewer TL, Watanabe IS, Gomes V (2015) Genotoxicity, potential cytotoxicity and cell uptake of titanium dioxide NPs in the marine fish *Trachinotus carolinus* (Linnaeus, 1766). AquatToxicol 158:218–229

Waghmode MS, Gunjal AB, Mulla JA, Patil NN, Nawani NN (2019) Studies on the titanium dioxide NPs: biosynthesis, applications and remediation. SN Appl Sci 1:310

Wang F, Liu X, Shi Z, Tong R, Adams CA, Shi X (2016) Arbuscular mycorrhizae alleviate negative effects of zinc oxide nanoparticle and zinc accumulation in maize plants—A soil microcosm experiment. Chemosphere 147:88–97

Wang S, Kurepa J, Smalle JA (2011) Ultra-small TiO$_2$ NPs disrupt microtubular networks in *Arabidopsis thaliana*. Plant Cell Environ 34:811–820

Wang S, Su R, Nie S, Sun M, Zhang J, Wu D,Moustaid-Moussa N (2014) Application of nanotechnology in improving bioavailability and bioactivity of diet-derived phytochemicals. J Nutr Biochem 25:363–376

Wang WN, Tarafdar JC, Biswas P (2013) Nanoparticle synthesis and delivery by an aerosol route for watermelon plant foliar uptake. J Nanoparticle Res 15:1417

Weir A, Westerhoff P, Fabricius L, Hristovski K, Von Goetz N (2012) Titanium dioxide NPs in food and personal care products. Environ Sci Technol 46:2242–2250

Wu B, Zhu L, Le XC (2017) Metabolomics analysis of TiO$_2$ NPs induced toxicological effects on rice (*Oryza sativa* L.). Environ Pollut 230:302–310

Yan A, Chen Z (2019) Impacts of silver NPs on plants: a focus on the phytotoxicity and underlying mechanism. Int J Mol Sci 20:1003

Yang F, Liu C, Gao F, Su M, Wu X, Zheng L, Hong F, Yang P (2007) The improvement of spinach growth by nano-anatase TiO$_2$ treatment is related to nitrogen photoreduction. Biol Trace Elem Res 119:77–88

Yang F, Hong F, You W, Liu C, Gao F, Wu C, Yang P (2006) Influences of nano-anatase TiO$_2$ on the nitrogen metabolism of growingspinach. Biol Trace Elem Res. 110(2):179–90. https://doi.org/10.1385/bter:110:2:179

Yang X, Cao C, Erickson L, Hohn K, Maghirang R, Klabunde K (2008) Synthesis of visible-light-active TiO -based photocatalysts bycarbon and nitrogen doping. J Catalysis 260(1):128–33

Yin JJ, Liu J, Ehrenshaft M, Roberts JE, Fu PP, Mason RP, Zhao B (2012) Phototoxicity of nano titanium dioxides in HaCaT keratinocytes—generation of reactive oxygen species and cell damage. Toxicol Appl Pharmacol 263:81–88

Zhang P, Cui H, Zhong X, Li L (2007) Effects of nano-TiO$_2$ semiconductor sol on prevention from plant diseases. Nanoscience 12:1–6

Zhang P, Ma Y, Zhang Z, He X, Zhang J, Guo Z, Tai R, Zhao Y, Chai Z (2012) Biotransformation of ceria NPs in cucumber plants. ACS Nano 6:9943–9950

Zhao L, Chen Y, Chen Y, Kong X, Hua Y (2016) Effects of pH on protein components of extracted oil bodies fromdiverse plant seeds and endogenous protease-induced oleosin hydrolysis. Food Chem. 1 (200):125–33. https://doi.org/10.1016/j.foodchem.2016.01.034

Zheng L, Hong F, Lu S, Liu C (2005) Effect of nano-TiO$_2$ on strength of naturally aged seeds and growth of spinach. Biol Trace Elem Res 104:83–91

Zhu X, Wang J, Zhang X, Chang Y, Chen Y (2010) Trophic transfer of TiO$_2$ NPs from daphnia to zebrafish in a simplified freshwater food chain. Chemosphere 79:928–933

Ziental D, Czarczynska-Goslinska B, Mlynarczyk DT, Glowacka-Sobotta A, Stanisz B, Goslinski T, Sobotta L (2020) Titanium Dioxide NPs: Prospects and Applications in Medicine. Nanomaterials 10:387

Interaction of Nano-TiO$_2$ with Plants: Preparation and Translocation

Kandasamy G. Moodley and Vasanthakumar Arumugam

Abstract

The application of nanotechnology is increasing at a rapid pace in the manufacture of products for both industrial and domestic markets. It is widely known that abundant nanoparticles of titanium dioxide are included in house products such as toothpaste and paints for improving a white colour as well as used as fillers in the products. Titanium nanoparticles are also added to food additives for a similar reason. Once their usefulness is over, these products will end up in water treatment plants or in solid waste disposal sites from where they may enter water bodies or arable land. If they enter into the edible plants, these nanoparticles may enter into the food chain. In the light of the above, this chapter focused on the methods of synthesis and characterization on nano-TiO$_2$ by various researchers and the results obtained with regard to the properties of nano-TiO$_2$. This was followed by summaries of studies on the interaction of nano-TiO$_2$ with various plants including the influence of these properties of nano-TiO$_2$ on the type and extent of the interactions.

Graphical Abstract

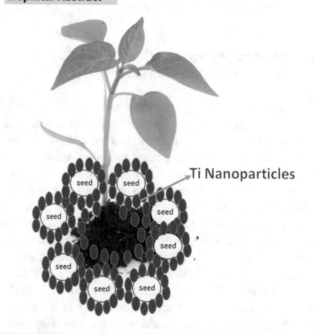

Keywords

Oxidative stress • Phytoextraction • Phytofiltration • Phytoremediation • Phytovolatilization • Shoot lengths • Sol–gel

1 Introduction

'Nanotechnology' is a term which is heard with increasing frequency and this will increase greatly in the future. To understand the term using a non-scientific concept, we may refer to the well-known practice of 'downsizing'. Before the early 1980s, televisions (TVs) were mostly big in size. Advances in technology have given rise to slim and lightweight 'smart' TVs. This trend in making products smaller is also noticed in the reduction in the size of motors cars.

K. G. Moodley (✉)
Department of Operations and Quality Management, Durban University of Technology, Durban, South Africa
e-mail: moodleykg@dut.ac.za

V. Arumugam
Key Laboratory of Ministry of Education for Advanced Materials in Tropical Island, Resources, Hainan University, No 58, Renmin Avenue, Haikou, 570228, China

© Springer Nature Switzerland AG 2021
P. Singh et al. (eds.), *Plant-Microbes-Engineered Nano-particles (PM-ENPs) Nexus in Agro-Ecosystems*,
Advances in Science, Technology & Innovation,
https://doi.org/10.1007/978-3-030-66956-0_5

However, it is no secret that there are advantages and disadvantages associated with very small cars. Likewise, nanotechnology has good and bad features as well. It is therefore pertinent to state that nanoparticles, which are very small versions of their precursors, may have desirable or undesirable properties. To earn the title of nanoparticle, it must be small enough that at least one dimension, but preferably two, should be 100 nm or less. Nanoparticles of titanium dioxide are present in many products which are used in common places. Nanotitanium dioxides (nano-TiO_2) are being used in several products (Bis and Wu 2005) such as toothpaste, personal care products, sunscreen, pigments and outdoor building materials like paving stones for decoration. The roles of nanotitanium dioxide in the above and other applications will become clear after the properties of titanium dioxide are described and discussed.

1.1 Background Information on the Source of Titanium Dioxide (TiO_2)

Titanium dioxide (TiO_2) is classified as a mineral, which is commonly known as titania. Just for clarity, we should note the difference between a mineral and an ore. A mineral occurs naturally in the earth's crust and has a definite range of formulae whereas an ore refers to a rock which is rich in minerals making extraction of the metal an economically viable process. The naturally occurring mineral forms of titanium dioxides are rutile, ilmenite and anatase. There are other crystal phases which are formed under specific conditions, especially in high pressures. The purpose in dwelling on the different crystalline forms of titanium dioxide is that the literature (Riaz and Naseem 2015; Tan et al. 2017) abounds with reports of titanium dioxide in a specific crystalline form. Also, titanium dioxide appears to be ubiquitous; so much, so that, it is even present in significant amounts, in beach sea sand (Shalini et al. 2020), in the rutile modification. The nanoforms of titanium dioxide were estimated (Robichaud et al. 2009), to amount 4 million metric tons per annum (MT $year^{-1}$). As a guesstimate, the present production of nano-TiO_2 could be double or more.

1.2 The Motivation for Focusing on Nano-TiO_2

Several personal care products, especially in most toothpastes contain significant amounts of nano-TiO_2. The end-of-use destiny of these items is wastewater treatment works (WWTW) where varying amounts of nano-TiO_2 are removed as sludge which is disposed on land. The water from WWTWs containing residual amounts nano-TiO_2 is invariably passed into rivers from the water may be utilized for agricultural purposes. If this scenario is currently in vogue, it implies that increasing amounts of nano-TiO_2 will be found in soil and water. Thus, there is a significant chance that these nanoparticles may be accumulated by edible plants. Therefore, the presence of nano-TiO_2 or its changed form may be an integral part of foods consumed by humans. Furthermore, nano-TiO_2 present in such soil and water may enter into the plants and affect them positively or negatively with regard to growth. If there is a need to deliberately reduce the amounts of nano-TiO_2 in the soils, the phenomenon of nanophytoremediation may be employed. Several reports covering both these cases involving interactions of nano-TiO_2 with plant life will be used to get an understanding of how these interactions occur and the results thereof in respect of the health of humans. The impact of factors such as crystal type, size, shape and possible pathways will be considered. This will be preceded by a brief report of the methods of generating nano-TiO_2, characterization of the products and very importantly the properties of nano-TiO_2. To accentuate the importance of the interactions of nano-TiO_2 with plants, brief notes on other applications of nano-TiO_2 will also be included.

2 Importance and Classification of the Methods of Synthesis of Nano-TiO_2

For describing the methods of synthesis, it has to be stated that the properties of nanoparticles such as size, shape and the reactivity of the faces of the crystal would be extremely useful. It is thus necessary to provide information on the various methods of synthesis of nanoparticles of TiO_2 with special reference to properties pertinent to the interaction of nano-TiO_2. To get a good perspective of the various methods used already, a general classification such as the following may be useful. Firstly, they could be physical or non-physical methods. Another classification may be 'green' and 'non-green' methods. The various methods may be driven by heat, electricity, sound, microwaves, solvents, chemicals and biochemicals. Green methods are taken to be driven by chemicals in plants; hence, biochemical for classification purposes. There may be methods which do not fall into these categories but the broad classification, given above, may have the benefits of highlighting the concepts involved. Synopses of some of the array of methods available to synthesize nanoparticles of nano-TiO_2 (Fig. 1) are given below.

Fig. 1 Series of methods for synthesizing nano-TiO₂

Green (Extraction) methods for synthesizing nano-TiO₂
- *From Trigella foenum graecum leaves*
- *From Moringa leaves*
- *From peels of fruit*

Physical methods for synthesizing nano-TiO₂
- *Ball Mill method*
- *Sol-Gel Method*

Thermal methods for synthesizing nano-TiO₂
- *Thermal Method*
- *Solvothermal Method*
- *Ultrasonic Method*

Chemical methods for synthesizing nano-TiO₂
Chemical Synthesis
Metal-Organic Chemical Vapour Deposition (MOCVD)

2.1 Green Methods for Synthesizing Nano-TiO₂

2.1.1 From Trigella Foenum-Graecum (Fenugreek) Plant Leaves

Green methods of synthesis are favoured for various reasons, with the chief among them being minimum pollution of the environment; for example, one of them involves the use of an extract of leaves (Subhapriya and Gomathipriya 2018) of a plant, namely *Trigella foenum-graecum*. Leaves from *Trigella foenum-graecum* were washed with deionized water, ground in a mortar and pestle. The resulting powder was added to water and stirred to yield an extract. The pH of the extract was adjusted to 8 with 1 M NaOH followed by the addition of titanium oxysulphate with stirring to give a precipitate of n-TiO₂. The latter require calcination at 700 °C for 3 h to give the final product. Extracts of sorgum roots have also been used to synthesize nano-TiO₂ (Dutta and Fulakar 2017). The phenol like compounds in these extracts was reported to function via reducing properties to produce nano-TiO₂. The nano-TiO₂ particles produced by this method were utilized to degrade methyl orange dye in wastewater samples.

2.1.2 From Moringa Leaves

A green technique for the synthesis of nanotitanium dioxide involves the leaves of *Moringa* which have been used for medicinal purposes over a very long period (Sivaranjani and Philominathan 2016). *Moringa oleifera* is a low-cost source of a precursor for the green synthesis of titanium dioxide nanoparticles which are being used for many interactions with plants.

2.1.3 From Peels of Fruit

A cheap, environment-friendly and eminently green method, to prepare titanium dioxide nanoparticles involved the exploitation of agricultural waste in the form of peels of fruits, namely plum, kiwi and peach (Naqvi et al. 2021). These cylindrical nanoparticles were found to have antibacterial activity. Extract of fruit peels was also used by Roopan et al. (2012) for photosynthesis of nanotitanium dioxide with rutile morphology. The resulting nanoparticles had spherical shapes and were in the size range from 23 to 25 nm.

2.2 Physical Methods for Synthesis of Nano-TiO₂

2.2.1 Ball Mill Method

i. *Method-1*

Physical methods for generating nanotitanium dioxide can also be classified as being eco-friendly; in that there is no need to remove solvent from the reaction for disposal elsewhere. A physical method was used to produce nanotitanium dioxide (Pavlovic et al. 2002). High purity titanium (IV) oxide powder was subjected to 10 mm zirconium (IV) oxide balls in a planetary ball mill for varying times. Distilled water as added to resulting powder and the mixture was put under ultrasonic radiation for 10 min. The solid was separated and dried at 100 °C for a day. The final powdery material was found to have antibacterial properties. Concerning the additional steps to an initial mechanical step, it is apparent that the synthesis of nanoparticles requires a combination of methods.

ii. *Method-2*

In a modified physical method for generating nanoparticles of titanium, Al-doped samples of TiO_2 were subjected to milling by a planetary type ball mill for 90 min (Nadeem et al. 2018). It has to be pointed out the reasons for doping are not connected to a physical method of synthesizing nanoparticles, but *inter alia*, to reduce the band gap, to increase photocatalytic activity and inhibit change of phase. Milling was followed by addition of deionized water, as drops, accompanied by concomitant stirring to yield a solution. The latter was ball-milled for 2 h followed by annealing, using stepwise increases in temperatures, between 773 and 1170 K. A similar procedure, without using Al-doping, gave pure nano-TiO_2.

2.2.2 Sol–Gel Methods

i. *Method 1*

Variations of the sol–gel technique have been used to prepare nanoparticles of titanium dioxide (Guo et al. 2016). In principle, this technique is fairly simple and starting material is generally a metal salt of a metal alkoxide which is subjected to hydrolysis resulting in the generation of a colloidal suspension. The latter is referred to as a sol. When polymerization sets in, the sol assumes a gelatin-like form which is referred to as a sol–gel (solid gel). If the sol–gel is placed in a mould and heated to drive out solvent, a substance with a porous nature called an aerogel is formed. Some examples will serve to the variations in the application of the sol–gel method. The precursor in the following variation was $TiCl_4$ with ethanol as solvent (Sabry et al. 2016). Controlled addition of $TiCl_4$ solution to ethanol gave a yellow solution. The formation of the sol–gel was monitored over periods from 1 to 5 days. After drying at 80 °C, the solid was subjected to calcination at 500 °C. It transpired that the size of the nanoparticles was dependent on the temperature used.

ii. *Method 2*

With tetra-n-butyl orthotitanate as a precursor and hydrochloric acid as solvent, nanoparticles of titanium dioxide in anatase and rutile phases obtained using the sol–gel method (Dalvandi and Ghasemi 2013). Here, the precursor and solvent were stirred well for 9 h to make a homogeneous mixture after which added liquid ammonia. An increase in calcination temperature led to a change in phase from anatase to rutile. As in the previous example, the size of the particle depended on the temperature at which the calcination was done.

iii. *Method 3 with combustion*

A combination of sol–gel with a combustion method gave pure anatase phase titanium dioxide nanoparticles (Jongprateep et al. 2015). The chosen precursor was either titanium isopropoxide (TTIP) or submicrometre sized particles of TiO_2 and method was sol–gel or combustion. The average size of powders by the combustion method using TTIP and submicroparticles as precursors, yielded particles with average sizes of 44 and 77 nm, respectively, while average sizes of particles using the sol–gel method for TTIP and submicrometre particles were found to be 48 and 85 nm, respectively.

2.3 Thermal Methods of Preparing Nanoparticles of Titanium Dioxide

2.3.1 Thermal Method

These techniques could be hydrothermal or solvothermal. The following examples should suffice to illustrate how the method is used. Nano-TiO_2 was produced using $TiCl_3$ as a precursor, by three different routes (Zeng and Zeng 2017). In the first route, $TiCl_3$ and NH_4F (as a mineralizer) were mixed in distilled water as the medium of reaction. As the dopant, $NiCl_2$ was added to the above mixture with vigorous stirring over 15-min of the period. In the second route, $TiCl_3$ was mixed with NH_4F only. For the third route, $TiCl_3$ was added to $NiCl_2$. All three reaction mixtures were subjected to heating in an autoclave at 180 °C for a whole day and night to give a precipitate and a solution phase. The solution phase was pipetted out to leave a precipitate containing the targeted material. The precipitate is washed with purified water and alcohol. The dry product was characterized by X-ray diffraction (XRD), scanning electron microscopy (SEM) and transmission electron microscopy (TEM). In summary, the results of these revealed that different types of crystals were formed from the three different routes used and that these researchers (Zeng and Zeng 2017) had succeeded in producing single crystals of anatase titanium dioxide which had one reactive crystal face.

2.3.2 Solvothermal Method

The solvothermal method was used to synthesize very fine crystals of TiO_2 targeted for photocatalytic uses (Kim et al. 2003). The precursor was chosen as titanium isopropoxide (TTIP). It was dissolved in dry toluene together with oleic acid, under an atmosphere of argon. Varying amounts of TIP were used to give 5:100; 10:100; 20:100 of TIP to solvent. The mixtures were stirred for 24 h, then placed in an autoclave and subjected to heat at the ramp rate of 4 °C per minute up to 250 °C. After cooling to ambient temperature, the addition of acetone yielded a precipitate which was isolated and dried in vacuole. XRD and TEM analysis showed particle sizes to be small; 4 and 6 nm, respectively. They were thus very suitable for photocatalytic applications.

2.3.3 Ultrasonic Methods

In a method using ultrasound, the precursor was titanium isopropoxide (TIP), in a beaker, to which propanol was added (Shirsath et al. 2013). The beaker was placed in a controlled temperature ultrasonic bath, and sonication was affected by placing a titanium horn in the beaker. Doping was done using five different concentrations of cerium nitrate added to the reaction after every 30 s after the initial addition of sodium hydroxide as well. After additions of solutions to the sonication beaker were completed, sonication was activated for a further 30 min. A time period was allowed for the formed precipitate to settle in the beaker. Thereafter, the precipitate was subjected to centrifugation, filtering and drying and finally calcining at 450 °C for 3 h. Undoped nano-TiO₂ was synthesized in the same way except that the doping agent was omitted. It was found that the catalytic activity of the nanotitanium dioxide derived via the sonication route was higher than obtained by the conventional route.

2.4 Chemical Method

In situ synthesis of nanotitanium dioxide on a piece of cotton fabric was performed by Sadr and Montazer (2014), who chose titanium tetra isopropoxide (TTIP) as the source of titanium. The other reagents were 100% glacial acetic acid, methylene blue, non-ionic detergents and distilled water while the fabric was bleached cotton of known waft, weave, yarn and density. Nanotitanium dioxide was prepared on the cloth by dispersing the TTIP on the cloth and then performing acid hydrolysis of TTIP. The sequence of steps whole process can be outlined as follows: preparation of aqueous acid solution in a glass beaker; immersion of sample of fabric in the above solution; reaction beaker placed in an ultrasonic bath and irradiated for 5 min; TTIP added dropwise to the reaction mixture at ambient temperature. The resulting mixture was sonicated for 4 h at room temperature and 75 °C for 2 h. The treated cloth was left in the beaker for 24 h at room temperature; thereafter, the cloth was washed and then dried at 70 °C for 15 min; a variety of tests was conducted on the fabric; the salient features of ones to note are: titanium dioxide nanoparticles were formed of the cotton fabric; the coating of nanoparticles afforded the fabric protection from ultraviolet (UV) ill-effects; the nanoparticles were formed at low temperature.

2.5 Metal–Organic Chemical Vapour Deposition (MOCVD)

In this method, titanium tetraisopropoxide (TTIP) was used (Pradhan et al. 2003) as the precursor which was placed in a stainless steel vessel maintained at 60 °C. Argon pressure was used to carry the precursor to a cold wall upright MOCVD reactor equipped with a susceptor to accommodate the tungsten carbide–cobalt substrate. The target substrate was cleaned by dipping in acetone, followed by ultrasonic treatment in a deionized water bath. Each deposition experiment was carried out for 1.5 h. The analysis revealed that the deposited material comprised TiO₂ nanorods in anatase phase (Pradhan et al. 2003).

3 Bioapplications of Nano-TiO₂

Manesh et al. (2018) investigated the biological aspects of the interaction of nanotitanium dioxide with plants and compared them with the role of nanotitanium in tandem with CdCl₂, a known toxin towards seedlings and plants. Radish seeds, commercially available nanotitanium dioxide and CdCl₂, were the principals in this interaction study. The nano-TiO₂ was the commercially available ones (namely Aeroxide P25 and Degussa Evonik) and radish seeds were from the species *Raphanus sativus* L. *parvus*. Radish seeds were treated with a series of nano-TiO₂ solutions/suspensions ranging from 1 to 1000 mg l⁻¹. In a separate experiment, radish seeds were subjected to solutions of CdCl₂ at concentrations in the range from 1 to 250 mg l⁻¹. In a third set-up, radish seeds were treated with a combination of nano-TiO₂ and CdCl₂. These experiments were followed by toxicity tests using the Organization for Economic Cooperation and Development (OECD) 208 protocols. Calculations of percentages of seed which germinated, germination index (GI) and root elongation were done. Other properties that were determined were cell morphology and oxidative stress after 5 days of treatment as described above. Furthermore, the Z-potential of nano-TiO₂ in Milli-Q water as exposure medium was also measured.

Dynamic light scattering experiments revealed that small aggregates of nano-TiO₂ had formed. The results showed that exposure to nano-TiO₂ to seeds had small effects on percentage of germination, germination index (GI) and root length compared to controls. By comparison, CdCl₂ caused a marked lowering of germination % and GI for control seeds and a concentration-dependent decrease on root length increase were observed. In summary, the data from the above experiments support the notion that the presence of nano-TiO₂ does not affect the toxicity arising from the presence of CdCl₂. Quite importantly changes in morphology, nuclei, vacuoles and shape of radish root cells were revealed even for a single exposure to CdCl$_2$. The presence or absence of nano-TiO$_2$ had no discernible effect on the toxicity of CdCl$_2$.

4 Interactions of Nano-TiO$_2$ with Plants

Although plants are the focus of attention in this chapter, one should be aware that it is one of the myriads of living species on earth. Prior to a brief consideration of the reported interactions of titanium dioxide nanoparticles with some of these other living species, knowledge of the estimated amounts of the global production of commonly used nano-materials would be both revealing and helpful in getting a grasp of the issues connected with the generation and applications of nanomaterials. In this regard, Piccinno et al. (2012) conducted an in-depth survey on the productions and applications of ten nanomaterials in Europe, USA and also in the whole world. They found the following order, from highest to lowest in terms of tons of nanomaterials:

TiO$_2$ > ZnO > SiO$_2$ > FeO$_x$ > AlO$_x$ > CeO$_x$ > CNT > Fullerenes > AgNPs > quantum dots

It is thus clear that the probability of finding nanoparticles in the environments of most countries would very high indeed. Furthermore, it is more likely to be present in soil and water than in air.

4.1 Uptake of Nano-TiO$_2$ from Water by Higher Species than Plants

As mentioned elsewhere in this chapter, nanotitanium dioxide present in water and soil. In this section, a brief account on effect of exposure to nanoparticles of TiO$_2$ to species higher than plants will be probed in view of the fact that nanoparticles are bound to enter into higher species from edible plants which are an integral part of the food chain for higher species. Crustaceans and fish are more highly evolved species than plants and are likely to have organs to deal with toxins which enter their systems. It was thus interesting that the effect of nanoparticles of titanium on a fish species which thrive in river water, namely rainbow trout, was assessed by Federici et al. (2007), in a 14-day exposure of the fish to low concentrations of the nanoparticles. No damage to organ tissues was reported but that the trout suffered oxidative stress.

In a similar study to the above, land-based crustaceans (isopods) were fed meals mixed with different amounts of titanium dioxide nanoparticles brushed over on pieces of hazelnut leaves of known mass (Valant et al. 2012). As controls, leaves without any nanoparticles were used. The experiment aimed to ascertain if the nanoparticles caused any damage to the cell membrane of the digestive glands of the crustaceans. It was found that there was no damage if the concentrations were 100 μg of the nanoparticles or less per gram of dry leaf for three days of feeding corresponding to the consumption of about 30 μg over three days.

The results for both the projects, summarized above, appear to indicate that animals like fish and crustaceans can tolerate or excrete small amounts of nanoparticles which enter their digestive systems. Rats are more highly evolved than fish. It would be of interest to know how nanoparticles affect them. It is common knowledge that trials on new drugs are invariably conducted on rats. In pursuance of this practice, there are no reports thus far, in the literature on experiments involving nanoparticles by humans. As anticipated, there is at least one study on the effect of nanotitanium dioxide on rats. This project (Long et al. 2007) was motivated by the danger that the very high presence of nano-TiO$_2$ in the environment and its well-established photoreactivity, nano-TiO$_2$ might interact negatively with biological targets such as the brains of animals such as rats. A commercially available nano-TiO$_2$ product with the trade name of Degussa P25 was utilized, as the source of the TiO$_2$ nanoparticles comprising 70% anatase and 30% rutile forms of TiO$_2$. From a mortified rat, microglia (BV2), rat dopaminergic (DA), neurons (N27) and culture of rat striatum were treated with P25 taken up in a physiological buffer medium. Measurement of physical properties of P25 was done under conditions that applied to biological specimens (Long et al. 2007). The results showed that the nanoparticles stimulated release of reactive oxygen species (ROS) from microglia and caused some damage to N27 neutrons. Even if one does not comprehend the technical terms involved, one can deduce, even tentatively, that nanoparticles of titanium dioxide are inimical to the well-being of rats (Long et al. 2007).

4.2 Uptake of Nano-TiO$_2$ by Plants

In the light of the brief background provided above, attention will be focused on the principal object of this chapter, namely to probe the interactions of nanotitanium dioxide with plants.

In terms of the main parts of plants, it is fair to state that nanotitanium dioxide should enter, through roots, stems and leaves. If the ingress of nanoparticles is through the roots and the nanoparticles are translocated to the leaves which are stripped from the plant for safe disposal, this would be classed as nanophytoremediation. The question which arises is: will plants, which have fewer body parts than more highly evolved species such as animals, be able to cope with nanoparticles which enter their systems? It is worth noting that a significant number of studies (Andersen et al. 2016) have focused on seedlings rather than on fully grown plants. Much of this has been driven by the expectation that nanoparticles would positively affect growth and growth rate (Faraji and Sepehri 2018).

5 Studies on the Effect of Nanoparticles on Germination and Growth of Seedlings

Several studies, some of which are described below, have been done on the effect of nanoparticles on seedlings in general and titanium dioxide nanoparticles in particular. Improvement in the germination of spinach seeds, using titanium dioxide nanoparticles up to a concentration of 400 μg ml^{-1}, was observed, but no improvement when the higher concentrations were tried (Zhang et al. 2020). An almost contrary effect was observed (Castiglione et al. 2011) after seeds of *Vicia narbonensis* L. and *Zea mays* L. were soaked in suspensions of titanium dioxide nanoparticles, having concentrations up to 4000 μg ml^{-1}.

A commercially available source of nanotitanium dioxide nanoparticles (Evonik P25) was used by Song et al. (2013), who found that soaking tomato seeds for 24 h in a suspension of the nanoparticles made no noticeable difference in germination periods. Evonik P25 titanium nanoparticles were tested by Larue et al. (2011) using a hydroponic technique, for growing three different seeds. Even after a 7-day treatment under hydroponic conditions, there was no decrease in germination time.

Based on their assessment that studies described briefly above were deficient with respect to methodology, Andersen et al. (2016) resorted to testing the effect of nanoparticles of TiO₂ and CeO₂ on ten different plant species. These researchers also used P25 Evonik Degussa on the basis that they can easily compare their results using the same product. An important difference was that they passed ultrasound radiation through the suspensions of the nanoparticles. They used the seeds of lettuce, cabbage, soybean, carrot, ryegrass, pineapple cucumber, oat, onion and corn. They sterilized the seeds with ethanol followed by 50% bleach. They made some changes to the test procedure whereby they replaced the sand with Petri dishes and performed the test under artificial light, to promote possible photocatalytic activity. They also increased the period over which the changes in germination were monitored. They did not detect any toxic effects on germination or growth of the seedlings.

5.1 A Study Involving Transplanting

In the following treatment of seedlings, there are differences from those described above. Firstly, two-year-old seedlings of *Ulmus elongata* were transplanted from one site in China to another site where the experiment was conducted by Gao et al. (2013). Secondly, the seeds were not the targets of the treatment; to wit, the leaves were. The nanoparticles used were anatase-TiO₂ which was synthesized from TiCl₄ as the precursor and benzyl alcohol via a sol–gel technique. Each experiment involved three leaves for spraying with three different concentrations (0.1, 0.2 and 0.3%) of the nanoparticle while a 4th leaf, a control, was sprayed with distilled water. Contrary to results given in the preceding example, treatment with nanoanatase stunted growth. At 0.4% concentration the leaves began to turn yellow, indicating a lack of chlorophyll. In the light of further experiments, it was concluded that carbohydrates and lipids were formed under toxicity caused by nanoparticles and that environmental factors play a very significant role in the interactions between nanoparticles and plants.

6 Effect of Nano-TiO₂ on Strength of Plants

Whether the structures of plants are strengthened, through interaction with titanium dioxide nanoparticles, can be revealed by analysis of the structures after such interactions have occurred. One study on this aspect was undertaken by McDaniel et al. (2013). Plants of tobacco, mustard and jalopeno were grown in potting soil under controlled conditions of temperature, humidity and light. They were then translocated to a hydroponic solution having a pH of 5.5 where it was left for 3–4 days for acclimatization in a soilless system. Thereafter, the hydroponic solution was replaced with suspensions of nanotitanium dioxide in water, containing 0, 75, 150 and 300 mg l^{-1} of the nanoparticles. After 7 days in this suspension, pieces of roots were excised from the plants, washed with water and treated, in order, as follows: with glutaraldehyde, osmium tetroxide, dehydrated with ethanol, dried, coated with gold by sputtering. The specimens were then analysed by SEM, attached to an (energy-dispersive X-ray spectroscopy) EDS detector. The results showed that titanium was detected on all three root specimens. The mustard root exposed to 300 mg l^{-1} had evidence of damage to epidermal cells whereas the roots of the same plants exposed to 75 and 150 mg l^{-1} showed growth. On the other hand, the roots of tobacco and jalapeno did not suffer any damage or experience growth. This may be taken to indicate that they are hyperaccumulators of TiO₂ or that they did not take up significant amounts of nanoparticles. In a study involving canola seeds and seedlings, Mahmoodzadeh et al. (2013) treated these seeds to nanoparticles of titanium dioxide at a very wide range of concentrations (10, 100, 1000, 1200, 1500, 1700 and 2000 mg l^{-1}). Concentrations of 2000 and 1500 mg l^{-1} gave the fastest and slowest germination rates, respectively. It was also found that the use of uniform size (20 nm mean size) nanoparticles increased both the rate of germination and robustness of the seedling.

All the results noted above, for different projects, reject the notion of '*one size fits all*' implied in statements which claim that nanoparticles are good for mankind because '*they help to increase agricultural production*'. The lesson to be learnt is that '*just as a few swallows do not make summer*', a few supportive research findings should not lead to euphoria in the minds of those who may benefit from such revelations. Corroboration of results using different methods may help to remove any doubts about the authenticity of the results. The above cautionary statements are not meant to demean or discredit the findings of many research groups but rather to encourage more research in a very crucial area, namely exploiting the benefits of nanotechnology for better living conditions for the world population of humans.

7 Phytoremediation and Nanophytoremediation of nano-TiO$_2$

Soil and water pollution by 'heavy metals' is a problem that has engaged and continues to engage the attention of many research organizations globally. Among the light metals, sodium is very much part of our daily lives in the form of sodium chloride or table salt. This ubiquitous white crystalline compound has beneficial and non-beneficial properties. The latter is often highlighted because 'high blood pressure' is attributed to high levels of sodium in the bloodstreams of persons having this condition. However, sufficient amounts of ions of sodium and chloride are also very useful as an ingredient of 'mouth wash and gargle' for reducing oral infections. Likewise 'heavy metals' also have positive and negative features concerning their usages. In the light of the above few remarks, about just one light element, it is noted that titanium is on the borderline between light and heavy metals in terms of the classification based on density and atomic number, namely that heavy metal should have a density of 5 g cm^{-3} and an atomic number greater than 23. Titanium has atomic number 22 and a density of 4.5 g cm^{-3}. On this basis, titanium is said to be on the borderline between light and heavy metals. In this chapter, titanium metal is not the subject of discussion; titanium dioxide in its nanoform is the focus. However, some aspects of titanium would be useful in understanding the properties of Ti(IV) which is the oxidation state of titanium in titanium dioxide.

The electronic configuration of the element Ti is: [Ar] $3d^2 4s^2$. On the other hand, Ti(IV) has the electronic configuration of [Ar]$3d^0 4s^0$. In electronic terms, this is a very stable electronic structure, being that of an inert element, namely argon. This implies that it would be energetically unfavourable to add or remove electrons from Ti in oxidation state IV. This, in turn, implies that oxidation or reduction will not occur readily. Stated differently, it could be expected that Ti(IV)oxide, that is, titanium dioxide should not be very reactive. The relatively large number of 'safe' uses or applications of titanium (IV) dioxide in macro- and nanoforms, may be taken as indirect testimony to this simplistic analysis. Two applications in particular, namely the addition of nanotitanium dioxide to foods (Ma et al. 2019) and its inclusion in denture products (Alirahlah et al. 2018) over many years without any reports of adverse effects, suggests that nanotitanium is not reactive in these instances. However, this is not to be interpreted that nanotitanium dioxide is not toxic to humans.

7.1 Effect of Concentration of Toxicity and the Role of Phytoremediation

One should be guided by the basic tenet of toxicology as enunciated some 500 years ago by Paracelsus, a Swiss physician and a chemist: '*All things are poison and nothing is without poison*'. Since then this has been paraphrased to read: '*The dose makes the poison*'. This implies that any substance can be harmful if consumed in very high doses. Even innocuous water, in exceedingly high doses, can be a problem for the human body.

Faraji and Sepehri (2018) investigated the effect of adding n-TiO$_2$ on stress caused by presence of Cd. They used n-TiO$_2$ at concentrations of 0, 500, 1000 and 2000 mg l^{-1} in the presence of sodium nitroprusside (SNP) to determine the effect of Cd (as 0, and 100 mM CdCl$_2$) on germination and growth of wheat seeds and seedlings. They found that the sue of the pair of n-TiO$_2$ and SNP working in concert reduced the stress due to Cd and thus promoted germination and growth of wheat seeds and seedlings, respectively.

In the light of the information of the background given above, it has to be conceded that the addition of large quantities of nanomaterial to soil may disrupt the microbial populations which serve desirable functions in soil conditioning. On the assumption that it is unnatural for high concentrations of nanoparticles to be in the soil as pollutants, plants have been used to uptake the nanoparticles of titanium dioxide employing a strategy which has been described as nanophytoremediation. Prior to the advent of nonophytoremediation, phytoremediation was in vogue some 50 years ago for removal of pollutants from soil using selected plants growing in the locality of the pollution (Judy et al. 2016) or invasive plants from outside the site of the pollution (Prabakaran et al. 2019). The latter are generally viewed as being destructive to established plants. Phytoremediation is a green technology for removing pollutants from contaminated soil. It has been classified, by those working in the field, into the following six types:

- phytoextraction (where pollutant moves from root to leaves which are generally removed and treated as waste (Eissa et al. 2014)
- phytovolatilization (in which transpiration of the undesirable material occurs through leaves (Islam et al. 2013),
- phytofiltration (filtration by biomass of the plant (Limmer and Burken 2016),
- phytostabilization (involves a reduction in movement of pollutants (Mendez and Maier 2008),
- phytodegradation (degradation using enzymes He et al. 2017),
- rhizome degradation (degradation by microbes in the rhizosphere Ouvrard et al. 2014).

The rhizosphere is the region of soil around the roots where microbes exist to promote growth.

7.2 Studies on Phytoremediation

On the basis that the presence of titanium dioxide in soil has not generally stunted growth of plants with some exceptions, plans are afoot to manufacture nanotitanium specifically for the protection of plants (Du et al. 2011). To test whether the new products would be taken up by plants, red clover was used in conjunction with a nitrogen-fixing bacterium (Moll et al. 2016). This combination was subjected to contact with two commercially available nanotitanium dioxides, namely P25 and E171. A control comprising macrotitanium dioxide was used as a control. It was found that E171 and the macrotitanium dioxide suppressed the growth rate of the bacterium while P25 produced no changes. The effect on red clover was that the shoot lengths decreased. The analysis showed that the nanoparticles had clustered to give units with dimensions greater than 100 nm and thus no longer nanoparticles (Moll et al. 2016).

As stated earlier, nanotitanium in the soil does not pose a serious problem to humans or lower life forms. On the contrary, nanotitanium aids in the uptake of heavy metals which are considered to be toxic to humans. The following study (Cai et al 2017) is a good illustration of this role of nanotitanium dioxide in plants. Since it is known that different types of nano-TiO$_2$ exhibit different properties, four types of nano-TiO$_2$ were chosen, namely anatase, pure rutile, hydrophilic or hydrophilic rutile and macrorutile. The study (Cai et al 2017) aimed to assess the effect of the presence of nanotitanium on the uptake and bioaccumulation of Pb by one variety rice, namely *Oriza sativa* from the soil which had a very high concentration of Pb which associated with toxic properties. The results showed that nano- and macroforms of titanium dioxide had the effect of reducing the uptake of Pb by rice seedlings by up to 80% depending on the actual type of titanium dioxide used. In another study in which oxides of Ti and Iron were involved, the effect of these in their positive and negative forms was assessed for their influence on the growth of roots of soybean and the associated soil microbes (Burke et al. 2015). Plants of soybean were grown in a greenhouse while receiving feeds of nanoparticles over six weeks. Then root growth and amount of nutrient were measured. For the root, the DNA method was applied to determine the extent of fungi and microbe formation. It was found that the charge of the nanoparticle played a significant part in growths investigated. Overall, nanoparticles of titanium dioxide proved to be less effective.

8 Summary and Conclusion

As noted in different sections of this chapter, the interaction of nanotitanium dioxide is influenced by several factors: the salient ones being a type of nanoparticle, the size, the shape, coating and microorganisms in the soil surrounding the roots of plants and toxicity of the nanoparticles towards plants. Furthermore, other factors such as ionic strength, pH and composition of the soil, including the presence of natural organic matter, on which the plants grow, should be taken into account when assessing the influence of nanoparticles of TiO$_2$ on the growth of plants. Regarding the type of nano-TiO$_2$, the source of the macrotitanium dioxide may be important, that whether it is a pure form of one of the crystal phases (rutile and anatase) in which TiO$_2$ exists or as a mixture of these two forms. The size and shape of the nanoparticles are crucial as they need to small enough as well as have a suitable shape for penetrating the cell walls of plants. It is worth noting that roots of plants have structures which allow ingress of nutrient matter but block what is detected to be 'invasive' particles. In a simplistic sense, the nanoparticles also have to contend with microorganisms which reside in the soil around the roots. It is clear from the foregoing points about the factors affecting interactions of nano-TiO$_2$, that there is much more to learn about the interactions of nanotitanium dioxide with plants than is known at present.

Acknowledgements The authors are grateful to the Natural Science Foundation of Hainan Province (2019RC166, 2019RC110) and National Natural Science Foundation of China (21965011) for financial support. Authors also acknowledge Durban University of Technology, South Africa, and Eskom Holdings, South Africa for financial support. Dr Vasanthakumar acknowledges the Postdoctoral Funding of Hainan Province, China.

References

Ajmal N, Saraswat K, Bakht MA, Riadi Y, Ahsan MJ, Noushad M (2019) Cost-effective and eco-friendly synthesis of titanium dioxide (TiO$_2$) nanoparticles using fruit's peel agro-waste extracts: characterization, in vitro antibacterial, antioxidant activities. Green Chem Lett Rev 12(3):244–254. https://doi.org/10.1080/17518253.2019.1629641

Andersen CP, King G, Plocher M, Storm M, Pokhrel LR, Johnson MG, Rygiewicz PT (2016) Germination and early plant development of ten plant species exposed to titanium dioxide and cerium oxide nanoparticles. Environ Toxicol Chem 35(9):2223–2229. https://doi.org/10.1002/etc.3374

Alirahlah A, Fouad H, Hashem M, Abdurahman A, Niazy AA, Abdulhakim, Al Badah A (2018) Titanium oxide (TiO$_2$)/poly-methylmethacrylate (PMMA) denture base nanocomposites: mechanical, viscoelastic and antibacterial behavior. Materials **11** (7):1096. https://doi.org/10.3390/ma11071096

Bis P, Wu CY (2005) Critical review: nanoparticles and the environment. J Air Waste Water Manag 55:708–746

Burke DJ, Pietrasiak N, Situ SF, Abenojar EC, Porche M, Kraj P, Samia ACS (2015) Iron oxide and titanium dioxide nanoparticle effects on plant performance and root associated microbes. Int J Mol Sci 16(10):23630–23650. https://doi.org/10.3390/ijms161023630

Cai F, Wu X, Zhang H, Shen X, Zhang M, Chen W, Gao Q, White JC, Tao S, Wang X (2017) Impact of TiO$_2$ nanoparticles on lead uptake and bioaccumulation in rice (*Oryza sativa* L). NanoImpact 5:101–108

Castiglione MR, Giorgetti L, Geri C, Cremonini R (2011) The effects of nano-TiO$_2$ on seed germination, development and mitosis of root tip cells of Vicia narbonensis L and Zea mays L. J Nanopart Res 13 (6):2443–2449. https://doi.org/10.1007/s11051-010-0135-8

Desireé M, Navas J, Sánchez-Coronilla A, Alcántara R, Fernández-Lorenzo C, Martín-Calleja J (2015) Highly Al-doped TiO$_2$ nanoparticles produced by Ball Mill method: structural and electronic characterization. Mater Res Bull 70:704–711. https://doi.org/10.1016/j.materresbull.2015.06.008

Du W, Sun Y, Ji R, Zhu J, Wu J, Guo H (2011) TiO$_2$ and ZnO nanoparticles negatively affect wheat growth and soil enzyme activities in agricultural soil. J Environ Monit 13(4):822–828. https://doi.org/10.1039/C0EM00611D

Eissa MA (2014) Phytoextraction of nickel, lead and cadmium from metal contaminated soils using different field. World Appl Sci J 32 (6):1045–1052. https://doi.org/10.5829/idosi.wasj.2014.32.06.912

Faraji J, Sepehri (2018) Titanium dioxide nanoparticles and sodium nitroprusside alleviate the adverse effects of cadmium stress on germination and seedling growth of wheat (*Triticum aestivum* L). Univ Sci 23(1):61–87

Federici G, Shaw BJ, Handy RD (2007) Toxicity of titanium dioxide nanoparticles to rainbow trout (*Oncorhynchus mykiss*): gill injury, oxidative stress, and other physiological effects. Aquat Toxicol 84 (4):415–430. https://doi.org/10.1016/j.aquatox.2007.07.009

Fulekar J, Dutta DP, Pathak B, Fulekar MH (2018) Novel microbial and root mediated green synthesis of TiO$_2$ nanoparticles and its application in wastewater remediation. J Chem Technol Biotechnol 93(3):736–743. https://doi.org/10.1002/jctb.5423

Gao J, Xu G, Qian H, Liu P, Zhao P, Hu Y (2013) Effects of nano-TiO$_2$ on photosynthetic characteristics of Ulmus elongata seedlings. Environ Pollut 176:63–70. https://doi.org/10.1016/j.envpol.2013.01.027

Guo X, Zhang Q, Ding X, Shen Q, Wu C, Zhang L, Yang H (2016) Synthesis and application of several sol-gel-derived materials via sol-gel process combining with other technologies: a review. J Sol-Gel Sci Technol 79(2):328–358. https://doi.org/10.1007/s10971-015-3935-6

He Y, Langenhoff AA, Sutton NB, Rijnaarts HH, Blokland MH, Chen F, Schröder P (2017) Metabolism of ibuprofen by Phragmites australis: uptake and phytodegradation. Environ Sci Technol 51 (8):4576–4584. https://doi.org/10.1021/acs.est.7b00458

Islam MS, Ueno Y, Sikder MT, Kurasaki M (2013) Phytofiltration of arsenic and cadmium from the water environment using *Micranthemum umbrosum* (JF Gmel) SF Blake as a hyper accumulator. Int J Phytorem 15(10):1010–1021. https://doi.org/10.1080/15226514.2012.751356

Jongprateep O, Puranasamriddhi R, Palomas J (2015) Nanoparticulate titanium dioxide synthesized by sol-gel and solution combustion techniques. Ceram Int 41:S169–S173. https://doi.org/10.1016/j.ceramint.2015.03.193

Judy JD , Kirby JK, Cavagnaro T, Paul M, Bertsch PM, (2016) Gold nanomaterial uptake from soil is not increased by arbuscular mycorrhizal colonization of *Solanum lycopersicum* (Tomato). Nanomaterials 6(68):1–9. https://doi.org/10.3390/nano6040068

Kim CS, Moon BK, Park JH, Choi BC, Seo HJ (2003) Solvothermal synthesis of nanocrystalline TiO$_2$ in toluene with surfactant. J Cryst Growth 257(3–4):309–315. https://doi.org/10.1016/S0022-0248(03)01468-4

Larue C, Khodja H, Herlin-Boime N, Brisset F, Flank AM, Fayard B, Carrière M (2011) Investigation of titanium dioxide nanoparticles toxicity and uptake by plants. J Phys: Conf Ser 304:012057. https://doi.org/10.1088/1742-6596/304/1/012057

Limmer M, Burken J (2016) Phytovolatilization of organic contaminants. Environ Sci Technol 50(13):6632–6643. https://doi.org/10.1021/acs.est.5b04113

Long TC, Tajuba J, Sama P, Saleh N, Swartz C, Parker J, Hester S, Lowry GV, Veronesi B (2007) Nanosize titanium dioxide stimulates reactive oxygen species in brain microglia and damages neurons in vitro. Environ Health Perspect 115:1631–1637. https://doi.org/10.1289/ehp.10216

Ma H, Lenz KA, Gao X, Li S, Wallis LK (2019) Comparative toxicity of a food additive TiO$_2$, a bulk TiO$_2$, and a nano-sized P$_{25}$ to a model organism the nematode *C. elegans*. Environ Sci Pollut Res Int 26:3556–3568. https://doi.org/10.1007/s11356-018-3810-4

Ma X, Yan J (2018) Plant uptake and accumulation of engineered metallic nanoparticles from lab to field conditions. Curr Opin Environ Sci Health 6:16–20. https://doi.org/10.1016/j.coesh.2018.07.008

Mahmoodzadeh H, Nabavi M, Kashefi H (2013) Effect of nanoscale titanium dioxide particles on the germination and growth of canola (*Brassica napus*) 25–32

Manesh RR, Grassi G, Bergami E, Marques-Santos LF, Faleri C, Liberatori G, Corsi I (2018) Co-exposure to titanium dioxide nanoparticles does not affect cadmium toxicity in radish seeds (*Raphanus sativus*). Ecotoxicol Environm Saf 148:359–366. https://doi.org/10.1016/j.ecoenv.2017.10.051

McDaniel E, Chen I, Balogh E, Yang Y, Ghoshroy S (2013) Structural analysis of plants exposed to titanium dioxide (TiO$_2$) nanoparticles. Microsc Microanal 19(S2):104–105. https://doi.org/10.1017/S1431927613002511

Mendez MO, Maier RM (2008) Phyto-stabilization of mine tailings in arid and semiarid environments an emerging remediation technology. Environ Health Perspect 116(3):278–283. https://doi.org/10.1289/ehp.10608

Moll J, Okupnik A, Gogos A, Knauer K, Bucheli TD, van der Heijden MGA (2016) Effects of titanium dioxide nanoparticles on red clover and its rhizobial symbiont. PLoS ONE 11(5):1–15. https://doi.org/10.1371/journal.pone.0155111

Nadeem M, Tungmunnithum D, Hano C, Abbasi BH, Hashmi SS, Ahmad W, Zahir A (2018) The current trends in the green syntheses of titanium oxide nanoparticles and their applications. Green Chem Lett Rev 11(4):492–502. https://doi.org/10.1080/17518253.2018.1538430

Naqvi S, Agarwal NB, Singh MP, Samim M (2021) Bio-engineered palladium nanoparticles: model for risk assessment study of automotive particulate pollution on macrophage cell lines. RSC Advances 11(3):1850–1861. https://doi.org/10.1039/D0RA09336J

Ouvrard S, Leglize P, Morel JL (2014) PAH phytoremediation: rhizodegradation or rhizoattenuation? Int J Phytorem 16(1):46–61. https://doi.org/10.1080/15226514.2012.759527

Pavlovic VB, Marinkovic V, Pavlovic VP, Nikolic Z, Stojanovic, Ristic MM (2002) Phase transformations and thermal effects of mechanically activated BaCO$_3$-TiO$_2$ system. Ferroelectrics 271 (1):391–396. https://doi.org/10.1080/713716214

Piccinno F, Gottschalk F, Seeger S, Nowack B (2012) Industrial production quantities and uses of ten engineered nanomaterials in Europe and the world. J Nanopart Res 14(9):1–11. https://doi.org/10.1007/s11051-012-1109-9

Prabakarana K, Lia J, Anandkumara A, Lenga Z, Chris BZ, Dua D (2019) Managing environmental contamination through phytoremediation by invasive plants: a review. Ecol Eng 138:28–37. https://doi.org/10.1016/j.ecoleng.2019.07.002

Pradhan SK, Reucroft RJ, Yang F, Dozier A (2003) Growth of TiO$_2$ nanorods by metalorganic chemical vapor deposition. J Cryst Growth 256:83–88. https://doi.org/10.1016/S0022-0248(03)01339-3

Riaz S, Naseem S (2015) Controlled nanostructuring of TiO$_2$ nanoparticles: a sol-gel approach. J Sol-Gel Sci Technol 74:299–309

Robichaud CO, Uyar AE, Darby MR, Zucker LG, Wiesner MR (2009) Estimates of upper bounds and trends in nano-TiO$_2$ production as a basis for exposure assessment. Env Sci Technol 12:4223–4233

Roopan SM, Bharathi A, Prabhakarn A, Rahuman AA, Velayutham K, Rajakumar G, Madhumitha G (2012) Efficient phyto-synthesis and structural characterization of rutile TiO$_2$ nanoparticles using *Annona squamosa* peel extract. Spectrochim Acta Part a Mol Biomol Spectrosc 98:86–90. https://doi.org/10.1016/j.saa.2012.08.055

Sabry RS, Al-Haidarie YK, Kudhier MA (2016) Synthesis and photocatalytic activity of TiO$_2$ nanoparticles prepared by sol-gel method. J Sol-Gel Sci Technol 78:299–306. https://doi.org/10.1007/s10971-015-3949-0

Sadr FA, Montazer M (2014) In situ sonosynthesis of nano TiO$_2$ on cotton fabric. Ultrason Sonochem 21(2):681–691. https://doi.org/10.1016/j.ultsonch.2013.09.018

Shirsath SR, Pinjari DV, Gogate PR, Sonawane SH, Pandit AB (2013) Ultrasound assisted synthesis of doped TiO$_2$ nano-particles: characterization and comparison of effectiveness for photocatalytic oxidation of dyestuff effluent. Ultrason Sonochem 20(1):277–286. https://doi.org/10.1016/j.ultsonch.2012.05.015

Shalini G, Hegde VS, Soumya M, Korkoppppa MM (2020) Provenence and implication of heavy minerals in the beach sands of India's Central West Coast. J Coastal Res 36:353–361

Sivaranjani V, Philominathan P (2016) Synthesize of Titanium dioxide nanoparticles using Moringa oleifera leaves and evaluation of wound healing activity. Wound Med 12:1–5. https://doi.org/10.1016/j.wndm.2015.11.002

Song U, Jun H, Waldman B, Roh J, Kim Y, Yi J, Lee EJ (2013) Functional analyses of nanoparticle toxicity: a comparative study of the effects of TiO$_2$ and Ag on tomatoes (Lycopersicon esculentum). Ecotoxicol Environ Saf 93:60–67. https://doi.org/10.1016/j.ecoenv.2013.03.033

Subhapriya S, Gomathipriya P (2018) Green synthesis of titanium dioxide (TiO$_2$) nanoparticles by *Trigonella foenum-graecum* extract and its antimicrobial properties. Microb Pathog 116:215–220. https://doi.org/10.1016/j.micpath.2018.01.027

Tan W, Peralta-Videa JR, Gardea-Torresday JL (2017) Interaction of titanium dioxide nanoparticles with soil components 1 and plants: current knowledge and future research need—a critical review. Environ Sci Nano. https://doi.org/10.1039/C7EN00985B

Valant J, Drobne D, Novak S (2012) Effect of ingested titanium dioxide nanoparticles on the digestive gland cell membrane of terrestrial isopods. Chemosphere 87(1):19–25. https://doi.org/10.1016/j.chemosphere.2011.11.047

Zeng B, Zeng W (2017) Hydrothermal synthesis and gas sensing property of titanium dioxide regular nano-polyhedron with reactive (001) facets. J Mater Sci Mater Electron 28(18):13821–13828. https://doi.org/10.1007/s10854-017-7228-4

Zhang K, Wang Y, Mao J, Chen B (2020) Effects of biochar nanoparticles on seed germination and seedling growth. Environ Pollut 256:113409. https://doi.org/10.1016/j.envpol.2019.113409

Plant Physiological Responses to Engineered Nanoparticles

Ahmed Abdul Haleem Khan

Abstract

Plants are reared in and around the world for a variety of purposes. The peasant community cultivates the same crop in continuous manner that result in more economic losses due to multiple reasons. The common among them may be diseases by pathogenic microorganisms and parasites (insect pests, nematodes), micronutrient deficiency and weeds. The green revolution introduced a number of agrochemicals to improve the crop yield, but prolonged use turned these complex organic compounds into recalcitrant and xenobiotics of ecosystem. The demand for sustainable agriculture has been introduced to improve the yield and meet the supplies as required. The rise in synthesis of particles of nanoscale with wide range of metal oxides such as Ag, Au, Al, Cd, Ce, Cu, Co, Fe, G, Ni, Mg, Pt, Pd, Mn, Ti, Zn was known to be beneficial in different fields. The nanoparticle synthesis is known to be done by different approaches like physical, chemical and biological (plant materials, bacteria and fungi). The nanoparticle applications in the field of agriculture are not as popular as compared to other allied aspects (medicine, pharmacy). There is progress in laboratory-level studies that could make nanoparticle-based products in agriculture as a substitute to agrochemicals. This chapter is intended to discuss the developments in the field of nanoparticles as a success story that proved the potential of nanoscale components as plant growth stimulants, fungicide, pest control, weedicides and micronutrient supply.

Keywords

Agrochemicals • Crop plants • Growth responses • Phytohormones • Phytotoxicity

A. A. H. Khan (✉)
Department of Botany, Telangana University, Nizamabad, 503322, (T.S.), India
e-mail: aahaleemkhan@gmail.com

1 Introduction

Plants are known to be autotrophic growing by acquiring solar energy and other requirements from substratum via the root system. The underground part root absorbs mineral nutrients required for growth and development from soil. The physiology and phenology of plants responds to the availability of mineral nutrients in the form of growth and reproduction. The deficiency and excess in minerals in plants expressed as symptoms, leads to several ill effects. Agriculture, an important source of food and feed around the globe, is facing challenge with increase in human population. The advent of the green revolution in 1970 changed the scenario to be profitable for the peasant community. The extensive use of synthetic chemicals developed an alarm to save the ecosystem. The series of approaches lead to go with smart/climate-resilient crops for reducing the threats for upcoming generations (Khan et al. 2017).

Nanoscale materials are natural or man-made/engineered forms that serve superior to bulk form. The nanoparticle (NPs) exist in size range <100 nm with spherical, tubular, irregular in shape found in single, fused or agglomerated forms in homologous or heterogeneous composition (Table 1). Nanotechnology has been applied in the field of farming systems to improve the yields of crop with minimizing the inputs. To overcome hazardous fertilizers, pesticides/insecticides, weedicides/herbicides, fungicides, and antibiotics nanoscale materials are exploited as a part of sustainable agriculture (Tables 2, 3, 4 and 5). The common fact is the applied fertilizers for nitrogen and phosphorus are absorbed by plants (30–50%) and the efficiency of absorption is low, and the possibility of interference with substratum increases and creates havoc. The nanoscale fertilizers increased the nutrient absorption efficiency with enhanced crop productivity. The important problem of agriculture is weed, pathogens (virus, fungi, bacteria, pests, nematodes) that reduce the crop yield.

Table 1 Engineered nanomaterials and their types

Engineered nanomaterials (ENMs)	Types
Carbon-based nanomaterials	Carbon nanotubes (CNTs): single walled carbon nanotube (SWCNT), multi walled carbon nanotube (MWCNT) and Graphene and fullerenes (C60 and C70)
Metal based nanomaterials (Inorganic)	Zero-valent metals (such as Au, Ag, and Fe ENMs), Metal oxides (nano-ZnO, TiO_2 and CeO_2), and Metal salts (such as nano silicates and ceramics)
Quantum dots	CdSe and CdTe
Nanosized polymers (Organic)	Dendrimers, liposomes and polystyrene
Organic—inorganic hybrids	Metal organic frameworks, covalent organic frameworks

Table 2 Engineered nanomaterials and their uptake by plants

Mode	Types	
Cuticle (size: 0.6–4.8 nm)	Lipophilic (non-polar) Hydrophilic (polar)	Leaves
Stomata (size: 20 nm)	Stomatal aperture through apoplast	
Lateral root junction	Apoplastic pathway	Roots
Cell–cell contact	Symplastic pathway	

Table 3 Impact of nanoparticles on plant productivity

Nanoparticles	Plant productivity	Reference
Nano-zinc oxide (nZnO) and nano-silicon (nSi)	Improved salt resistance in plant, load of annual crop and quality of mango fruit	Elsheery et al. (2020a, b)
Urea doped hydroxyapatite nanoparticles (Ur@HANP)	Proved alternative for N and P fertilizers	Pradhan et al. (2020)
Titanium NPs (Ti-NPs)	Alleviated As-induced toxic responses in *Vigna radiata* L	Katiyar et al. (2020)
Molybdenum oxide nanoparticles (MoO_3-NPs)	Effective on the productivity of common bean plant	Osman et al. (2020)
Calcium tetraborate nanocrystals-Boron (B) nano-fertilizer (NF) - lettuce (*Lactuca sativa*) and zucchini (*Cucurbita pepo*) growth and physiology	Effective at plant productivity on B-limited soils	Meier et al. (2020)
Iron (III) oxide (Fe_2O_3) NMs applied to wheat plants in a hydroponics	Efficient in plant growth	Al-Amri et al. (2020)
Zinc oxide nanoparticles (ZnONPs) and zinc ions (Zn^{2+})	Reduced total As in rice	Ma et al. (2020)
Zinc oxide nanoparticles (ZnO-NPs)	Reduce toxicity of Cd in tomato plants	Faizan et al. (2020)
Fe_3O_4 nanoparticles (NPs)	Protective role of NPs for microbes and plant (maize) roots	Yan et al. (2020)
Manganese (III) oxide nanoparticles (MnNPs)	Alleviate salinity stress in *Capsicum annuum* L	Ye et al. (2020)

Table 4 Impact of nanoparticles formulation with herbicides

Nanoparticles	Impacts	Reference
Poly(lactic-co-glycolic-acid) (PLGA) with atrazine on potato	Herbicide was effective and alternative to inhibit weed growth	Schnoor et al. (2018)
Fullerenol nanoparticles (FNP) with paraquat on honey bee (*Apis mellifera carnica*)	Antioxidative effects with protection against oxidative stress	Kojic et al. (2020)
Different NMs (Fe, Mn_3O_4, SiO_2, Ag, and MoS_2) on spinach	Proved to enhance photosynthesis and potential as nanofertilizer	Wang et al. (2020a, b)
Spinach, apple and corn leaves Glyphosate (Gly) on cysteamine-modified gold nanoparticles (AuNPs-Cys)	Evaluated the Gly distribution on plant tissues	Tu et al. (2019)
Triazine + ZnO-NPs on Corn	Determination of traces of herbicide	Li et al. (2017)
Poly(ε-caprolactone) nanocapsules with neem oil	Environmentally friendly formulation for applications in agriculture	Pasquoto-Stigliani et al. (2017)
Herbicides: imazapic and imazapyr Alginate/chitosan and chitosan/tripolyphosphate nanoparticles	Encapsulation of herbicides improved mode of action and reduced toxicity	Maruyama et al. (2016)
Gold nanoparticles (AuNPs) and nanosheets - triazine herbicides (prometryn, atrazine, terbumeton and secbumeton) in spiked maize	Au/LDH nanohybrids can also be applied to extract other analytes	Li et al. (2018)
Paraquat on novel nanoparticles of pectin, chitosan, and sodium tripolyphosphate (PEC/CS/TPP)	Efficient and formulation of NPs showed herbicide activity in maize/mustard	Rashidipour et al. (2019)
Metolachlor water-based mPEG − PLGA nanoparticle formulation	Polymeric nanoparticles served pesticide carrier on *O. sativa*, *Digitaria sanguinalis* with low environmental impact	Tong et al. (2017)
2,4-dichlorophenoxy acetic acid (2,4-D) mesoporous silica nanoparticles (MSNs)	Nanoformulation showed good bioactivity on target plant cucumber (*C. sativus* L.) and wheat (*T. aestivum* L.)	Cao et al. (2018)
Poly(ε-caprolactone) (PCL) nanocapsules containing atrazine	Nanocapsules potentiated the post-emergence control of *Amaranthus viridis* (slender amaranth) and *Bidens pilosa* (hairy beggarticks)	Sousa et al. (2018)
Polycaprolactone nanocapsules (PCL) containing pretilachlor	Barnyard grass found cytotoxicity and rice-no toxic effect	Diyanat et al. (2019)
Plant virus nanoparticles (VNPs) and virus-like particles (VLPs): tobacco mild green mosaic virus (TMGMV), cowpea mosaic virus (CPMV), Physalis mosaic virus (PhMV), mesoporous silica nanoparticles (MSNPs) and poly(lactic-co-glycolic acid) (PLGA) formulation	Plant viruses were superior to synthetic mesoporous silica nanoparticles and poly(lactic-co-glycolic acid) for the delivery and controlled release of pesticides	Chariou et al. (2019)
Mesoporous silica nanoparticles (MSNs) - Diquat dibromide (DQ)	Exhibited herbicidal activity against *Datura stramonium* L	Shan et al. (2019)

The synthetic chemicals are applied to reduce the burden of diseases, and in turn, the chemical forms target the beneficial agents. There are several reports on nanomaterials as effective control efficacies and improved crop yields (Khan et al. 2015a, b, 2019). The efficiency of engineered nanoparticles (ENPs) is focused to highlight the merits on plant physiology that increase the crop yields with multifarious applications. The need of sustainable farming to protect the environment with high yields from different crop plants is investigated to feed the growing human population. The nanoscale particles are reported for alternatives to reduce the burden of agrochemicals. The reports on use of ENPs to boost plant physiological responses and enhance the yield and quality of crop, medicinal and ornamental plants (herbs, shrubs and trees) were evaluated to prove the

efficiency in different conditions (Figs. 1 and 2). In this chapter, the role of ENPs in test plants is presented in different categories, viz.: (1) Plant growth responses, (2) Fertilizer effects in different plants, (3) Nano-harvest, (4) Phytoaccumulation, and (5) Toxicity effects.

2 Plant growth responses to engineered nanoparticles

Improved metabolic profile

The soil amended with cadmium sulfide nanoparticles (CdS-NPs) was used for broad bean (*Vicia faba* L.) plant cultivation and evaluation of the phenotypic, biochemical

Table 5 Impact of nanoparticles formulation for insecticides

Nanoparticles	Insecticidal activity	References
Chitosan and agrochemical loaded chitosan (spinosad and permethrin) nanoparticles	More effective with a lasting residual effect on *Drosophila melanogaster*	Sharma et al. (2019)
Pesticide (ferbam)-gold nanoparticles (AuNPs)	NP's served as carrier for delivering pesticides	Hou et al. (2016)
Carboxylic multiwall carbon nanotubes (CMNTs) as adsorbent to remove fenvalerate	Showed stability and non-aggregatable as adsorbent	Naeimi et al. (2016)
Carboxymethyl chitosan modified carbon nanoparticles (CMC@CNP), as carrier for emamectin benzoate (EB)	Test NP performance based on pH-responsive controlled release. The release of EB was sustained, steady and prolonged persistence time on maize with *Mythimna separate*	Song et al. (2019)
Nanoformulation (NF) of thiamethoxam (TMX) - cellulose nanocrystals (CNCs)	Insecticidal activity against *Phenacoccus solenopsis*	Elabasy et al. (2020)
Copper-based nanopesticide Kocide 3000	Effective on genes related to detoxification and reproductive system of *Daphnia magna* (water flea)	Aksakal and Arslan (2020)
Zinc oxide nanoparticles (ZnO NPs) and silica nanoparticles (SiO$_2$ NPs) against: adults of rice weevil (*Sitophilus oryzae* L.); red flour beetle (*Tribolium castaneum* Herbst.) and cowpea beetle (*Callosobruchus maculatus* F.)	Proved potential as stored seed protectant	Haroun et al. (2020)
Silica nanoparticles (SiO$_2$-NPs) against *Sitophilus oryzae, Rhizopertha dominica, Tribolium castaneum,* and *Orizaephilus surinamenisis*	NPs were effective than conventional pesticides	El-Naggar et al. (2020)
Silver nanoparticles (AgNPs) from leaf extract of *Holostemma ada-kodien*	Toxic against *Anopheles stephensi, Aedes aegypti,* and *Culex quinquefasciatus* and Antimicrobial activity	Alyahya et al. (2018)
Fe$_2$O$_3$NPs on Bt-transgenic scotton	Increased the Bt-toxin in leaves and roots	Nhan et al. (2016)
Neem oil-loaded zein nanoparticles	Mortality effects on *Acanthoscelides obtectus, Bemisia tabaci* and *Tetranychus urticae*	Pascoli et al. (2020)

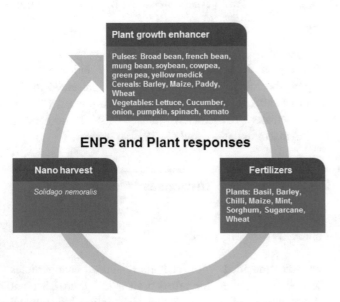

Fig. 1 Impact of engineered nanoparticles (ENPs) in different plants

and metabolic responses. The findings of study proved the alleviation of toxicity by CdS-NP in soil and without change in phenotypic effects in broad bean plant and upregulation of antioxidative metabolic profiles of the leaves (Tian et al. 2020). Another study investigated the role of iron oxide nanoparticles (Fe$_3$O$_4$-NPs) on phloem-sap metabolite composition in pumpkin (*Cucurbita maxima* L.) plants. The results showed that the test NPs were translocated to the aerial parts of plants with increased metabolites in phloem sap and improved oil composition of the plant (Tombuloglu et al. 2020). The reports of studies on plants like broad bean and pumpkin proved treatment with test NPs enhanced metabolites that alleviate toxicity and improved oil composition.

Recovery from drought

The study explored the impact of silicon nanoparticles (Si-NPs) on seedlings of barley (*Hordeum vulgare*) treated

Fig. 2 Physiological responses to engineered nanoparticles (ENPs) by test plants

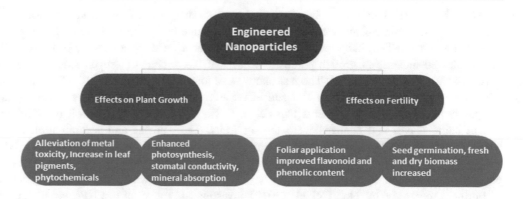

with different drought intensities and recovery from drought stress. The findings showed NP aggregates formation in plant tissues, pores of large size in roots and stomata in leaves were closed rapidly. There was an increase in total chlorophyll and carotenoid content of leaves in the test plants. The plants showed changes in antioxidant enzymes, cell injury, osmolyte, metabolite profile and the membrane stability indices. The study proved that application of NPs directly in soil was suitable for post-drought recovery of barley plants (Ghorbanpour et al. 2020).

Counteract membrane damage

The effects of engineered nanoparticles such as Ag, Co, Ni (metals) and CeO_2, Fe_3O_4, SnO_2, TiO_2 (metal oxides) on seedlings of basil (*Ocimum basilicum* L.) grown in mix of 20% sandy soil and 80% peat were investigated by Antisari et al. (2018). The results indicated that the test metal-NPs were accumulated in the roots and the selected NPs, i.e., Ag, Co, CeO_2 and Ni, were translocated from the root to shoot, leaves and then to edible part of the test plant. The relative short exposure accumulated Ca in roots that counteracted the membrane damage by nanoparticles (Antisari et al. 2018).

Induced root formation

The study carried out by Ahmad et al. (2020) on ZnO and CuO-ENPs application on in vitro formation of root, antioxidant (non-enzymatic) activities and steviol glycosides (SGs) in Candyleaf (*Stevia rebaudiana*) regenerants. The results of NP treatment showed that the percentage of rooting and SGs—rebaudioside A and stevioside—were increased. The phytochemical studies (flavonoid content, phenolic content) and 2,2-diphenyl-1-picryl hydrazyl (DPPH)-free radical scavenging activity were high in regenerants. The biochemical and morpho-physiological responses of candyleaf were proved to elicit defense against test ENPs (Ahmad et al. 2020).

Increased in vitro regeneration

The study by Zia et al. (2020) analyzed the effect of silver nanoparticles (AgNPs) on in vitro regeneration of carnation cultivars cv. *Noblessa*, cv. *Antigua* and cv. *Mariposa*. The number of shoots/explant of cv. *Noblesse* and cv. *Antigua* and cv. *Mariposa* showed the highest regeneration rate. The study concluded that test nanoparticles were effective for increasing in vitro shoot multiplication and regeneration of plants (Zia et al. 2020).

Improved yield and nutritional quality

The effects of $nCeO_2$ and nCuO on yield and nutritional quality of cucumber by foliar application was studied by Wang et al. (2020a, b) in three week-old cucumber seedlings grown in soil. The test plants were evaluated for parameters such as Ce, Cu and other nutritional elements, stomatal conductance (Gs), transpiration rate (E), net photosynthesis rate (Pn), yield, fruit size, weight and firmness. The results showed increase in the fresh weight and reduced Zn content in test plant fruits treated with nanoparticles (Hong et al. 2016). Another study by Wang et al. (2020a, b) on Chinese scallion (*Allium fistulosum*) plants from soil amended with CuO particles [nano (nCuO), bulk (bCuO) and $CuSO_4$] grown in greenhouse conditions. These plants were used to evaluate the allicin content, nutrient element and enzymatic antioxidants. The test plants showed enhanced nutrient and allicin contents in scallion by nCuO treatment and suggested the use of nanofertilizer for onion crop (Wang et al. 2020a, b).

Further, the bulbs of *Allium cepa* were assessed for mitotic index (MI) and chromosomal aberrations (CAs) after treatment with TiO_2 and ZnO-NPs, and their mixtures (1:1) by Fadoju et al. (2020). The results of recovery test in treated bulbs showed transient CAs induced by both NPs and the frequency of aberrations was high. The finding proved the potential of tested NPs to induce mutation in somatic cells of bulbs of *A. cepa* (Fadoju et al. 2020). In a study by Bakshi et al. (2019), the tomato plants were grown in sewage sludge amended with nano-TiO_2 in agricultural soil. The NP-treated plants showed increase in growth parameters like leaf biomass (142%), fruit yield (102%). The test plants were found with decrease in tannin, lignins of leaf and increase in carbohydrate levels, change in elements like Fe, B, P, Na and Mn in stem, leaves and changes were less in fruits. The tomato fruits showed no significant Ti enrichment, and TiO_2-NPs proved safe in improving growth and biochemical parameters (Bakshi et al. 2019).

Further, the study by Noori et al. (2020) investigated tomato seedlings (*Lycopersicon esculentum*) for physiological and molecular responses upon exposure to AgNPs and silver nitrate ($AgNO_3$) in hydroponic media for 7 days. The results showed 2–7 times decrease in growth rate and increase in H_2O_2 and malondialdehyde in exposed plants than control. There was decrease in enzymatic antioxidant (50%) and upregulation of genes of ethylene-inducing xylanase (EIX), peroxidase (POX) and phenylalanine ammonia-lyase in test plants (Noori et al. 2020).

Improved photosynthesis

The bean (*Phaseolus vulgaris* L.) plants were sprayed and applied in soil with CeO_2-NPs by Salehi et al. (2018). The results showed absorption of test NPs in dose-dependent manner; the uptake and translocation by plants was through both roots and leaves that increased Ce content. The spraying lowered stomatal density and increased stomatal length, and alteration in photosynthesis and the electron transport chain. The increase in Ce content induced accumulation of osmolytes (proline), phytosiderophores (muconate and mugineate) and proteins involved in folding or turnover. The spray mode of NPs in bean plants was effective compared to soil application (Salehi et al. 2018).

Amelioration of Cd toxicity

The cowpea plants were evaluated by Ogunkunle et al. (2020) for Cd toxicity after foliar treatment of nano-TiO_2 in six episodes. The results showed that NPs promoted parameters in roots and leaves like chlorophyll b, total chlorophyll content and stress enzymes. The seeds were with increase in Zn, Mn and Co levels and the roots, shoots and grains showed decrease in Cd levels in NP-treated plants after Cd stress. The nano-TiO_2 foliar application proved its ameliorative potential in cowpea plants for Cd toxicity (Ogunkunle et al. 2020).

Lian et al. (2020) investigated maize (*Zea mays* L.) crop production in Cd-contaminated soils by application of TiO_2-NPs. The results showed that root exposure enhanced Cd uptake and created phytotoxicity in the test plant. The foliar exposure decreased shoot Cd content and alleviated Cd-induced toxicity by test NPs through increase in activities of superoxide dismutase (SOD) and glutathione S-transferase (GST) and upregulation of citrate cycle, galactose, alanine, aspartate, glutamate, glycine, serine and threonine metabolism. Sharifan et al. (2020) investigated the uptake and accumulation of surface charge of ENP cerium oxide nanoparticles (CeO_2-NPs) and cadmium (Cd) in the presence of inorganic phosphorous in soybean seedlings grown in hydroponic system. The results showed mutual effect of test NPs and Cd-affected phosphate level in treated plants.

Increase in seed number

The green pea (*Pisum sativum*) plants were grown in soil by Ochoa et al. (2017) amended with nano-CuO, bulk-CuO (bCuO) and $CuCl_2$ and indole-3-acetic acid (IAA). The results showed that NPs reduced the number of plants and pod biomass by about 50%. The results suggested that the nutritional quality of test pea pods was improved by use of nano-CuO and bCuO (Ochoa et al. 2017). The effect on seed germination and growth performance of pea (*P. sativum*) by treatment by poly(vinylpyrrolidone) (PVP) stabilized on platinum nanoparticles (Pt:PVP) was investigated. The germination rate was decreased, and the other parameters like dormancy period, arbuscular mycorrhizal fungi and rhizobial colonization in treated plants were decreased. The results proved that the average number of seeds was increased to 163.5% and the average seed weight was decreased by 66.7% (Rahman et al. 2020).

Decreased fertilizer use

The greenhouse experiments by Pandorf et al. (2020) were conducted to study the growth of romaine lettuce (*Lactuca sativa*) and nitrate leaching through soil by using 2D graphite carbon nanoparticles (CNPs). Then, the NPs were combined with fertilizer and the effect on yield, nitrate leaching and nutrient uptake by lettuce plant was evaluated by applying in the soil. The results showed that test NP lowered fertilizer dose and decreased nitrate infiltration through the soil (Pandorf et al. 2020).

Improved germination

The germination of *Vigna radiata* was studied by Jung et al. (2020) of single and binary mixtures of CdO, CuO nanoparticles under humidity 70–80%. The results showed the germination rate of bean was high at 80% humidity and less with single metal NP exposure in both humidity levels. The metal accumulation rate was high with the nCuO than nCdO in treated bean plants (Jung et al. 2020). Pariona et al.

(2017) reported the germination and early growth of oak (*Quercus macdougallii*) by application of citrate coated two types of Fe_3O_4-NPs. The germination was increased 33%, and the growth, dry biomass and chlorophyll concentration were enhanced. The study suggested that NPs treatments could improve reforestation of threatened forestry species (Pariona et al. 2017).

The investigation of ZnO and TiO_2 nanomaterial synthesis was attempted to increase the rate of transplant production in eggplant, pepper and tomato. The results showed the effect of nanomaterials gel-coated seedlings on parameters like mean germination time, and germination coefficient of variation was reduced. The performance of growing transplants was efficient for the safer production of transplants by gel-coated nanomaterials on test plants (Younes et al. 2020).

Increased plant biomass

The effects on hydroponically grown *Nigella arvensis* L. by application of engineered aluminum and nickel oxide (Al_2O_3 and NiO-NPs) nanoparticles on parameters like growth, oxidative stress and activities of antioxidants were investigated by Chahardoli et al. (2020). The less concentrations increased plant biomass, and the high levels of the test nanoparticles decreased *N. arvensis* biomass. There was an increase of enzymatic antioxidants such as ascorbate peroxidase (APX), catalase (CAT), superoxide dismutase (SOD) and peroxidase (POD) in roots and shoots. The parameters like scavenging activity by 2,2-diphenyl-1-picryl hydrazyl (DPPH), capacity of total antioxidant, reducing power, iridoids, saponin and phenols in treated plants were increased by test NPs. The application of NiO-NPs on test plants inhibited the antioxidant activities, secondary metabolites formation, total antioxidant capacity, scavenging activity by DPPH and total saponin content (Chahardoli et al. 2020).

Effect on plant microbiota

Vitali et al. (2019) explored the poplar (*Populus nigra*) plants for the effect of silver nanoparticles (AgNPs) application. The test plant microbiota levels in leaf and root were evaluated after NP treatment. The results showed increase in bacteria and fungi in the treated leaf and reduced the bacterial and fungal biodiversity in the root. The study showed the phyllosphere and rhizosphere poplar-associated microbiota of a tree species from a polluted environment (Vitali et al. 2019).

Increase in plant root and shoot

The sludge conditioned in soil with single and binary mixture of nanoparticles (Ag_2O, TiO_2) was investigated by Singh and Kumar (2020a, b) for effect on spinach plants grown in pot experiments. The Ag_2O NPs treated plants showed no growth effects. The root length and fresh weight were increased in spinach at high concentration of single and binary mixture of TiO_2NPs. The binary mixture and TiO_2 increased total chlorophyll content and decreased with higher tendency of root surface adsorption by Ag_2O. The study of single and binary mixture of NPs reported no acute toxicity in the treated spinach. The spinach leaves grown in sludge enriched with NPs were found unsafe for consumption due to accumulation of Ag and Ti metals (Singh and Kumar 2020a, b).

The study assessed the role of nano-zerovalent iron (FeO nanoparticles) on sunflower (*Helianthus annuus*) plants cultivated in soil with hexavalent chromium (Cr IV). The amelioration of Cr toxicity in sunflower plants by application of test NPs was evaluated by Mohammadi et al. (2020). The results revealed that the higher concentration of test nanoparticles increased plant morphological and physiological parameters and decrease in Cr uptake. The factors like bioaccumulation (BAF) and translocation (TF) in root and shoot tissues were reduced. The results of NP-treated plants under Cr toxicity were found through reduced Cr uptake and increased activity by SOD, CAT, POX and APX detoxification enzymes (Mohammadi et al. 2020).

Mitigation of chilling stress

The effects on sugarcane leaves treated with NPs of silicon dioxide ($nSiO_2$), zinc oxide (nZnO), selenium (nSe), graphene nanoribbons (GNRs) as foliar sprays were investigated by Elsheery et al. (2020a, b). The ameliorative effects of NPs against chilling stress in test plants for photosynthesis and photoprotection were evaluated. The results of NPs application reduced the adverse chilling effects in treated seedlings by the increased PS-II photochemical efficiency (*Fv/F*m), maximum photo-oxidizable PS-I (*P*m), photosynthetic gas exchange and the chlorophyll and carotenoid content. It was proved that among the tested NPs, $nSiO_2$ showed higher amelioration effects to mitigate chilling stress in sugarcane (Elsheery et al. 2020a, b).

Improved defenses

The foliar spray of Fe_3O_4-NPs on *Nicotiana benthamiana* plant was reported for the increased (both dry and fresh) weights, activation of antioxidants and upregulation of salicylic acid (SA) synthesis. The accumulation of endogenous SA in test plants conferred the plant resistance against *Tobacco Mosaic Virus* (TMV) infection (Cai et al. 2020). The exposure of tobacco (*Nicotiana tabacum*) seedlings to AgNPs and ionic silver was investigated by Stefanic et al. (2020) for the physiological effects, changes in ultrastructure and proteomics. The results revealed high toxicity in treated seedlings due to oxidative stress parameters by ionic Ag than nanosilver. The root cells showed the presence of silver in the nanoparticle form. The leaf chloroplasts of treated plantlets were changed that altered rate of photosynthesis. The majority of primary metabolism proteins was up-regulated that helped to cope with silver-induced toxicity through enhanced energy production and reinforced defense in treated plants (Stefanic et al. 2020).

Kokina et al. (2020) demonstrated that the five-weeks-old yellow medick (*Medicago falcata* L.) plants grown using hydroponics with Fe_3O_4-NPs. The results of treatment induced increase in root length, chlorophyll a fluorescence in yellow medick. The parameters that conferred resistance to powdery mildew disease (fungal) were reduced genome instability, genotoxicity and expression of miR159c (Kokina et al. 2020).

Responses of nanoparticles treated paddy

Peng et al. (2020) examined the effects on paddy soil and rice plants under flooded condition by ZnO, CuO and CeO_2 nanoparticles (NPs) for the bioavailability and translocation. The results showed that test NPs enhanced redox potential of paddy soil. The NPs induced the elements (Cu and Ce) accumulation in rice roots. The Zn concentration in shoots was high by ZnO-NPs with translocation factor—1.5. The root cortex accumulated Zn and Cu was accumulated in the root exodermis in the NPs treated plants (Peng et al. 2020).

Wu et al. (2020) reported hydroponic cultivation of rice (*Oryza sativa* L.) seedlings for comparative evaluation of the metallic (AgO-NPs) and sulfidized (Ag_2S-NPs) silver nanoparticles. The test NPs were investigated for iron plaque formation and effects of silver uptake in seedlings. The results revealed iron plaque in seedlings and the AgO and Ag_2S-NPs bioavailability. The study alarmed concern for the wetland plants for Ag_2S-NPs exposure. The high Fe levels facilitate bioavailability of Ag_2S-NP in Fe-rich environments (Wu et al. 2020). The rice seedlings in a hydroponic were investigated for arsenite (As(III)) or arsenate (As(V)) and CuO-NPs or Cu(II) accumulation by Wang et al. (2019). Cu in both forms were found to reduce the total As accumulation, and Cu(II) was more effective than CuO-NPs in rice tissues. The results proved that nano-enabled agrichemicals were alternative to conventional metal salts in agriculture for safe application (Wang et al. 2019).

Zhang et al. (2020a, b) reported on the heavy metals chemical speciation and micronutrient bioavailability in paddy soil by TiO_2-NPs, ZnO-NPs and CuO-NPs by flooding–drying simulation. The results showed that the NPs addition increased pH, Eh and electrical conductivity (EC) in soil. The acid-soluble fraction showed increase in the Zn and Cu concentrations that led to enhanced bioavailability of test

metals in the soil. The NPs treated soil showed decrease in Cd bioavailability with the TiO_2-NPs and increase by ZnO and CuO-NPs (Zhang et al. 2020a, b). The ZnO-NPs toxicity in rice seedlings by using sodium nitroprusside (SNP, a NO donor) was investigated for the regulatory mechanisms of nitric oxide (NO) in counteracting test NPs toxicity by Chen et al. (2015). The results showed reduced accumulation of Zn, production of reactive oxygen species and lipid peroxidation. The test seedlings showed increase in reduced glutathione and activities by peroxidase, catalase and ascorbate peroxidase. The study provided evidence for NO in amelioration of test NPs phytotoxicity in rice seedlings (Chen et al. 2015).

The germination and growth of rice (*O. sativa* L., cv. Swarna) seedlings were evaluated by Gupta et al. (2018) for phytostimulatory effect by silver nanoparticles (AgNPs). The results showed that tested concentrations of NPs promoted both the shoot and root growth and increased the length and biomass, phenolic metabolites, chlorophyll-a and carotenoid contents of seedlings. The study showed changes in activities of catalase (CAT), ascorbate peroxidase (APX) and glutathione reductase (GR) and gene expression of antioxidative enzymes in seedlings (Gupta et al. 2018).

The study by Zhang et al. (2020a, b) evaluated the phytotoxicity and cadmium (Cd) migration in *O. sativa* by TiO_2-NPs in the soil–rice system. The high Cd content decreased the height and biomass of test plants and metal enrichment in paddy soil. The increase in height, biomass and the total chlorophyll in the leaves was reported at tillering stage. The booting stage showed reduction of malondialdehyde (MDA) by 15–32% and the peroxidase (POD) activity 24–48%. The leaves (booting and heading stage) and the catalase (CAT) activity in the tillering stage were reduced. The results suggested that Cd migration was found promoted by TiO_2-NPs in the soil–rice system (Zhang et al. 2020a, b).

Responses of nanoparticle-treated wheat plants

The wheat (*Triticum aestivum* L.) plants were investigated for bioaccumulation and translocation of NPs, growth, photosynthesis and gas exchange by application of biochar supplemented with cerium oxide nanoparticles (CeO_2NPs) by Abbas et al. (2020). The results indicated that CeO_2NPs promoted the plant growth by triggering photosynthesis, transpiration and stomatal conductance in dose-dependent manner. The biochar amendment with CeO_2NPs reduced the accumulation of Ce and alleviated the phytotoxic effects on wheat plant growth. The findings proved that NPs bioavailability to plants could be inhibited by supplementation of biochar (Abbas et al. 2020).

Khan et al. (2020a, b) investigated the wheat (*T. aestivum* L.) plant growth and uptake of Cd grown in pot under ambient conditions in Cd-contaminated soil by Si-NPs at different water levels. The results showed that NP application improved the wheat plant growth and photosynthesis, reduced the Cd levels in wheat grains and the oxidative stress in leaves. The different parameters like hydrogen peroxide production, leakage of electrolytes and malondialdehyde were reduced and superoxide dismutase and peroxidase activities by NPs application on wheat plants. The test NPs improved in wheat plant growth and reduced oxidative stress and Cd in tissues in dosage-dependent manner (Khan et al. 2020a, b).

The study by Zhang et al. (2018) examined the effect of nCu exposure on the root morphology, physiology and gene transcription levels of wheat (*T. aestivum* L.). The results showed decrease in relative growth rate of roots and the formation of lateral roots. The nitrogen uptake was increased, and auxin was accumulated in lateral roots of test plants. The antioxidant (proline) was induced that scavenged excess reactive oxygen species and alleviation of Cu phytotoxicity (Zhang et al. 2018). In a study by Rico et al. (2020), two generations of wheat plants were exposed to low or high nitrogen soil amended with CeO_2-NPs. The results of NP treatment showed change in DNA/RNA metabolites, i.e., thymidine, uracil, guanosine, deoxyguanosine, adenosine monophosphate in test plants. The wheat grains exhibited decrease in Fe concentration by 13–16% (Rico et al. 2020).

The mesoporous silica nanoparticles (MSNs) effects were evaluated in wheat and lupin plants by Sun et al. (2016) for the growth and development. The results of NP application increased rate of germination and plant biomass. The growth of test plants was accompanied with enhanced total protein, chlorophyll and rate of photosynthesis was increased (Sun et al. 2016). The study of wheat (*T. aestivum* L.) in a greenhouse under drought and non-drought conditions was evaluated by use of urea coated with ZnO-NPs or bulk ZnO by Dimkpa et al. (2020a, b). The drought treatment and NP application on wheat plants affected parameters like time of panicle initiation was increased, grain yield reduced, and uptake of Zn, nitrogen (N), and phosphorus (P) was inhibited. The drought-treated plants with ZnO-NPs reduced panicle initiation, and bulk ZnO showed no effect on panicle initiation. The NPs coated urea increased grain yield by 51% and uncoated urea enhanced to 39%. The coated ZnO-NPs increased Zn uptake to 24% in plants and 8% with uncoated ZnO. The coated bulk ZnO applied to test plants enhanced Zn uptake to 78% and uncoated increased to 10%. The Zn treatment to plants of without drought showed no change in time for panicle initiation. The findings demonstrated that NPs coated urea increases the performance of treated plants and accumulation of Zn. This study suggested application of nanoscale micronutrients as an approach for better crop yield (Dimkpa et al. 2020a, b).

3 Engineered nanoparticles as fertilizers in different plants

The factorial-based randomized design was applied to study the morphology and biochemistry of basil (*Ocimum basilicum* L.) plants by Abbasifar et al. (2020) through treatment with Zn and Cu-NPs. The results of nutrient treatments (4000 ppm) Zn-NPs and (2000 ppm) Cu-NPs improved the plant morphology. The leaves of treated basil plants showed increase in chlorophyll-a, b, total chlorophyll and carotenoid. The total phenolic and flavonoid content and antioxidant activity was improved by test NPs. The foliar application of the Zn and Cu-NPs improved the quantity and quality in basil (Abbasifar et al. 2020).

Linares et al. (2020) investigated the seedlings growth of *Hordeum vulgare* in soil with AgNPs. The shoot and root tissues of barley were evaluated for Ag bioconcentration and distribution after exposure to test NPs. The bioconcentration values of Ag were high in the plants grown in soil from OECD than the Delacour. The morphological changes in barley seedlings were small shoots and short, thick roots after exposure to NPs. It was concluded that early diagnosis of test NP exposure was plant structural responses in seedlings in biosolid-amended soils (Linares et al. 2020).

A study on the application of bulk and nanoparticles—zinc oxide, titanium oxide and silver on chilli seeds cv. PKM 1 by using template-free aqueous solution was performed by Kumar et al. (2020). The nanoparticles effects were analyzed by parameters like electrical conductivity, antioxidant enzymes, i.e., catalase and lipid peroxidase, germination (%), shoot, root length and seedling vigor. The results of 1000 mg kg^{-1} ZnO-NPs treated chilli seeds showed the increase in germination and seedling vigor (Kumar et al. 2020).

Kubavat et al. (2020) performed a study based on chitosan-nanoparticle (CN) prepared and incorporated with potassium (CNK) to tested pot trials of *Zea mays* plant. The different doses of K-formulation were investigated on NP-treated maize plants. The accumulation was increased in fresh (51%) and dry biomass (47%) in amended soils with reduced potassium rates (75% CNK). The CNK improved root growth by enhancing porosity, water conductivity and friability of soil. The nano-formulation and the treatment showed no deleterious effects on test plant but improved carbon-cycling activity (Kubavat et al. 2020). The composites of microcrystalline cellulose, chitosan and alginate biopolymers along with ZnO nanoparticles were tested by Martins et al. (2020) for their potential for controlled release of Zn. The study was reported by growing the maize plants in four agriculture soils with distinct pH and organic matter. The conventional Zn salts applied was leached from the soil, and Zn was less labile for ZnO-NPs. The ZnO-biopolymers supplied Zn better than forms applied to test plants. The plants grown in acidic soil with poor Zn and ZnO-NPs/alginate beads resulted in steady Zn concentration. The results indicated avoidance of early stage Zn toxicity and the Zn requirement of maize plant were done by ZnO-NPs/alginate beads (Martins et al. 2020).

The investigation of germination and development of seedling in corn after seed priming with ZnO-NPs, bulk ZnO and $ZnCl_2$ were evaluated by Neto et al. (2020). The seed priming promoted germination, root length, dry biomass, seedling growth 25% by NP and 12%-$ZnCl_2$ than control. The bulk ZnO seed priming showed similar growth with the control. The NP seed priming was an alternative that supported the delivery of essential micronutrient (Zn) to seedlings of corn (Neto et al. 2020). The peppermint (*Mentha piperita* L.) plants were treated with different fertilizers, i.e., control, chemical fertilizer, arbuscular mycorrhiza fungus, 50% chemical fertilizer + arbuscular mycorrhiza fungus - *Glomus mosseae*, nano-chelated fertilizer, 50% chemical fertilizer + nano-chelated fertilizer, nano-chelated fertilizer + arbuscular mycorrhiza fungus to evaluate desirable essential oil production and reduce chemical inputs by Ostadi et al. (2020). The results showed the impacts of growth parameters, i.e., plant height, number of lateral branches per plant and leaf greenness with increased N, P, K and Fe contents in test plant. The increase of peppermint dry matter, essential oil content and yield revealed the use of integrative chemical fertilizers with nanofertilizers as an alternative and eco-friendly approach (Ostadi et al. 2020).

Gomaa et al. (2020) has carried the field experiments on growth, yield of sorghum by addition of mineral, nano-fertilization and different weed control. The application of NPK mineral and NPK nanoparticles fertilizers revealed the high yield of sorghum was achieved by hand hoeing one time with herbicide (Gomaa et al. 2020). The study performed by Alimohammadi et al. (2020) evaluated the yield of sugarcane (*Saccharum officinarum*) by application of urea and nano-nitrogen chelate (NNC) fertilizers and nitrate leaching from soil. The results showed that nitrate leaching was high with urea and low for NNC. The sugarcane stem height was increased by application of both fertilizers in increased doses (Alimohammadi et al. 2020).

The foliar application of ZnO-NPs and $ZnSO_4$ on winter wheat (*T. aestivum* L.) was evaluated by Sun et al. (2020) for increasing the Zn content in the grain. ZnO-NPs increased the Zn in the wheat grain was in limit for human consumption. The results demonstrated that ZnO-NPs fertilizer increased Zn in wheat grain and contributed for improved human nutrition (Sun et al. 2020). The effect of zinc nanofertilizer was evaluated by Prajapati et al. (2018) for growth and yield of wheat (*T. aestivum* L.). The experiments were seed treatment, foliar application and seed treatment + foliar application of bulk Zn and nano-Zn. The results showed that the seed treatment followed by three foliar sprays of ZnO-NPs after sowing proved to enhance the height, number of effective tillers, length of spike, test weight, yields of grain and straw along with grain and straw zinc content and uptake by grain and straw (Prajapati et al. 2018).

The integrative effects of wheat and nutrient acquisition were evaluated by Dimkpa et al. (2020a, b) in soil under treatments, i.e., drought, organic fertilizer (OF) and nano- vs. bulk ZnO particles. The drought effect reduced chlorophyll levels, delayed panicle emergence, reduced grain yield treatment of nano- and bulk ZnO reported to alleviate stress with increase in chlorophyll, accelerated panicle emergence under drought, increased grain yield and OF also increased chlorophyll levels, increased yield under drought and counteracted with Zn. The results of the study demonstrated that drought effects in food crops were alleviated by ZnO particles and Zn-rich OF and found potential mitigation strategies for sustaining food production (Dimkpa et al. 2020a, b).

4 Nano-harvest with engineered mesoporous silica nanoparticles (MSNPs).

Solidago nemoralis hairy root cultures were performed for harvesting of polyphenolic flavonoids using engineered mesoporous silica nanoparticles (MSNPs) functionalized with both titanium dioxide (TiO_2) and amines (NH_2) to promote cellular internalization. The results of the study demonstrated continuous isolation of biomolecules from living and functioning plant cultures (Khan et al. 2020a, b).

5 Phytoaccumulation of Engineered Nanoparticles (ENPs) in Plants

The different behaviors of leaf samples from *Dittrichia viscosa* and *Cichorium intybus* for the phytoaccumulator characteristics were studied by Abdallah et al. (2020) to evidence sequestration of heavy metals as nanoparticles from autogenous environment, i.e., steel manufacturing company. The results showed different behaviors of phytoaccumulation in *Dittrichia viscosa* and nanoparticle composition. The levels of heavy metals NPs estimated from the nearby industries and *Cichorium intybus* plants were similar (Abdallah et al. 2020).

6 Phytotoxic Effects of Engineered Nanoparticles (ENPs) in Different Plants

The results of study performed by Falco et al. (2020) on leaves of broad bean (*Vicia faba*) exposed to silver nanoparticles (AgNPs) revealed that the photochemical efficiency of photosystem II (PS-II) was reduced. The NPs increased the non-photochemical quenching and decrease in stomatal conductance (*Gs*) and CO_2 assimilation with overproduction of reactive oxygen species (ROS). The photosynthesis process was affected negatively by accumulation of NPs in the leaves of bean plants (Falco et al. 2020).

A study performed by Mylona et al. (2020) to test the impacts of Ag-NP for sensitive responses and toxicity on the seagrass (*Cymodocea nodosa*). The results showed changes in the cytoskeleton, endoplasmic reticulum, ultrastructure of seagrass treated with test NPs. The function of photosystem II, markers of oxidative stress and cell viability were altered in test plants. The leaf, rhizome, root elongation and protein content in seagrass were decreased, and antioxidant enzyme activity was increased (Mylona et al. 2020).

The *in-vitro* grown seedlings of *Abelmoschus esculentus* (okra) were investigated by Baskar et al. (2020) for phytotoxic effects by metal oxide NPs such as nickel oxide (NiO), copper oxide (CuO) and zinc oxide (ZnO). The tested NPs suppressed plant growth in a concentration-dependent manner. The results showed decrease in chlorophyll content, length of shoot and root, enhanced ROS and malondialdehyde (MDA), altered anthocyanin, total phenols and flavonoids in the NP-treated seedlings of *A. esculentus*. Among the tested Ni-NPs toxicity was high than CuO and ZnO-NPs in the treated seedlings (Baskar et al. 2020).

Yang et al. (2020) performed study on the rice (*O. sativa* L.) plants grown under hydroponic condition to assess for phytotoxicity of copper oxide nanoparticle (CuO-NPs) for seven days of exposure. The treated plants were found with suppressed growth rate, increased malondialdehyde (MDA) content and electrical conductivity in shoots. The leaf chlorophyll-a, b, carotenoid, catalase and superoxide dismutase were decreased. The results of the study reported effective CuO-NPs concentration that affected the growth and development of rice seedlings through oxidative damage and decrease in chlorophyll and carotenoid synthesis (Yang et al. 2020).

Priester et al. (2017) studied growth of soybean (*Glycine max*) in soil enriched with $nCeO_2$ or $nZnO$. The results showed increase in lipid peroxidation and ROS and decrease in total chlorophyll that damaged leaf. The quantum efficiency of PS-II and seed protein remained unchanged by test NP on soybean plants. The NPs generated stress and damage in soybean leaves (Priester et al. 2017). The seed yield of *Glycine max* (cv. Kowsar) grown in soil with N-fixing bacteria (*Rhizobium japonicum*) inoculant was evaluated by Yusefi-Tanha et al. (2020) for CuO-NPs (25, 50 and 250 nm) phytotoxicity. The results showed the differential alteration of antioxidant enzymes such as APX, CAT, POX, SOD and MDA dependent on the type, concentration and interactions of copper compound (Yusefi-Tanha et al. 2020).

The study performed by Singh and Kumar (2020a, b) showed growth of *Spinacia oleracea* by the treatment of single and binary mixture of CuO and ZnO-NPs in the soil. The results revealed the adverse effects of test NPs on spinach plant biomass and fresh weight (Singh and Kumar 2020a, b). The physiology and biochemistry of spinach plants after foliar application of lead oxide nanoparticles (PbO-NPs) on lead (Pb) accumulation and associated health risks were evaluated by Natasha et al. (2020). The results showed accumulation of Pb decreased in leaf pigments; dry weight and the activities of catalase and peroxidase were increased. The translocation was limited toward root tissues in test plants by NPs. The foliar deposition of metal-enriched particles (PM) affects growth of spinach and ingestion of metal-contaminated vegetables results in health issues (Natasha et al. 2020).

7 Conclusion

Nanoparticles are reported to replace the bulk forms in coming generations as the concern to get enhanced outputs from agriculture to feed increasing human population. The nanoparticles efficacy is dosage-dependent manner and required in small quantities to get benefits. The abiotic/biotic stresses are important hurdles of the present agroecosystem around the planet; the nanoparticles are important tools to boost yields. The engineered nanoparticles (ENPs) attracted the community of researchers to investigate the effects in variety of plant habits. The findings support the nanoscale particles improve the variety of physiological aspects such as germination, in vitro regeneration, metabolic profile, leaf, shoot, fruit and root growth, antioxidant (enzymatic and non-enzymatic) levels, nutrient uptake, colonization of microbiota, defenses against diseases, essential oil and amelioration of stress (drought, chilling and metal). The nanomaterials were applied to plants either sole or in combination with biochar, AM fungi, chemical fertilizer that enhanced efficiency of plant uptake resulted in high yields. The market for nanofertilizers, nanopesticides, nano-herbicides and other nano-agrochemicals is near to conquer and revolutionize the yields.

Acknowlegements The author expresses deep gratitude toward Hon'ble Vice-chancellor, Prof. Naseem (Hon'ble Registrar) and members, Department of Botany, Telangana University. Thanks to editors and anonymous reviewers for enlighten and encourage during the endeavor.

References

Abbas Q, Liu G, Yousaf B, Ali MU, Ullah H, Munir MAM, Ahmed R, Rehman A (2020) Biochar-assisted transformation of engineered-cerium oxide nanoparticles: Effect on wheat growth, photosynthetic traits and cerium accumulation. Ecotoxicol Environ Saf 187:109845

Abbasifar A, Fatemeh Shahrabadi F, ValizadehKaji B (2020) Effects of green synthesized zinc and copper nano-fertilizers on the morphological and biochemical attributes of basil plant. J Plant Nutr 43 (8):1104–1118

Abdallah BB, Zhang X, Andreu I, Gates BD, Mokni RE, Rubino S, Landoulsi A, Chatti A (2020) Differentiation of nanoparticles isolated from distinct plant species naturally growing in a heavy metal polluted site. J Hazard Mater 386:121644

Ahmad MA, Javed R, Adeel M, Rizwan M, Ao Q, Yang Y (2020) Engineered ZnO and CuO nanoparticles ameliorate morphological and biochemical response in tissue culture regenerants of Candyleaf (*Stevia rebaudiana*). Molecules 25:1356

Alimohammadi M, Panahpour E, Naseri A (2020) Assessing the effects of urea and nano-nitrogen chelate fertilizers on sugarcane yield and dynamic of nitrate in soil. Soil Science and Plant Nutrition. https://doi.org/10.1080/00380768.2020.1727298

Al-Amri N, Tombuloglu H, Slimani Y, Akhtar S, Barghouthi M, Almessiere M, Alshammari T, Baykal A, Sabit H, Ercan I, Ozcelik S (2020) Size effect of iron (III) oxide nanomaterials on the growth, and their uptake and translocation in common wheat (*Triticum aestivum* L.) Ecotoxicol Environ Saf 194:110377

Alyahya SA, Govindarajan M, Alharbi NS, Kadaikunnan S, Khaled JM, Mothana RA, Al-anbr MN, Vaseeharan B, Ishwarya R, Yazhiniprabha M, Benelli G (2018) Swift fabrication of Ag nanostructures using a colloidal solution of *Holostemma ada-kodien* (Apocynaceae) – antibiofilm potential, insecticidal activity against mosquitoes and non-target impact on water bugs. J Photochem Photobiol, B 181:70–79

Antisari LV, Carbone S, Bosi S, Gatti A, Dinelli G (2018) Engineered nanoparticles effects in soil-plant system: Basil (*Ocimum basilicum* L.) study case. Appl Soil Ecol 123:551–560

Aksakal FI, Arslan H (2020) Detoxification and reproductive system-related gene expression following exposure to Cu(OH)$_2$ nanopesticide in water flea (*Daphnia magna* Straus 1820). Environ Sci Pollut Res 27(6):6103–6111

Bakshi M, Line C, Bedolla DE, Stein RJ, Kaegi R, Sarret G, Real AEPd, Castillo-Michel H, Abhilash PC, Larue C (2019) Assessing the impacts of sewage sludge amendment containing nano-TiO2 on tomato plants: A life cycle study. J Hazardous Mater 369:191–198

Baskar V, Safia N, Preethy KS, Dhivya S, Thiruvengadam M, Sathishkumar R (2020) A comparative study of phytotoxic effects of metal oxide (CuO, ZnO and NiO) nanoparticles on *in-vitro* grown *Abelmoschus esculentus*. Plant Biosystems - an international journal dealing with all Aspects of Plant Biology DOI: https://doi.org/10.1080/11263504.2020.1753843

Cai L, Cai L, Jia H, Liu C, Wang D, Sun X (2020) Foliar exposure of Fe3O4 nanoparticles on *Nicotiana benthamiana*: Evidence for nanoparticles uptake, plant growth promoter and defense response elicitor against plant virus. J Hazard Mater 393:122415

Cao L, Zhou Z, Niu S, Cao C, Li X, Shan Y, Huang Q (2018) Positive-charge functionalized mesoporous silica nanoparticles as nanocarriers for controlled 2,4-dichlorophenoxy acetic acid sodium salt release. J Agric Food Chem 66:6594–6603

Chahardoli A, Karimi N, Ma X, Qalekhani F (2020) Effects of engineered aluminum and nickel oxide nanoparticles on the growth and antioxidant defense systems of *Nigella arvensis* L. Sci Rep 10:3847

Chariou PL, Dogan AB, Welsh AG, Saidel GM, Baskaran H, Steinmetz NF (2019) Soil mobility of synthetic and virus-based model nanoparticles. Nat Nanotechnol 14:712–718

Chen J, Liu X, Wang C, Yin S-S, Li X-L, Hu W-J, Simon M, Shen Z-J, Xiao Q, Chu C-C, Peng X-X, Zheng H-L (2015) Nitric oxide ameliorates zinc oxide nanoparticles-induced phytotoxicity in rice seedlings. J Hazard Mater 297:173–182

Dimkpa CO, Andrews J, Fugice J, Singh U, Bindraban PS, Elmer WH, Gardea-Torresdey JL, White JC (2020a) Facile coating of urea with low-dose ZnO nanoparticles promotes wheat performance and enhances Zn uptake under drought stress. Front Plant Sci 11:168

Dimkpa CO, Andrews J, Sanabria J, Bindraban PS, Singh U, Elmer WH, Gardea-Torresdey JL, White JC (2020b) Interactive effects of drought, organic fertilizer, and zinc oxide nanoscale and bulk particles on wheat performance and grain nutrient accumulation. Sci Total Environ 722:137808

Diyanat M, Saeidian H, Baziar S, Mirjafary Z (2019) Preparation and characterization of polycaprolactone nanocapsules containing pretilachlor as a herbicide nanocarrier. Environ Sci Pollut Res 26:21579–21588

Elabasy A, Shoaib A, Waqas M, Shi Z, Jiang M (2020) Cellulose nanocrystals loaded with thiamethoxam: fabrication, characterization, and evaluation of insecticidal activity against *Phenacoccus solenopsis* Tinsley (Hemiptera: Pseudococcidae). Nanomaterials 10:788

El-Naggar ME, Abdelsalam NR, Fouda MMG, Mackled MI, Al-Jaddadi MAM, Ali HM, Siddiqui MH, Kandil EE (2020) Soil application of nano silica on maize yield and its insecticidal activity against some stored insects after the post-harvest. Nanomaterials 10:739

Elsheery NI, Helaly MN, El-Hoseiny HM, Alam-Eldein SM (2020a) Zinc oxide and silicone nanoparticles to improve the resistance mechanism and annual productivity of salt-stressed mango trees. Agronomy 10:558

Elsheery N, Sunoj VSJ, Wen Y, Zhu JJ, Muralidharan G, Cao KF (2020b) Foliar application of nanoparticles mitigates the chilling effect on photosynthesis and photoprotection in sugarcane. Plant Physiol Biochem 149:50–60

Fadoju OM, Osinowo OA, Ogunsuyi OI, Oyeyemi IT, Alabi OA, Alimba CG, Bakare AA (2020) Interaction of titanium dioxide and zinc oxide nanoparticles induced cytogenotoxicity in *Allium cepa*. Nucleus. https://doi.org/10.1007/s13237-020-00308-1

Faizan M, Faraz A, Mir AR, Hayat S (2020) Role of zinc oxide nanoparticles in countering negative effects generated by cadmium in *Lycopersicon esculentum*. J Plant Growth Regul. https://doi.org/10.1007/s00344-019-10059-2

Falco WF, Scherer MD, Oliveira SL, Wender H, Colbeck I, Lawson T, Caires ARL (2020) Phytotoxicity of silver nanoparticles on *Vicia faba*: evaluation of particle size effects on photosynthetic performance and leaf gas exchange. Sci Total Environ 701:134816

Ghorbanpour M, Mohammadi H, Kariman K (2020) Nanosilicon-based recovery of barley (*Hordeum vulgare*) plants subjected to drought stress. Environ. Sci.: Nano 7: 443–461

Gomaa MA, Rehab IF, Kordy AM, Bilkess, Salim MA Top of Form (2020) Assessment of Sorghum (*Sorghum bicolor* L.) productivity under different weed control methods, mineral and nano fertilization. Egypt Acad J Biolog Sci 11(1): s1–11

Gupta SD, Agarwal A, Pradhan S (2018) Phytostimulatory effect of silver nanoparticles (AgNPs) on rice seedling growth: An insight from antioxidative enzyme activities and gene expression patterns. Ecotoxicol Environ Saf 161:624–633

Haroun SA, Elnaggar ME, Zein DM, Gad RI (2020) Insecticidal efficiency and safety of zinc oxide and hydrophilic silica nanoparticles against some stored seed insects. Journal of Plant Protection Research 60(1):77–85

Hong J, Wang L, Sun Y, Zhao L, Niu G, Tan W, Rico CM, Peralta-Videa JR, Gardea-Torresdey JL (2016) Foliar applied nanoscale and microscale CeO2 and CuO alter cucumber (*Cucumis sativus*) fruit quality. Sci Total Environ 563–564:904–911

Hou R, Zhang Z, Pang S, Yang T, Clark JM, He L (2016) Alteration of the non systemic behavior of the pesticide ferbam on tea leaves by engineered gold nanoparticles. Environ Sci Technol 50:6216–6223

Jung ES, Sivakumar S, Hong S-C, Yi P-I, Jang S-H, Suh J-M (2020) Influence of relative humidity on germination and metal accumulation in *Vigna radiata* exposed to metal-based nanoparticles. Sustainability 12:1347

Katiyar P, Yadu B, Korram J, Satnami ML, Kumar M, Keshavkant S (2020) Titanium nanoparticles attenuates arsenic toxicity by up-regulating expressions of defensive genes in *Vigna radiata* L. Journal of Environmental Sciences 92:18–27

Khan MA, Wallace WT, Sambi J, Rogers DT, Littleton JM, Rankin SE, Knutson BL (2020a) Nanoharvesting of bioactive materials from living plant cultures using engineered silica nanoparticles. Mater Sci Eng C Mater Biol Appl 106:110190

Khan ZS, Rizwan M, Hafeez M, Ali S, Adrees M, Qayyum MF, Khalid S, Rehman MZ, Sarwar MA (2020b) Effects of silicon nanoparticles on growth and physiology of wheat in cadmium contaminated soil under different soil moisture levels. Environ Sci Poll Int 27(5):4958–4968

Khan AAH (2019) Cytotoxic potential of plant nanoparticles. In: Abd-Elsalam K and R. Prasad. (eds) Nanobiotechnology applications in plant protection. Nanotechnology in the life sciences. Springer, Cham. pp. 241–265.

Khan, AAH, Naseem, Vardhini BV (2017) Resource-conserving agriculture and role of microbes. In: Prasad R. and N. Kumar. Microbes & sustainable agriculture. IK International Publishing House, New Delhi. pp 117–152

Khan AAH, Naseem VBV (2015) Synthesis of nanoparticles from plant extracts. International Journal of Modern Chemistry and Applied Science 2(3):195–203

Khan AAH, Naseem, Vardhini BV (2015b) Fungus mediated synthesis of metal nanoparticles. In: Sivaramaiah, G. (ed) Proceedings of national seminar on new trends on advanced materials (NTAM). Paramount Publishing House, Hyderabad. pp. 82–88.

Kokina I, Plaksenkova I, Jermaļonoka M, Petrova A (2020) Impact of iron oxide nanoparticles on yellow medick (*Medicago falcata* L.) plants. Journal of Plant Interactions 15(1): 1–7

Kojic D, Purac J, Celic TV, Jovic D, Vukasinovic EL, Pihler I, Borisev I, Djordjevic A (2020) Effect of fullerenol nanoparticles on oxidative stress induced by paraquat in honey bees. Environ Sci Pollut Res 27:6603–6612

Kubavat D, Trivedi K, Vaghela P, Prasad K, Anand KGV, Trivedi H, Patidar R, Chaudhary J, Andhariya B, Ghosh A (2020) Characterization of chitosan based sustained release nano-fertilizer formulation as a soil conditioner whilst improving biomass production of *Zea mays* L. Land Degrad Dev. https://doi.org/10.1002/ldr.3629

Kumar GD, Raja K, Natarajan N, Govindaraju K, Subramanian KS (2020) Invigouration treatment of metal and metal oxide nanoparticles for improving the seed quality of aged chilli seeds (*Capsicum annum* L.), Materials Chemistry and Physics 242: 122492

Lian J, Zhao L, Wu J, Xiong H, Bao Y, Zeb A, Tang J, Liu W (2020) Foliar spray of TiO2 nanoparticles prevails over root application in reducing Cd accumulation and mitigating Cd-induced phytotoxicity in maize (*Zea mays* L.). Chemosphere 239: 124794

Linares MG, Jia Y, Sunahara GI, Whalen JK (2020) Barley (*Hordeum vulgare*) seedling growth declines with increasing exposure to silver nanoparticles in biosolid-amended soils. Can J Soil Sci. https://doi.org/10.1139/cjss-2019-0135

Li X, Sun Y, Yuan L, Liang L, Jiang Y, Piao H, Song D, Yu A, Wang X (2018) Packed hybrids of gold nanoparticles and layered double hydroxide nanosheets for microextraction of triazine herbicides from maize. Microchim Acta 185:36

Li X, Sun Y, Sun Q, Liang L, Piao H, Jiang Y, Yu A, Song D, Wang X (2017) Ionic-liquid-functionalized zinc oxide nanoparticles for the solid-phase extraction of triazine herbicides in corn prior to high-performance liquid chromatography analysis. J Sep Sci 40 (14):2992–2998

Ma X, Sharifan H, Dou F, Sun W (2020) Simultaneous reduction of arsenic (As) and cadmium (Cd) accumulation in rice by zinc oxide nanoparticles. Chem Eng J 384:123802

Martins NCT, Avellan A, Rodrigues S, Salvador D, Rodrigues SM, Trindade T (2020) Composites of biopolymers and ZnO NPs for controlled release of zinc in agricultural soils and timed delivery for Maize. ACS Applied Nano Materials 3(3):2134–2148

Meier S, Morales FMA, Gonzalez ME, Seguel A, Merino-Gergichevich C, Rubilar O, Cumming J, Aponte H, Alarcon D, Mejias J (2020) Synthesis of calcium borate nanoparticles and its use as a potential foliar fertilizer in lettuce (*Lactuca sativa*) and zucchini (*Cucurbita pepo*). Plant Physiol Biochem 151:673–680

Mohammadi H, Amani-Ghadim AR, Matin AA, Ghorbanpour M (2020) Fe^0 nanoparticles improve physiological and antioxidative attributes of sunflower (*Helianthus annuus*) plants grown in soil spiked with hexavalent chromium. 3 Biotech 10(1):19

Mylona Z, Panteris E, Moustakas M, Kevrekidis T, Malea P (2020) Physiological, structural and ultrastructural impacts of silver nanoparticles on the seagrass *Cymodocea nodosa*. Chemosphere 248:126066

Maruyama CR, Guilger M, Pascoli M, Bileshy-Jose N, Abhilash PC, Fraceto, de Lima R, (2016) Nanoparticles based on chitosan as carriers for the combined herbicides imazapic and imazapyr. Sci Rep 6:19768

Natasha SM, Farooq ABU, Rabbani F, Khalid S, Dumat C (2020) Risk assessment and biophysiochemical responses of spinach to foliar application of lead oxide nanoparticles: A multivariate analysis. Chemosphere 245:125605

Naeimi A, Saeidi M, Baroumand N (2016) Carboxylated carbon nanotubes as an efficient and cost-effective adsorbent for sustainable removal of insecticide fenvalerate from contaminated solutions. Int Nano Lett 6:265–271

Nhan LV, Ma C, Rui Y, Cao W, Deng Y, Liu L, Xing B (2016) The effects of Fe2O3 nanoparticles on physiology and insecticide activity in non-transgenic and Bt-transgenic cotton. Front Plant Sci 6:1263

Neto ME, Britt DW, Lara LM, Cartwright A, dos Santos RF, Inoue TT, Batista MA (2020) Initial development of corn seedlings after seed priming with nanoscale synthetic Zinc oxide. Agronomy 10:307

Noori A, Donnelly T, Colbert J, Cai W, Newman LA, White JC (2020) Exposure of tomato (*Lycopersicon esculentum*) to silver nanoparticles and silver nitrate: physiological and molecular response. Int J Phytoremediation 22(1):40–51

Ogunkunle CO, Odulaja DA, Akande FO, Varun M, Vishwakarma V, Fatoba PO (2020) Cadmium toxicity in cowpea plant: Effect of foliar intervention of nano-TiO2 on tissue Cd bioaccumulation, stress enzymes and potential dietary health risk. J Biotechnol 310:54–61

Ochoa L, Medina-Velo IA, Barrios AC, Bonilla-Bird NJ, Hernandez-Viezcas JA, Peralta-Videa JR, Gardea-Torresdey JL (2017) Modulation of CuO nanoparticles toxicity to green pea (*Pisum sativum* Fabaceae) by the phytohormone indole-3-acetic acid. Sci Total Environ 598:513–524

Osman SA, Salama DM, El-Aziz MEA, Shaaban EA, Elwahed MSA (2020) The influence of MoO3-NPs on agro-morphological criteria, genomic stability of DNA, biochemical assay, and production of common dry bean (*Phaseolus vulgaris* L.). Plant Physiol Biochem 151:77–87

Ostadi A, Javanmarda A, Machiani MA, Morshedloo MR, Nouraein M, Rasouli F, Maggi F (2020) Effect of different fertilizer sources and harvesting time on the growth characteristics, nutrient uptakes, essential oil productivity and composition of *Mentha* x *piperita* L. Ind Crops Prod 148:112290

Pascoli M, de Albuquerque FP, Calzavara AK, Tinoco-Nunes B, Oliveira WHC, Gonçalves KC, Polanczyk RA, Vechia JFD, de Matos STS, de Andrade DJ, Oliveira HC, Souza-Neto JA, de Lima R, Fraceto LF (2020) The potential of nanobiopesticide based on zein nanoparticles and neem oil for enhanced control of agricultural pests. J Pest Sci 93:793–806

Pariona N, Martínez AI, Hernandez-Flores H, Clark-Tapia R (2017) Effect of magnetite nanoparticles on the germination and early growth of *Quercus macdougallii*. Sci Total Environ 575:869–875

Pandorf M, Pourzahedi L, Gilbertson L, Lowry GV, Herckes P, Westerhoff P (2020) Graphite nanoparticle addition to fertilizers reduces nitrate leaching in growth of lettuce (*Lactuca sativa*). Environ. Sci.: Nano 7: 127–138

Pasquoto-Stigliani T, Campos EVR, Oliveira JL, Silva CMG, Bilesky-Jose N, Guilger M, Troost J, Oliveira HC, Stolf-Moreira R, Fraceto LF, de Lima R (2017) Nanocapsules containing neem (*Azadirachta indica*) oil: development, characterization, and toxicity evaluation. Sci Rep 7:5929

Peng C, Tong H, Shen C, Sun L, Yuan P, He M, Shi J (2020) Bioavailability and translocation of metal oxide nanoparticles in the soil-rice plant system. Sci Total Environ 713:136662

Pradhan S, Durgam M, Mailapalli DR (2020) Urea loaded hydroxyapatite nanocarrier for efficient delivery of plant nutrients in rice. Archives of Agronomy and Soil Science. https://doi.org/10.1080/03650340.2020.1732940

Prajapati BJ, Patel SB, Patel RP, Ramani VP (2018) Effect of Zinc nano-fertilizer on growth and yield of wheat (*Triticum aestivum* L.) under saline irrigation condition. Agropedology 28(01): 31–37

Priester JH, Moritz SC, Espinosa K, Ge Y, Wang Y, Nisbet RM, Schimel JP, Goggi AS, Torresdey JL, Holden PA (2017) Damage assessment for soybean cultivated in soil with either CeO(2) or ZnO manufactured nanomaterials. Sci Total Environ 579:1756–1768

Rahman MS, Chakraborty A, Mazumdar S, Nandi NC, Bhuiyan MNI, Alauddin SM, Khan IA, Hossain MJ (2020) Effects of poly (vinylpyrrolidone) protected platinum nanoparticles on seed germination and growth performance of *Pisum sativum*. Nano-Structures & Nano-Objects 21:100408

Rashidipour M, Maleki A, Kordi S, Birjandi M, Pajouhi N, Mohammadi E, Heydari R, Rezaee R, Rasoulian B, Davari B (2019) Pectin/chitosan/tripolyphosphate nanoparticles: efficient carriers for reducing soil sorption, cytotoxicity, and mutagenicity of paraquat and enhancing its herbicide activity. J Agric Food Chem 67:5736–5745

Rico CM, Wagner D, Abolade O, Lottes B, Coates K (2020) Metabolomics of wheat grains generationally-exposed to cerium oxide nanoparticles. Sci Total Environ 712:136487

Salehi H, Chehregani A, Lucini L, Majd A, Gholam M (2018) Morphological, proteomic and metabolomic insight into the effect of cerium dioxide nanoparticles to *Phaseolus vulgaris* L. under soil or foliar application. Sci Total Environ 616–617:1540–1551

Stefanic PP, Cvjetko P, Biba R, Domijan A-M, Letofsky-Papst I, Tkalec M, Sikic S, Cindric M, Balen B (2020) Physiological, ultrastructural and proteomic responses of tobacco seedlings exposed to silver nanoparticles and silver nitrate. Chemosphere 209:640–653

Sharifan H, Wang X, Ma X (2020) Impact of nanoparticle surface charge and phosphate on the uptake of coexisting cerium oxide nanoparticles and cadmium by soybean (*Glycine max.* (L.) *merr.*). Int J Phytoremediation 22(3): 305–312

Shan Y, Cao L, Xu C, Zhao P, Cao C, Li F, Xu B, Huang Q (2019) Sulfonate-functionalized mesoporous silica nanoparticles as carriers for controlled herbicide diquat dibromide release through electrostatic interaction. Int J Mol Sci 20:1330

Sharma A, Sood K, Kaur J, Khatri M (2019) Agrochemical loaded biocompatible chitosan nanoparticles for insect pest management. Biocatal Agric Biotechnol 18:101079

Singh D, Kumar A (2020a) Binary mixture of nanoparticles in sewage sludge: Impact on spinach growth. Chemosphere 254:126794

Schnoor B, Elhendawy A, Joseph S, Putman M, Chacon-Cerdas R, Flores-Mora D, Bravo-Moraga F, Gonzalez Nilo F, Salvador-Morales C (2018) Engineering Atrazine loaded poly (lactic-co-glycolic Acid) nanoparticles to ameliorate environmental challenges. J Agric Food Chem 66(30):7889–7898

Singh D, Kumar A (2020b) Quantification of metal uptake in *Spinacia oleracea* irrigated with water containing a mixture of CuO and ZnO nanoparticles. Chemosphere 243:125239

Song S, Wang Y, Xie J, Sun B, Zhou N, Shen H, Shen J (2019) Carboxymethyl chitosan modified carbon nanoparticle for controlled emamectin benzoate delivery: improved solubility, pH responsive release, and sustainable pest control. ACS Appl Mater Interfaces 11(37):34258–34267

Sousa GFM, Gomes DG, Campos EVR, Oliveira JL, Fraceto LF, Stolf-Moreira R, Oliveira HC (2018) Post-emergence herbicidal activity of nanoatrazine against susceptible weeds. Front Environ Sci 6:12

Sun D, Hussain HI, Yi Z, Rookes JE, Kong L, Cahill DM (2016) Mesoporous silica nanoparticles enhance seedling growth and photosynthesis in wheat and lupin. Chemosphere 152:81–91

Sun H, Du W, Peng Q, Lv Z, Mao H, Kopittke PM (2020) Development of ZnO nanoparticles as an efficient Zn fertilizer: using synchrotron-based techniques and laser ablation to examine elemental distribution in wheat grain. J Agric Food Chem 68 (18):5068–5075

Tian L, Zhang H, Zhao X, Gu X, White JC, Zhao L, Ji R (2020) CdS nanoparticles in soil induce metabolic reprogramming in broad bean (*Vicia faba* L.) roots and leaves. Environ. Sci.: Nano 7: 93–104

Tombuloglu H, Anıl I, Akhtar S, Turumtay H, Sabit H, Slimani Y, Almessiere M, Baykal A (2020) Iron oxide nanoparticles translocate in pumpkin and alter the phloem sap metabolites related to oil metabolism. Sci Hortic 265:109223

Tong Y, Wu Y, Zhao C, Xu Y, Lu J, Xiang S, Zong F, Wu X (2017) Polymeric nanoparticles as a metolachlor carrier: water-based formulation for hydrophobic pesticides and absorption by plants. J Agric Food Chem 65:7371–7378

Tu Q, Yang T, Qu Y, Gao S, Zhang Z, Zhang Q, Wang Y, Wang J, He L (2019) *In situ* colorimetric detection of glyphosate on plant tissues using cysteamine-modified gold nanoparticles. Analyst 144 (6):2017–2025

Vitali F, Raio A, Sebastiani F, Cherubini P, Cavalieri D, Cocozza C (2019) Environmental pollution effects on plant microbiota: the case study of poplar bacterial-fungal response to silver nanoparticles. Appl Microbiol Biotechnol 103(19):8215–8227

Wang Y, Deng C, Cota-Ruiz K, Peralta-Videa JR, Sun Y, Rawat S, Tan W, Reyes A, Hernandez-Viezcas JA, Niu G, Li C, Gardea-Torresdey JL (2020a) Improvement of nutrient elements and allicin content in green onion (*Allium fistulosum*) plants exposed to CuO nanoparticles. Sci Total Environ 725:138387

Wang A, Jin Q, Xu X, Miao A, White JC, Gardea-Torresdey JL, Ji R, Zhao L (2020b) High-throughput screening for engineered nanoparticles that enhance photosynthesis using mesophyll protoplasts. J Agric Food Chem 68(11):3382–3389

Wang X, Sun W, Ma X (2019) Differential impacts of copper oxide nanoparticles and Copper (II) ions on the uptake and accumulation of arsenic in rice (*Oryza sativa*). Environ Pollut 252:967–973

Wu Y, Yang L, Gong H, Dang F, Zhou DM (2020) Contrasting effects of iron plaque on the bioavailability of metallic and sulfidized silver nanoparticles to rice. Environ Pollut 260:113969

Yan L, Li P, Zhao X, Ji R, Zhao L (2020) Physiological and metabolic responses of maize (*Zea mays*) plants to Fe_3O_4 nanoparticles. Sci Total Environ 718:137400

Yang Z, Xiao Y, Jiao T, Zhang Y, Chen J, Gao Y (2020) Effects of copper oxide nanoparticles on the growth of rice (*Oryza sativa* L.) seedlings and the relevant physiological responses. Int J Environ Res Public Health 17: 1260

Ye Y, Cota-Ruiz K, Hernandez-Viezcas JA, Valdes C, Medina-Velo IA, Turley RS, Peralta-Videa JR, Gardea-Torresdey JL (2020) Manganese nanoparticles control salinity-modulated molecular responses in *Capsicum annuum* L. through priming: a sustainable approach for agriculture. ACS Sustainable Chemistry & Engineering 8(3):1427–1436

Younes NA, Hassan HS, Elkady MF, Hamed AM, Dawood MFA (2020) Impact of synthesized metal oxide nanomaterials on seedlings production of three Solanaceae crops. Heliyon 6:e03188

Yusefi-Tanha E, Fallah S, Rostamnejadi A, Pokhrel LR (2020) Particle size and concentration dependent toxicity of copper oxide nanoparticles (CuONPs) on seed yield and antioxidant defense system in soil grown soybean (*Glycine max* cv. Kowsar). Sci Total Environ 715:136994

Zhang W, Long J, Li J, Zhang M, Ye X, Chang W, Zeng H (2020a) Effect of metal oxide nanoparticles on the chemical speciation of heavy metals and micronutrient bioavailability in paddy soil. Int J Environ Res Public Health 17:2482

Zhang Z, Ke M, Qu Q, Peijnenburg WJGM, Lu T, Zhang Q, Ye Y, Xu P, Du B, Sun L, Qian H (2018) Impact of copper nanoparticles and ionic copper exposure on wheat (*Triticum aestivum* L.) root morphology and antioxidant response. Environ Pollut 239:689–697

Zhang W, Long J, Geng J, Li J, Wei Z (2020b) Impact of Titanium dioxide nanoparticles on Cd phytotoxicity and bioaccumulation in rice (*Oryza sativa* L.). Int J Environ Res Public Health 17: 2979

Zia M, Yaqoob K, Mannan A, Nisa S, Raza G, Rehman R (2020) Regeneration response of carnation cultivars in response of silver nanoparticles under in vitro conditions. Vegetos 33:11–20

Engineered Nanoparticles and Soil Health

Engineered Nanoparticles in Agro-ecosystems: Implications on the Soil Health

Disha Mishra, Versha Pandey, and Puja Khare

Abstract

Soil health has been considered as one of the important factors for maintaining ecosystem boundaries, balanced biogeocycles, sustaining plant growth, support habitat, and balanced environmental functions. However, along with the presence of persistence xenobiotics, the entry of newer engineered nanoparticles (ENPs) to the agro-ecosystem has directly influenced the soil health. ENPs are now having tremendous potential to shape the global economy and thus their production has increased deliberately. They are refined from bulk materials to offer unprecedented interactions with small-scale molecules or naturally occurring compounds that are produced on a scale of ∼1–100 nm. These nano-architects are chiefly employed for controlled delivery of fertilizers, pesticides, hormones, genetic material, nano-sensors, and rebuilding of soil structure in agro-ecosystem. However, they undergo various transformations like aggregation, sorption, dissolution, decomposition, dispersion, and transportation in soil environment which directly affects the soil health. Thus, their exposure has resulted in various implications like disturbed soil microflora, impeded decomposition of organic matter, lowered nutrient and carbon reserves, and additionally toxicity to soil microbial communities. The scientific communities have widely reviewed major concerns about their origin, interaction, distribution, toxicity, and mitigation in the soil ecosystem. However, the unethical and uncontrolled liberation of ENPs to the environment always made it a matter of concern. Therefore, strong regulation, risk assessment, and mitigation strategies are required for the sustainable use of ENPs. Here, we have attempted to review the structures, properties, mobility, interaction with soil components, impact on soil health, toxicological profile, effects on soil microbial communities, and assessment methods. This will provide valuable approaches to tackle the challenges associated with ENPs and directions for future research.

Keywords

Nutrient content • Persistent • Risk assessment • Soil microflora • Toxicity

1 Introduction

Engineered nanoparticles (ENPs) are the artificially derived nanometer-scale components (1–100 nm in dimension) which are produced for advanced nanomaterial construction in smart applications (Auffan et al. 2009). ENPs are composed of two-layer, i.e., the surface layer having small doped molecules like metal ions, polymer, surfactants, and the inner core referring nanoparticle itself (Raliya 2019). Currently, many smart application of ENPs in soil nanotechnology has been documented, including as nanobiosensors, delivery of nutrients, growth hormones, pesticides, food additives, and genetic improvement of plants (Jampílek and Kráľová 2017; Dar et al. 2020; Saxena et al. 2020). Therefore, to keep harmony in the soil functioning such as sustained growth of microbial species, nutrient bioavailability, plant growth, crop yield, and application of engineered nanoparticles (ENPs) have been considered as an emerging and potentially viable technique for agricultural practices. An increasing number of studies have suggested massive production and liberation of ENPs in the ecosystem has raised the question about their transformation, toxicity, risk, and uncertainty during application. They are widely accepted by the scientific community for the application in agricultural purposes; however, accidentally soil becomes a major sink of ENPs through different exposure routes (Kumar et al. 2012).

D. Mishra (✉) · V. Pandey · P. Khare
Central Institute of Medicinal and Aromatic Plants, Lucknow, 226015, India
e-mail: mishra.disha1@gmail.com

© Springer Nature Switzerland AG 2021
P. Singh et al. (eds.), *Plant-Microbes-Engineered Nano-particles (PM-ENPs) Nexus in Agro-Ecosystems*,
Advances in Science, Technology & Innovation,
https://doi.org/10.1007/978-3-030-66956-0_7

Modeling studies have suggested that soil receives a higher amount of ENPs than air or water. They easily adapt different electrical, magnetic, optical, properties than its bulk material and influence soil physico-chemical properties which manipulates soil texture, particle size, soil pH, microbial population, and simultaneously causes potential toxicity. The release of ENPs to the soil could be from a point or diffuse sources including liberation directly through primary particles, or transformation after reactions like agglomeration, aggregation, association with soil matrix, or dissociation. Direct exposure pathways of ENPs to soils occur when ENPs are used for delivery of fertilizer, pesticides, for remediation purposes, or via an accidental release. However, the deliberate entry of ENPs to the soil environment may lead to bioaccumulation, expanded toxicity, loss of organic matter, alteration in soil biodiversity, and altered soil physico-chemical structures. The application of ENPs to soil has varied according to application and includes two major categories, i.e., (1) In the organic forms (carbon nanotubes, fullerenes) and (2) inorganic forms (metal nanoparticles, silica-based, and quantum dots). However, the absence of proper monitoring methods and complex heterogeneous environment of soil turns it difficult to measure than in any other environment. The fate and travel of ENPs are usually governed by soil properties, like pH, texture, organic matter, water regime, and ionic strength. It is often argued that during travel into soil components the transformation of ENPs occurs which makes it difficult to extrapolate in a realistic scenario.

The detailed study about the effect of ENPs in soil ecosystem has outlined a clear sketch about modification in microbial enzymatic activities due to metals and metal oxides, alteration in soil pollutant mobility, toxicity in the plant (De La Rosa et al. 2011), accumulation in plant tissues, soil and sediments (Cornelis et al. 2014), control of plant insects (Debnath et al. 2012). However, the behavior and fate of the ENPs in the soil will be determined by a complex set of factors of both ENPs and soil. The chapter will provide an overview of the ENPs in the agro-ecosystem considering their synthesis, mobility, transformation, interaction, accumulation, toxicity, assessment methods, and impact on soil health briefly. This would further be designed as a framework and provide relevant information for futuristic studies regarding ENPs in soil environment.

2 Synthesis and Types of Engineered Nanoparticles

Generally, two different synthesis approaches have been employed for the construction of ENPs namely top-down and bottom-up methods. The top-down is a destructive approach, which involves the division of larger particles into smaller particles ENPs. The top-down methods are usually destructive methods that are costly, time-consuming, and not suitable for large-scale production. Various methods like mechanical milling, nanolithography, laser ablation, sputtering, and thermal decomposition chemical methods, photo-lithography are suggested for this purpose (Dhand et al. 2015). While in bottom-up known as building up approach, the ENPs are formed from relatively simpler substances. The bottom-up or constructive method is made of material from atom to clusters to ENPs through sol–gel, spinning, chemical vapor deposition, and pyrolysis processes (Khan et al. 2019).

Between organic and inorganic types they are mainly applied in the form of metal/metal oxides, carbon-based, silica-based, quantum dots, and dendrimers. Their application in the form of metal oxides such as ZnO, TiO_2, CeO_2, CrO_2, Fe_3O_4, and binary oxides was frequently noticed (Bhatt and Tripathi 2011).

To create silica ENPs, the covalent grafting of polymers was carried out through the use of various polymers such as polystyrene and polyacrylamide (Adams 2018). Nano-SiO_2 was reported to promote seed germination and stimulated the antioxidant system, promote plant, and root growth (Gul et al. 2014). Mesoporous silica was also used for delivery of nutrients, fertilizers, drugs, gene, and DNA to the plant cell due to their high surface area, pore-volume, stability, and tunable structure. Thus, various nanodevices with tremendous effects could offer novel insights for the safer use of these nanoparticles.

Carbon nanotubes (CNT) are cylindrical layers of graphene with single or multiwalled designed as open and closed ends. Various nanodevices for application in agricultural purpose and pollutant remediation were constructed due to their unique conductive, optical, and thermal properties. CNT can also make soil nutrient-rich and enhance its biota as well as chemical and physical properties. CNT provides adsorption sites due to the cylindrical structure and a wide range of toxic compounds can easily absorb on it. While the use of CNT-based nanosponges was identified as a great tool for remediation of xenobiotics from soil (Manjunatha et al. 2016). Unfortunately, CNTs have shown potential toxicity in human cells due to their penetrability and accumulation in the cytoplasm (Prasad et al. 2017).

Nanocrystal quantum dots are semiconducting heterostructured materials such as cadmium selenide (CdSe), indium phosphide (InP), or zinc selenide (ZnSe) with controlled optical and electrical properties (Xiaoli et al. 2020). Quantum dots (QDs) can be employed for live imaging in plant tissue for retrieving information about physiological processes. Sometimes, ENPs were made by dendrimers, and these are normally organic-based ENPs which are complex, multifunctional polymers with their size range between 1 and 10 nm diameters. They are having branched asymmetric structures with nanospheres or nanocapsules shape (Ishtiaq et al. 2020). The unique structures with a solid center and surrounded spherical surface of dendrimers have shown tremendous capability in the field of sensors development and also as a sorbent for contaminants (Zhang et al. 2017). In the future, more focused research toward the development of application-specific nanoparticles through controlling reaction parameters, shape, size, and morphology should be done.

3 Exposure of Engineered Nanoparticles in Soil

The rapidly evolving synthesis of ENPs has provided maximum chances of ENPs to enter in soil compartment during traveling. As far as the concern of their entry to the soil ecosystem, they can enter through point or nonpoint sources. The direct exposure route consists mainly of the application of nanofertilizers, nanopesticides, seed treatment preparation, agrofilms, or for remediation of contaminated land or groundwater. However, the accidental liberation was primarily from diffuse emission, ENPs containing products, solid waste disposal, landfilling, and incineration or mishandling of those during transportation (Walden and Zhang 2016). The product matrix has severely affected the ENPs physical and chemical characteristics and thus long-term application of ENPs resulted in bioaccumulation in soil. Nonetheless, after liberation, they get interacted with the heterogeneous structure of the soil. Soil provides a suitable habitat for the retention of ENPs, as they can adsorb on the soil pores, forms aggregates with organic matter, or establish electrostatic interaction, ligand exchanges networking with the soil-solid matrix. The surface chemistry of ENPs plays an important role in deciding its mobility, stability with inorganic and organic soil colloidal suspension (Alimi et al. 2018). The release of silver nanoparticles was found more in presence of natural organic matter as without that in the soil the reduction in negative surface potential of ENPs would cause more aggregation in soil (Li et al. 2013).

Application of wastewater sludge enriched with Zn and Ag nanoparticles in soil final concentration of 1400 and 140 mg/kg for Zn and Ag, respectively, has shown a reduction in the fungal community in soil (Durenkamp et al. 2016). The calculated risk assessment of ENPs released through personal care products has suggested that about 43% of it ends up in landfills, 0.8% directly goes to the soil, and 32% in water bodies. The uprising concentration of ENPs was mainly due to the usage of sunscreen, facial moisturizer, hair coloring agents, body wash, toothpaste, and shampoo (Keller et al. 2014). After entering the ecosystem, it is very easy to enter the soil either via wastewater sludge, landfilling, or atmospheric deposition. However, the fate of ENPs after liberation has been described in detail in the later section. The impetus of the application of ENPs in the soil through key drivers is being summarized in Table 1. This would provide insight into their application in various forms and possible impacts on the soil ecosystem.

4 The Fate of Engineered Nanoparticles in the Soil

The modification in the physico-chemical characteristics took place due to major transformation reactions occurring between soil matrix and ENPs. The soil reaction occurring inside soil pore or soil solution resulted in either their retention or mobilization. The surface area, size, charge, density, and shape of ENPs play a major role in determining their fate in the soil matrix. The complex and heterogeneous environment of soil leads to aggregation, sedimentation, dissolution, the transformation of ENPs. Furthermore, their bioavailability in the soil is mainly influenced by soil chemistry and soil microorganism (Dwivedi et al. 2015). It is well predicted that the residence time of ENPs is more in soil and sediments than in aquatic system. Based on their biodegradation potential, they eventually build up in the soil and thus become bioavailable for plants and terrestrial organisms. Although many theories have been suggested to the fate of ENPs in soil, however, clear mechanisms of fate remain unclear due to the heterogeneous surface of the soil. The interaction between these processes and the ENPs transfer determines the fate and finally the ecotoxicological potential of ENPs in the soil matrix. The upcoming section will summarize the different fate behavior of ENPs according to the consensus of various scientific theories.

4.1 Engineered Nanoparticles and Colloids

Many processes inside the soil matrix are generally governing their fate in soil. Regarding this, colloids of the soil (diameter <1 μm) play a magnificent role in the interaction chemistry of ENPs. These fractions of soil are very mobile and active components with high surface area and often turn as carriers for different contaminants and nutrients in the soil. These colloid particles govern the transport of various engineered nanoparticles and then impart to environmental pollution (Pan and Xing 2012). The Derjaguin–Landau–Verwey–Overbeek (DLVO) theory and the colloid filtration theory both have explained their transport in the soil porous media due to the structural similarities between natural

Table 1 Application mode of different types of engineered nanoparticles

Classes	Types	Application	References
Metallic	Manganese and copper nanoparticle	Act as micronutrient nanofertilizer, reduction in the rate of release of micronutrients to plants and help in N-fixation	Kopittke et al. (2019), Zahra et al. (2015)
	Iron and magnesium nanoparticle	Reduce the concentration of polychlorinated biphenyls in soils up to 56%	Olson et al. (2014)
	Iron sulfide nanoparticles with carboxymethylcellulose	Immobilizes Hg in soils up to 65–91%	Gao et al. (2013)
	Zero-valent iron nanoparticle	Act as excellent phosphate ion absorbent (90–98%)	Lin and Xing (2007)
		Degrade polybrominated diphenyl ethers up to 67%	Qiu et al. (2011), Xie et al. (2016)
		Removal of Cr (VI) up to 56–98%	Yang et al. (2019)
		Removes nitrates from soils, water, and sediments	Liu and Wang (2019)
		Degraded molinate (a carbothionate herbicide)	Joo et al. (2005)
Metallic oxide	Nano-titanium oxide, iron oxide	Enhances rhizopheric phosphorus content when applied on *Lactua sativa*	Zahra et al. (2015)
	TiO$_2$	Helps in the bioremediation of various organic compounds such as phenol, p-nitrophenol, salicylic acid, and benzene	Zhang et al. (2010)
		Extensively used as photocatalyst for waste treatment	Li et al. (2008)
	CeO$_2$	Improved plant growth, biomass yield, grain yield in *Triticum aestivum* L	Rico et al. (2014)
	ZnO	Act as nanofertilizer to boost the yield and growth of food crops	Sabir et al. (2014)
		Removal of Cr by 45–53%	Ahmed and Yusuf (2015)
Carbon	Graphene oxide	Act as suitable amendment to immobilize copper in polluted soil 65%	Baragaño et al. (2020)
		Sorption of volatile organic compounds, pesticides, heavy metals, and pharmaceuticals	Gao et al. (2013), Deng et al. (2017)
	Carbon nanotubes	Sorption of metals (Cu, Ni, Cd, Pb, Ag, Zn)	Khin et al. (2012)
		Adsorbed cationic dyes up to 97.2%	Li et al. (2003)
	Fullerenes	Sorption of organic compounds (e.g., naphthalene)	Cheng et al. (2004)
		Used for remediation of organometallic compounds	Ballesteros et al. (2000)
Silica	SiO$_2$ nanoparticle	Used for bioremediation of polycyclic aromatic hydrocarbons pyrene efficiency of 75–102%	Topuz et al. (2011)
	Silica nanoparticles	Removal of cationic dyes (86%)	Tsai et al. (2016)
Polymeric nanoparticles		Helps in removal of hydrophobic pollutants from soils (e.g., phenanthrene) by 85.2%	Tungittiplakorn et al. (2005)
Dendrimers		Removal of copper (II) from sandy soil up to 85%	Xu and Zhao (2005), Zou et al. (2016)

colloids and ENPs. The organic (<30 nm) and inorganic particles (>20–30 nm) of iron or aluminum oxide or clay oxides, larger colloids of soil minerals (>100 nm) play important role in adhering and act as the carrier of ENPs. Wang et al. (2015) have described the main key factors governing the transport of ENPs in soil porous media. The interaction between silver nanoparticles and natural soil colloids has shown deposition of nanoparticles followed by hetero-aggregation and hence confirming their reduced mobility in soil solution (Cornelis et al. 2013). The transport of nanoscale zero-valent iron (nZVI) was also promoted by

soil-colloids interaction behavior suggesting the highest mobility in quartz while least in diatomite (Zhang et al. 2019). The transport of ENPs in the soil column was mainly affected by texture, charge, porosity, and adsorption capacity of colloids fraction.

4.2 Aggregations

The word "aggregate" is known as the clusters of ENPs in different shapes. This phrasing is aggravated passing through the potency of adhesion of the majority of ENPs to both and other particles, by which ENPs frequently agglomerates to form particle clusters (Cornelis et al. 2014). Aggregation reduces the specific exterior area of particles and interfacial energy. In aggregation presence of the natural organic matter commonly limits their movement in the soil. The actions of soil organisms manipulate carbon maintenance time and return in soil, which revolves change carbon stabilization, aggregation, and yield. Aggregation favors the movement and deposition of ENPs in soil solution. It can be homoaggregation (between ENPs) or hetroaggregation (between ENPs and other soil components such as clay, minerals, or oxides). The collision among the ENPs resulted in the formation of aggregates and thus the establishment of weaker van der Walls forces or strong chemical bonds takes place. The presence of natural organic nanoparticles in the soil porous media further facilitates the hetero-aggregation among them, thus, severely affects their bioavailability, toxicity, and transport across porous media. Aggregation of metal-based ENPs (Ti, Cu, Au, Ag, Ni) was already noted in soil (Cornelis et al. 2014). Hetero-aggregation most likely occurs than homoaggregation rates in soil pores and appears to vary depending on the soil colloid and ENPs nature. The formation of large hetero-aggregates may also reduce the translocation and uptake of ENPs through plant cells and membranes, and thus decreasing their biological availability (Gogos 2015). The loss of ENPs polymeric coatings under sunlight catalyzed redox reactions may stimulate instability and favors hetero-aggregation. The soil pH, clay content, organic matter, and particle characteristics play an important role in ENPs transport and retention (Abbas et al. 2020). The movement of ENPs across the soil solution was likely to be influenced by zeta potential, surface coating of ENPs, however, the use of emulsifier during processing helps in stabilizing while capping agents prevent degradation and transformation (Sajid et al. 2015).

4.3 Deposition

Engineered nanoparticles (ENPs) undergo deposition in the soil as a result of collisions and bonding with the surface. The Brownian diffusion, interception, or gravitational settling inside soil pore wall are responsible and thus deposition took at a place (Cornelis et al. 2014), while hydrodynamic drag forces further allow their travel to the collector sites (Torkzaban et al. 2007). The absence of repulsion due to the presence of similar charges on the surface and high collision efficiency always promotes their deposition. The large aspect ratio of CNT has resulted in higher deposition than other colloids or ENPs (Lin et al. 2010), due to the coiling features of CNT around soil particles (Canady and Kuhlbusch 2014). The deposition can be considered as analogous to the

aggregation of relatively small ENPs with a much larger colloid, and as dominates as aggregation in the soil matrix. Li et al. (2020) have demonstrated relatively high mobility of silver nanoparticles (Ag NPs) in the loamy sand than in silty soil under low ionic strength and higher flow rates. Further, the transport of the Ag NPs in loamy sand was slowed at a low flow rate, due to the dominance of diffusion and depositions after compression of the electrical double layer of Ag NPs and soil surface (Braun et al. 2015). The effect of input concentration, size, and surface coating of Ag NPs for the transport was also studied and it was stated that migration was less in ultisols due to high surface area and retention sites. The increased concentration, lower particle size, and surface coating of Ag NPs have promoted the transport (He et al. 2019). However, the transport and deposition of ENPs are a complex process that is jointly affected by several factors such as physico-chemical properties of soil, pore-water solution, ENPs features as well as hydrodynamic behavior. The transport and retention of CuO nanoparticles in soil subsurface environment was also affected by soil pH, ionic strength, and humic acid (Fig. 1).

However, the establishment of van der Walls forces between nanoparticles and collector surface was repulsive, promoting an unfavorable deposition due to interaction energy, collision, and aggregation in soil (Ma et al. 2018; Wu et al. 2020).

4.4 Oxidation/dissolution

The ENPs generally undergoes different oxidation process, followed by their complexation with organic matter and chelating agents and finally adsorbs on the colloidal surface. The dissolution, oxidation–reduction reactions largely depend upon the structure of ENPs such as soft metal cations Mg, Ag, Zn, and Cu are susceptible to these reactions (Boxall et al. 2007). However, the reason for the increasing trend of ENPs use has exposed the soil with a rising concentration of ENPs. The speed of dissolution of ENPs in soils can be explained by the type, texture, and source material of ENPs (Rodrigues et al. 2016). The dissolution of metal-based ENPs is affected by their chemical properties and soil conditions like pH, organic carbon, texture, and size (Arora et al. 2012). Still, the kinetics of oxidation of the different ENPs in a complex soil matrix, and the relevant controlling factors are unexplained. According to the reports in soil systems, the dissolution of ENPs allows the liberation of free ions in soil solution which further transforms by reacting with organic matter, soil chelators, or adsorb on soil particles/minerals followed by precipitation of non-soluble reactive counterparts. In non-saturated aerated soil, i.e., low pH and high oxygen contents and the rate of oxidation enhances. In contrast, the presence of organic coatings

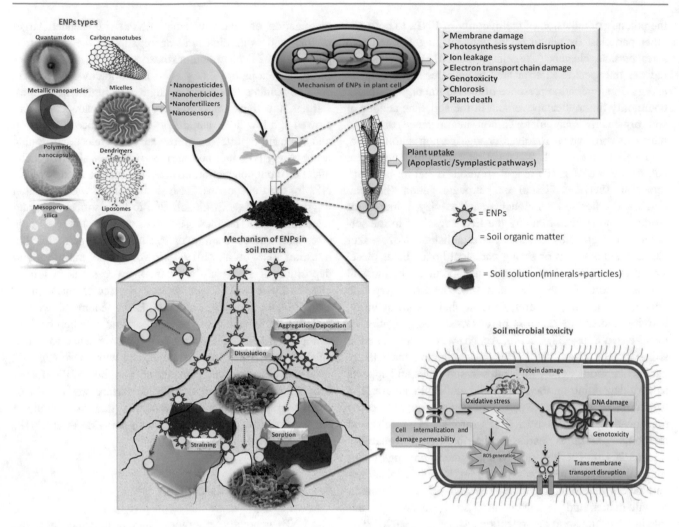

Fig. 1 Fate of engineered nanoparticles (ENPs) in the soil matrix and associated toxicity (modified from Santiago et al. 2016)

impedes dissolution while sometimes their quick degradation promotes dissolution (Chen 2018). The dissolution of Ag NPs was reported due to chemical reactions in the soil (Benoit et al. 2013) while Au nanoparticles were impervious to oxidative dissolution due to its instability of oxidized Au (Au^{+3}) hence readily reduced in soil.

The oxidation and dissolution of CeO_2 nanoparticles were enhanced by complex formation between chelating agents and Ce^{+3} in soil, thus, lowering the bioavailability and transport of nanoparticles (Zhang et al. 2017; Rodrigues et al. 2016). In contrast, sometimes, the dissolution process is inhibited by certain factors in soil subsurface likewise the dissolution of Ag NPs was hampered by iron oxides due to the formation of electrostatic attraction followed by hetero-aggregation (Wang et al. 2019). Thus, the dissolution kinetics of nanoparticles is a very complex process and therefore the factors affecting the dissolution should be critically evaluated. Overall future study should incorporate

detailed inspection of the in vitro fate behavior of ENPs for better identification of risk associated with ENPs application in soil.

5 Factors Affecting Transport of Engineered Nanoparticles in Soil

As discussed above, the aggregation, transport, and deposition of ENPs have been affected by several parameters like size, surface area, zeta potential, hydrophobicity, structure, and synthesis route. These properties may interact with soil solution and it was proposed that ENP size between 1 and 30 nm behave differently for the aforementioned processes (Santiago-Martín et al. 2016). Due to their smaller size, they start to aggregate in soil, however, they can keep on changing the properties in soil solution depend on the particle size. The high surface area further boosts its activity and therefore the aggregation with soil particles amplified. However, the particle size becomes a major parameter as surface atoms increase with a decrease in particle size (Alan et al. 2020). Ag NPs of diameters around 10 nm showed higher penetration capacity into the cell than of particle size of 20–100 nm (Ivask et al. 2014; Goswami et al. 2017). The surface charge of ENPs often governs their binding pattern with clay or minerals in soil solution likewise the electro-static interaction of negatively charge cerium oxide nanoparticles was increased with clay edges. Similarly, the low affinity of cerium oxide nanoparticles with the surface of kaolinite suggested strong electrostatic interactions between them. The charge-dependent aggregation was due to the variation in hydrodynamic size and surface charge (Guo et al. 2019). Further, the coarse surface of some clay minerals also provides binding sites for positively charged ENPs (Ghorbanpour et al. 2020).

The retention of functionally stabilized Ag NPs was increased with the interaction of iron and clay minerals (Hoppe et al. 2014). Furthermore, the retention in the soil largely depends on the ionic strength, mass concentration, and particle numbers which contributes to the filling of retention site and concentration-dependent mass transfer in soil solution (Alan et al. 2020). However, the increase in the magnitude of electrical double layer forces between the charged colloids and minerals leads to the release of ENPs. Likewise, Zn nanoparticles and carbon nanotubes have shown low mobility in different ionic strength of clay minerals soil and natural soil, respectively (Zhao et al. 2012). It might be related to their shape, aspect ratio, surface charge, size distribution, and interconnected soil pores. In addition to this, the surface coating also responsible for their fate in the soil like a coating of polyvinylpyrrolidone and citrate has boosted the transport and reduced the retention of Ag NPs in soil, which related to obstruction in the solid phase retention sites in soil (Kanel et al. 2015). Consequently, ENPs surface modification generates electrostatic, steric, or repulsive forces which more likely to reduce the aggregation and thus enhance their transport and bioavailability (Goswami et al. 2017). Similarly, the uncoated Ag NPs were more bioavailable than citrate-coated nanoparticles due to a rise in their stabilization after coating (Cornelis et al. 2014). The transport of sodium dodecylbenzene sulfonate surfactant on the transport of Ag NPs and CNTs in saturated porous media has shown high mobility in soil column and they exhibited similar transport patterns as of natural clay soil (Tian et al. 2010).

Soil is the complex mixture with heterogeneous features thus extrapolation of the effect of any one characteristic of the ENPs cannot be described perspicuously. Nevertheless, the complex mechanism simultaneously occurring in the soil system helps to solve the question related to their abundance, mobility, bioavailability, transformation, and toxicity.

6 Effect of Engineered Nanoparticles on the Soil Properties

Being the natural sink of ENPs, the soil environment has been critically affected by their presence. Most of the studies have suggested the complex reactions occurring in soil media have been actively mediated by both soil components and ENPs (Pradhan and Mailapalli 2017; Abbas 2020). In this view, the discussion about the effect of the ENPs on the soil properties and edaphic biota has been elaborated here.

6.1 Effect on Physico-Chemical Properties

Soil pH is a major governing parameter that directly influences soil health, indicates its nutrient status, and also about ionic strength of soil solution. Somehow, it plays a major role in the ionization of various organic/inorganic compounds and changes their solubility and responsible for the sorption of many compounds. It is pragmatic that variation in the pH of the soil is sometimes mediated through the accumulation of the different ENPs mentioned above (Schultz et al. 2015). These variations in pH further lead to toxicity in soil fauna and led to metal ion solubility, nutrient availability, plant growth, and clay dispersion (Zhang et al. 2018; Tarafdar and Adhikari 2015). In another study, the change in soil pH from 5.9 to 6.8 has shown no changes in solubility of copper nanoparticles, however, it was positively correlated with the change in the organic matter content of soil (Gao et al. 2019). In contrast, the solubility of Zn was related to negatively correlate with soil pH, due to its retention and adsorption on the clay particles (García-Gómez et al. 2018). In addition to this the agglomeration, discharge, oxidation, and release of nanoparticles are highly dependent on soil pH (Nowack et al. 2012). Likewise, the impact of pH on ZnO ENPs breakdown has caused danger on the population of *Folsomia candida* and *Eisenia fetida* in soil (Kool et al. 2011). The gravity-driven transport of ENPs like TiO_2, CeO_2, and $Cu(OH)_2$ owing to their effect on soil pH and nutrient release in unsaturated soils has determined and small changes in the soil pH were detected due to the release of natural ions (Mg^{2+}, H^+) through substitution suggesting the high retention of ENPs in soil (Conway and Keller, 2016). Furthermore, the interaction of ENPs with dissolved organic matter would greatly alter the magnitude of fate, transport, binding, and bioavailability of ENPs in soil.

The ubiquitous organic matter is generally composed of heterogeneous and different molecular compounds and thus multiple interaction mechanisms such as hydrogen bonding, electrostatic interaction, hydrophobic binding, π-π interaction, cation bridging, and adsorption take place between ENPs and organic matter surface. These interaction leads to

Fig. 2 Interaction of engineered
nanoparticles (ENPs) in
agro-ecosystem

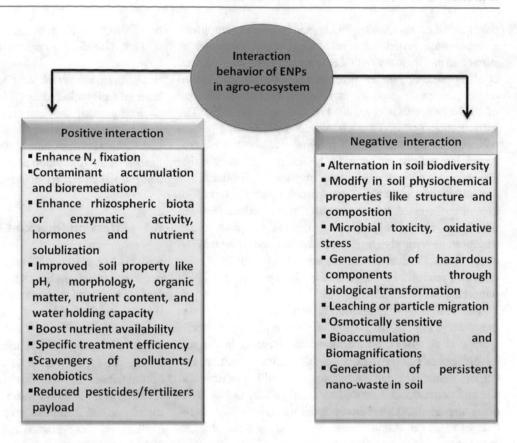

aggregation, sedimentation, dissolution, oxidation, reduction, and deposition of ENPs and eventually modify the bioavailability and toxicity for soil environment. The release of metal ions from certain ENPs and metal ion complexes causes toxicity to microorganism and the dissolved organic matter could significantly alter their release by blocking ENPs oxidation sites. Recent research about this has clearly stated that the reduction of Ag^+ through dissolved organic matter has diminished the acute toxicity in *Daphnia magna* (Zhang et al. 2016). Also, in different natural organic macromolecules types, such as humic acid, fulvic acid, alginic acid, and tannic acid have collectively alleviated the ZnO induced antimicrobial activity in *Bacillus subtilis* due to their binding of Zn^{+2} with natural organic matter (Ma et al. 2013). Similar findings were also observed by Nie et al. (2020) who showed that soil organic matter reduces Ag^+ to Ag NPs which was mediated through free organic radicals and reducing surface groups of organic matter. The interaction of Ag^+ with soil organic matter has helped in elucidating the formation of silver nanoparticles (Nie et al. 2020). Different interaction behaviors of ENPs are presented in Fig. 2.

The natural organic matter often used as a stabilizer for ENPs production (Grillo et al. 2015), as a controlling agent for the stability of nanoparticles in the environment for control of toxicity (Omar et al. 2014) and remediation of

media contaminated with heavy metals (Karnib et al. 2014) or organic compounds (Tang et al. 2014; Grillo et al. 2015) has critically reviewed the emphasis of natural organic matter for the interaction and stability of nanoparticles. The effect of C60 on soil microbial activity was also suggested due to the strong binding of C60 to soil organic matter (Patra et al. 2016). The adsorption of ENPs on the surface of soil organic matter has curtailed their mobility which ultimately changed their impact on soil properties.

The impact of Ag ENPs on the five different soils with varying physico-chemical properties has demonstrated that its toxicity was positively correlated with clay content and pH of the soil while the organic matter has not shown any relation with toxicity (Schlich and Hund-Rinke 2015). Surface adsorption and diminished actions of humic and fulvic acid in soil were responsible for the restraining of Ag ENPs disintegration (Javed et al. 2019).

Including these properties, the physical structure and hydraulic properties of soil were also affected upon exposure with ENPs. Besides soil texture, nutrient content, soil ionic strength, and pH of the soil solution also significantly impact ENPs transfer in soil (Patra et al. 2016). It was found that the concentration of valence of soil salt cation was also changed providing greater stability of divalent cation calcium (Ca^{2+}) than monovalent cation potassium (K^+) (Makselon et al. 2018). It has been observed that there is 30%, 45%, and 2%

reduction in hydraulic conductivity when the soil is treated with 2% Al_2O_3, ZnO, and CuO ENPs, respectively as compared to untreated clay owing to pore-clogging (Tan 2017). In support of this study, it was found that reduction in hydraulic conductivity by clogging soil pores when there is an addition of nanosized material (montmorillonite) to fine-grained soil. Similar results also observed when metal oxide ENPs are added to sands (Braun et al. 2015). The application of nZVI in the soil at concentrations of 1 and 4 g l^{-1} showed no consequences on the hydraulic conductivity of natural soil (Reginatto et al. 2020).

6.2 Effect on Biological Properties of Soil and Phytotoxicity

The increased liberation of ENPs in soil has resulted in inevitable accumulation in soil and thus becomes a serious threat to the soil microbial community. Due to the dynamic nature of ENPs, they produce certain toxicity to soil microorganisms either directly or indirectly through interacting with other organic compounds. But due to the dynamic features of ENPs, it is always under discussion and newer theories have been developed day by day. The chemicals present in the root exudates in the rhizosphere greatly affect the physical properties of soil such as pH, cation exchange capacity (CEC), and salinity which ultimately influence ENPs aggregation and dissolution. Moreover, the rhizospheric microbiome also produces biomolecules which affect the ENPs fate. For instance, amino acids such as cysteine have shown to fasten ENPs aggregation rates but not long-term aggregation to a larger size (Hsieh 2010).

The impact of ENPs on the plant growth-promoting rhizobacteria (PGPR) like *P. aeruginosa, P. putida, P. fluorescens, B. subtilis*, soil nitrifying bacteria, and phosphate solubilizing bacteria was visible with retarded growth in culture conditions (Kumar 2018). Metal ENPs are generally toxic to microorganisms thus these ENPs damages plant-fungi and plant-bacteria association. Nano-ZnO and nano-TiO_2 were reported toxic against *B. subtilis, E. coli*, and *V. fischeri* (Li et al. 2011). Metal oxide ENPs of Cu is found to be toxic against PGPR such as *K. pneumoniae, P. aeruginosa, S. paratyphi*, and *Shigella* strains due to antibacterial property of Cu (Mahapatra et al. 2008). Iron and copper-based ENPs are observed to react with peroxides present in the soil, releasing free radicals which have a toxic effect on microorganisms (Saliba et al. 2006). Javed et al. (2019) showed that TiO_2 and CuO ENPs reduced the microbial biomass of the paddy soil due to their chemical characteristics. Similar outcomes were contemplated for the impact of ZnO, TiO_2, CeO_2, and Fe_3O_4 ENPs for lowering bacterial communities in saline/black soil and it was possibly

responsible for the decreased enzymatic activities of invertase, urease, catalase, and phosphatase in the soil (You et al. 2018). Pérez-Hernández et al. (2020) have also summarized the impact of various ENPs on the soil microbiota and suggested that they were responsible for the reduction in the population of mesofauna and microfauna in soil.

Arbuscular mycorrhizal fungi (AMF) are known to show symbiotic association with plants and thus largely affect plant growth. However, it has been reported that AMF diversity decreased after exposure to Fe_3O_4 NPs at the concentration of 10 mg kg^{-1} (Cao et al. 2017). On the other hand, AMF remediates the toxicity when ENPs are exposed to plants. A study revealed the AMF inoculation eliminated the negative effect of ZnO ENPs on maize by increasing plant growth and nutrient uptake. Similar effects were reported in tomato plants when inoculated with AMF, the plant showed a reduction in Ag uptake up to 12% after exposure of Ag NPs (Noori et al. 2017). AMF restricts the uptake of ENPs by plants through the discharge of glycoprotein known as glomalin which acts as a chelator in the rhizospheric region (Siani et al. 2017).

The exposure of copper oxide nanoparticles (1000 mg/kg) prevail suppressed immune response followed by the mortality of the earthworm species (*Metaphire posthuma*) which was related to phagocytosis, production of cytotoxic molecules, stress enzymes, and loss of total protein of coelomocytes (Gautam et al. 2018). In another study, CuO ENPs were found to limit the life span of another invertebrate species *Enchytraeus crypticus* (Gonçalves et al. 2017). More susceptibility of ENPs was recorded in the case of juvenile species of *Lumbricus rubellus* than adult populations after exposure with C-60 ENPs (Van Der Ploeg et al. 2013). The ENPs remain attached to soil colloids, invertebrates internalize the ENPs by ingestion and eventually get transferred to the gut epithelium. The ENPs toxicity toward soil invertebrates through hindering ribosomal and histone activity, by disrupting sugar, protein, and lipid metabolism (Novo et al. 2015). Moreover, NMR studies have shown the amino acid such as leucine, valine, isoleucine, and sugars such as glucose and maltose are potential bioindicators of ENPs toxicity in invertebrates (Liang et al. 2017). The negative impact of Ag ENPs on reproduction ability in *E. andrei* (Velicogna et al. 2017) and multiwalled carbon nanotubes (MWCNTs) on *E. fetida* population were also reported (Zhang et al. 2014) and it might be due to the inhibition of antioxidant enzymes and restriction on metabolic pathways.

The phytotoxicity of ENPs mainly depends on its size, shape, chemical properties, and chemical subcellular sites where it is accumulated. Depending upon the chemical and physical nature of plant cell wall, ENPs act as a carrier or modulator which interacts with cellular processes. ENPs when interacts physically with plant cells, it mainly clogs the cellular structures mechanically while chemically it influences specific groups such as sulfhydryl and carbonyl groups, thereby imparting oxidative stress by hindering the cellular homeostasis. Particle size and surface properties greatly influence the plant-ENPs interaction. Depending upon the physical properties of ENPs, these can either act as nanofertilizers or as phytotoxic agents (Pradhan and Mailapalli 2017). The metal ENPs impart toxic effects in plants through three mechanisms; first: the ENPs release specific ions which might be toxic to plants. For instance, Ag^+ ions released from Ag ENPs transport through the plasma membrane hinder cellular respiration which ultimately results in cell death. Second, their chemical interactions with cellular components may produce chemical radicals to generate oxidative stress in plants. Thirdly, ENPs can directly interact with plant cells and disrupt membrane integrity. Since metal-based ENPs are sparingly soluble, impart a detrimental effect on plants (Verma et al. 2018). The outcomes of toxicity are recorded as inhibition in seed germination, lowered photosynthetic rate, plant growth, and development, fruit production, followed by reactive oxygen species (ROS) generation, and hampering in the synthesis of major biomolecules for cell growth (Pullagurala et al. 2019). The co-exposure of TiO_2 NPs (0, 100, 250 mg l^{-1}) and Cd (0, 50 mM) determined that root exposure has shown a more prominent effect rather than foliar exposure on hydroponic culture study in maize (*Zea mays* L.). In addition to this, the accumulation of TiO_2 was also reported (increased by 1.61 and 4.29 times) upon root exposure and foliar spray of TiO_2 nanoparticles helped in reducing Cd accumulation and further lowering Cd-induced phytotoxicity in maize than root exposure (Lian et al. 2020).

A similar type of study was also performed to evaluate the effect of TiO_2 on the phytotoxicity of Cd in *Oryza sativa* L. and TiO_2 was found to lower the toxicity and accumulation of Cd in booting and tillering stages of plants (Zhang et al. 2020). Furthermore, the impact of PbS nanoparticles on the *Zea mays* L. was also studied at various hydroponic treatments and it was concluded that it exerts potential toxicity to plant, seed germination, and root elongation. The STEM-EDS mapping has suggested the presence of PbS inside cortical cells, cytoplasm, and intracellular space suggesting its translocation and accumulation (Ullah et al. 2020). In this view, Tripathi et al. (2017) have summarized the potential mechanism of phytotoxicity, anatomical, physiological, biochemical, and molecular damages due to ENPs. They have also pointed out the defense and detoxification mechanism led by plants owing to the accumulation of ENPs inside plant cells (Tripathi et al. 2017). Another study has highlighted the Ag NPs-driven changes in photochemical efficiency of *Vicia faba* through leaves injection at the concentration of 100 ppm and it has the prominent repercussion of decreasing photosystem II efficiency and increases non-photochemical quenching, affecting stomatal conductance, and assimilation of carbon dioxide (Falco et al. 2020).

7 Monitoring Methods for Engineered Nanoparticles

The exposure of ENPs results in adsorption, aggregation, deposition, or accumulation in the soil–plant system, and thus valid analytical methods are required to monitor their quantification, traveling, imaging in the ecosystem. To address the characteristic features of ENPs which mainly influence their interaction in the environment, the broad range of methods are available for their recognition and categorization, jointly with microscopy, spectroscopy, chromatography, synchrotron radiation-based methods, and also including dynamic light scattering (DLS), voltammetry, isotopic methods, size partition, and sensor-based methods.

The shape, morphology, particle size, size distribution, and aggregation, accumulation can easily be acquired by using different microscopy-based techniques. For this scanning/transmission electron microscopy (SEM/TEM), scanning transmission electron microscopy (STEM), X-ray fluorescence microscopy (XRF), atomic force microscopy (AFM), transmission X-ray microscopy (TXM), confocal laser scanning microscopy (CLSM), and hyperspectral microscopy are currently available modern spectroscopy techniques. They can easily imaging the ENPs up to nanometer size, helps to acquire 2D/3D images, in the detection limit µg-ng/g without using any external standard in imaging. Many researchers have previously used these kinds of imaging techniques to obtain different monographs of ENPs which are useful to their kinetic investigation, in vivo toxicity, translocation, accumulation, and particle behavior (Jampílek and Kráľová 2017). However, these methods have certain limitations like tedious sample preparation, whole sample representation, geometry, mechanical tips, and loss of material during staining which counteracts for inaccuracy in data acquisition.

Furthermore, spectroscopy methods like UV–Vis spectroscopy, X-ray diffraction (XRD), Fourier transform infrared (FT-IR) spectroscopy, and energy-dispersive X-ray spectroscopy (EDX) are frequently used due to their easy handling, fast sample preparation, user-friendly, minimum aggregation, direct analysis, and economically viable. To investigate the fluorescent-labeled ENPs matrix-assisted laser desorption/ionization (MALDI), laser-induced fluorescence (LIF), or iron-trap (IT) mass spectrometry has been previously conducted (Shrivastava et al. 2019). In this view, the information about the particular nanoparticle, nanoparticle aggregation state, and average particle size, functional characteristics, and presence of outer coating can be obtained through these methods. Both NMR and IR spectroscopy were employed to detect the surface functionality of ENPs (Jiang et al. 2012). Mostly, FT-IR data was utilized to study humic substance adsorption onto silica and magnetite ENPs (Ma et al. 2018). These types of analysis are quite helpful in quantifying the quality of ENPs and also which type of functional group are present. It also detect surface charge present on ENPs and address different types of ENPs and their suitability to various type of soil for nutrient arability and deficiency to enhance soil quality. Mostly, XRD techniques are acquired to determine the crystalline nature, phase, and grain size of the nanoparticles (Jorge et al. 2013).

Dynamic light scattering (DLS) is the most commonly used technique to measure the aggregation rate/kinetics of ENPs through the measurement of zeta potential (Peijnenburg et al. 2015). Zeta potential helps to study particle aggregation or particle behavior of ENPs in the environment. Another is the mass spectrometry techniques, which generally consist of inductively coupled plasma-mass spectroscopy (ICP-MS), matrix-assisted laser desorption/ionization (MALDI), laser-induced fluorescence (LIF), or iron-trap (IT) mass spectrometry. Total metal concentrations in metallic nanoparticles can easily analyzed by aqua regia digestion followed by ICP–OES and ICP-MS measurements. Commonly used separation methods based on filtration, centrifugation, chromatography, and electrophoresis techniques are the conventional available strategies for the detection of size, shape, and charge of ENPs. Among that the most popularly high-performance liquid chromatography (HPLC), field-flow fractionation (FFF), size-exclusion chromatography (SEC), and capillary electrophoresis (CE) are used (Luo et al. 2014). They were well applied to study the various features of multiwalled carbon nanotubes, silica nanoparticles, and metal nanoparticles due to faster separation, high efficiency, low sample volume, and high sensitivity (Navratilova et al. 2015). Their combination with appropriate techniques like ICP-MS, UV–visible spectroscopy, nephelometry, and static light scattering (SLS) can extend their applications due to their broad size separation ranges and relatively moderate sample disruption (Choi et al. 2007). While the size-exclusion chromatography has higher partition efficiency, it can undergo as of fixed-phase interactions. FFF and HDC are having high detection limits (as per detector) but non-ideal samples of ENPs require additional pre-fractionation steps during sample preparation (Pornwilard and Siripinyanond 2014).

In the advancement of techniques, few methods like small-angle X-ray scattering (SAXS), small-angle neutron scattering (SANS), X-ray reflectometry (XR), and neutron reflectometry (NR) are advantageous due to fast sample analysis and data extraction. The knowledge about the surface properties, organic coatings, and crystallographic behavior can excogitate by application of X-ray-based methods such as X-ray absorption (XAS), fluorescence

(XRF), and photoelectron spectroscopy (XPS) as well as diffraction (XRD) as they are non-destructive, flexible, and relatively less expensive (Nurmi et al. 2005). Extensive applications of these techniques include the measurement of percentage crystallinity, detection of fine-grained minerals such as nanoparticles, nano-clays, and mix layer identifications. The XAS technique is often preferred due to its non-destructive nature, collection of wet samples (soil, sediments, and tissue) with absorption spectra but high metal concentration often creates hindrances in measurement (Tiede et al. 2008).

As emerging techniques currently small-angle neutron scattering (SANS), small-angle X-ray scattering (SAXS), and static light scattering (SLS) are utilized to study the presence of ENPs in the solid and liquid phase (Polte et al. 2010). To study the atomic, molecular, and structural features mostly RAMAN and laser-induced fluorescent spectroscopy were utilized. As an advanced mechanism, synchrotron radiation-based techniques were being adapted for localization and speciation of ENPs as they are non-destructive, higher spatial resolution, and higher detection limits (Castillo-Michel et al. 2017). These are often combined with XAS, photothermal flow cytometry (PTFC), and photoacoustic flow cytometry PAFC to study the ENPs in live plant tissue (Nedosekin et al. 2010). Also, the radioactive stable isotopes based methods such as autoradiography and positron emission tomography are combined with SEM and TEM to track and visualize ENPs in the in vivo soil–plant system. Stable isotopes provide as a tracer with no harm to radiation and have a long half-life. Although rapidly evolving sophisticated analytical techniques have shown its immense potential in testing the localization, speciation, uptake, availability, biotransformation, and toxicity in the soil–plant system. But in the future, the focus should be more on isotopic and sensors based methods with the right synchronization of technologies to understand the exposure risk of ENPs.

8 Conclusions and Future Directions

The exposure, transfer, accumulation, and transformation of ENPs in the agro-ecosystem have now become an imperative subject to address their environmental fate and risk. Engineered nanoparticles (ENPs) have been enacted as sophisticated technology with conceivable consequences for sustainable agriculture. A significant contribution of ENPs for crop management, delivery agents, sensing material, disease control, and soil conservation is well known. But the unregulated exposure of ENPs in soil has shown serious implications on soil health like damage of soil structure, loss of soil fertility, and toxicity in soil microflora. Therefore, the key focus should be on the development of biodegradable nanoparticles to avoid the nano-waste-related toxicity in specific soil microorganism, plant, and long-term persistent in soil. The risk assessment with a systematic study of their production, acceptable limit, degradation, and multicentre field trial should not be overlooked in foreseeable future.

Considering all the points, a more comprehensive in vivo, in vitro study, elaborated ENPs interaction mechanism, database modeling for their bioavailability, biomagnifications, on-field monitoring through highly sophisticated techniques are the crucial points of care. Also, the strict implementation of regulatory affairs regarding their production, application exposure, and disposal, identification of exposure source, and fate pathways should be implemented to overcome the challenges and risk of ENPs. Moreover, the developing nano-era of ENPs has created a revolution in sustainable agriculture; nonetheless, the hazards associated with their continuous application cannot be ignored. Thus, the knowledge of ENPs presented here could pave a way for future research to minimize the detrimental impacts of ENPs in the soil environment through designing sustainable, green, and more efficient ENPs.

References

Abbas Q, Yousaf B, Ali M, Munir MA, MEl-Naggar, A Rinklebe J, Naushad M, (2020) Transformation pathways and fate of engineered nanoparticles (ENPs) in distinct interactive environmental compartments. Rev Environ Int 138:105646. https://doi.org//10.1016/j.envint.2020.105646

Adams ML (2018) Release of bioactive agents from mesoporous silica nanoparticles for biological applications. Arthur Lakes Library. 11124/172826

Ahmed NA, Yousef NS (2015) Synthesis and characterization of zinc oxide Nano particles for the removal of Cr (VI). Int J Eng Res 1235–1243. ISSN 2229–5518

Alan BO, Barisik M, Ozcelik HG (2020) Roughness effects on the surface charge properties of silica nanoparticles. J Phys Chem C 124:7274–7286. https://doi.org//10.1021/acs.jpcc.0c00120

Alimi OS, Farner Budarz J, Hernandez LMT, N, (2018) Microplastics and nanoplastics in aquatic environments aggregation deposition and enhanced contaminant transport. Environ Sci Techno 52:1704–1724. https://doi.org/10.1021/acs.est.7b05559

Arora S, Rajwade JM, Paknikar KM (2012) Nanotoxicology and in vitro studies the need of the hour. Toxicol Appl Pharmacol 258:151–165. https://doi.org/10.1016/j.taap.2011.11.010

Auffan M, Rose J, Bottero JY, Lowry GV, Jolivet JP, Wiesner MR (2009) Towards a definition of inorganic nanoparticles from an environmental health and safety perspective. Nat Nanotechnol 4:634–641. https://doi.org/10.1038/nnano.2009.242

Ballesteros E, Gallego M, Valcarcel M (2000) Analytical potential of fullerene as adsorbent for organic and organometallic compounds from aqueous solutions. J Chromatogr A 869:101–110. https://doi.org/10.1016/S0021-9673(99)01050-X

Baragaño D, Forján R, Welte L, Gallego JLR (2020) Nanoremediation of As and metals polluted soils by means of graphene oxide nanoparticles. Sci Rep 10:1–10. https://doi.org/10.1038/s41598-020-58852-4

Benoit R, Wilkinson KJ, Sauvé S (2013) Partitioning of silver and chemical speciation of free Ag in soils amended with nanoparticles. Chem Cent J 7:75. https://doi.org/10.1186/1752-153X-7-75

Bhatt I, Tripathi BN (2011) Interaction of engineered nanoparticles with various components of the environment and possible strategies for their risk assessment. Chemosphere 82:308–317. https://doi.org/10.1016/j

Boxall AB, Tiede K, Chaudhry Q (2007) Engineered nanomaterials in soils and water: how do they behave and could they pose a risk to human health. Nanomed Lond 2:919–927. https://doi.org//10.2217/17435889.2.6.919

Braun A, Klumpp E, Azzam R, Neukum C (2015) Transport and deposition of stabilized engineered silver nanoparticles in water saturated loamy sand and silty loam. Sci Total Environ 535:102–112. https://doi.org//10.1016/j.scitotenv.2014.12.023

Canady R, Kuhlbusch T (2014) The life cycle of conductive plastics based on carbon nanotubes. In: Wohlleben W, Kuhlbusch T, Schnekenburger J, Lehr CM (eds) Safety of nanomaterials along their lifecycle release exposure and human hazards. CRC Press, pp 399–415

Cao J, Feng Y, Lin X, Wang J, Xie X (2017) Iron oxide magnetic nanoparticles deteriorate the mutual interaction between arbuscular mycorrhizal fungi and plant. J Soils Sedim 17:841–851. https://doi.org/10.1007/s11368-016-15618

Castillo Michel HA, Larue C, del Real AEP, Cotte M, Sarret G (2017) Practical review on the use of synchrotron based micro-and nano-X-ray fluorescence mapping and X-ray absorption spectroscopy to investigate the interactions between plants and engineered nanomaterials. Plant Physiol Biochem 110:13–32. https://doi.org//10.1016/j.plaphy.2016.07.018

Chen H (2018) Metalbased nanoparticles in agricultural system behavior transport and interaction with plants. Chem SpeciatBioavailab 30:123134. https://doi.org//10.1080/09542299.2018.1520050

Cheng X, Kan AT, Tomson MB (2004) Naphthalene adsorption and desorption from aqueous C60 fullerene. J Chem Eng Data 49:675–683. https://doi.org/10.1021/je030247

Choi WK Li, L Chew HG, Zheng F (2007) Synthesis and structural characterization of germanium nanowires from glancing angle deposition. Nanotechnology 18: 385302. https://doi.org/10.1088/0957-4484/18/38/385302

Conway JR, Keller AA (2016) Gravity-driven transport of three engineered nanomaterials in unsaturated soils and their effects on soil pH and nutrient release. Water Res 98:250–260. https://doi.org//10.1016/j.watres.2016.04.021

Cornelis G, Hund-Rinke K, Kuhlbusch V, den Brink N, Nickel C (2014) Fate and bioavailability of engineered nanoparticles in soils. A review. Crit Rev Environ Sci Technol 44:2720–2764. https://doi.org//10.1080/10643389.2013.829767

Cornelis G, Pang L, Doolette C, Kirby JK, McLaughlin MJ (2013) Transport of silver nanoparticles in saturated columns of natural soils. Sci Total Environ 463:120–130. https://doi.org//10.1016/j.scitotenv.2013.05.089

Dar FA, Qazi G, Pirzadah TB (2020) Nano-biosensors nextgen diagnostic tools in agriculture. In: Hakeem K, Pirzadah T (eds) Nanobiotechnology in agriculture. Nanotechnology in the life sciences. Springer Cham, pp 129–144. https://doi.org/https://doi.org/10.1007/978-3-030-39978-8_7

De La Rosa G, Lopez-Moreno ML, Hernandez-Viezcas JA, Montes MO, Peralta-Videa J, Gardea-Torresdey J (2011) Toxicity and biotransformation of ZnO nanoparticles in the desert plants *Prosopis juliflora*-velutina *Salsola tragus* and *Parkinsonia florida*. Int J Nanotechnol 8:6–7. https://doi.org//10.1504/IJNT.2011.04019

Debnath N, Mitra Das S, Goswami A (2012) Synthesis of surface functionalized silica nanoparticles and their use as entomotoxic nanocides. Powder Technol 221: 252–256. https://doi.org/10.1016/j.powtec.2012.01.009

Deng CH, Gong JL, Zhang P, Zeng GM, Song B, Liu HY (2017) Preparation of melamine sponge decorated with silver nanoparticles-modified graphene for water disinfection. J Colloid Interface Sci 488:26–38. https://doi.org//10.1016/j.jcis.2016.10.078

Dhand C, Dwivedi N, Loh XJ, Ying ANJ, Verma NK, Beuerman RW, Ramakrishna S (2015) Methods and strategies for the synthesis of diverse nanoparticles and their applications: a comprehensive overview. RSC Adv 5:105003–105037. https://doi.org//10.1039/C5RA19388E

Durenkamp M Pawlett, M Ritz K, Harris JA, Neal AL, McGrath SP (2016) Nanoparticles within WWTP sludges have minimal impact on leachate quality and soil microbial community structure and function. Environ Pollut 211:399–405. https://doi.org/10.1016/j.envpol.2015.12.063

Dwivedi AD, Dubey SP, Sillanpää M, Kwon YN, Lee C, Varma RS (2015) Fate of engineered nanoparticles: implications in the environment. Coord Chem Rev 287:64–78. https://doi.org/10.1016/j.ccr.2014.12.014

Falco WF, Scherer MD, Oliveira SL, Wender H, Colbeck I Lawson, T Caires AR (2020) Phytotoxicity of silver nanoparticles on *Vicia faba*: evaluation of particle size effects on photosynthetic performance and leaf gas exchange. Sci Total Environ 701:134816. https://doi.org/10.1016/j.scitotenv.2019.134816

Gao P, Ng K, Sun DD (2013) Sulfonated graphene oxide–ZnO–Ag photocatalyst for fast photodegradation and disinfection under visible light. J Hazard Mater 262:826–835. https://doi.org//10.1016/j.jhazmat.2013.09.05

Gao X, Rodrigues SM, Spielman-Sun E, Lopes S, Rodrigues S, Zhang Y, Lowry GV (2019) Effect of soil organic matter soil pH and moisture content on solubility and dissolution rate of CuO NPs in soil. Environ Sci Technol 53:4959–4967. https://doi.org//10.1021/acs.est.8b07243

García-Gómez C, Fernández MD, García S Obrador, AF Letón M, Babín M (2018) Soil pH effects on the toxicity of zinc oxide nanoparticles to soil microbial community. Environ Sci Pollut Res 25: 28140–28152. https://doi.org/10.1007/s11356-018-2833-1

Gautam A, Ray A, Mukherjee S, Das S, Pal K, Das S, Ray S (2018) Immunotoxicity of copper nanoparticle and copper sulfate in a common Indian earthworm. Ecotoxicol Environ Saf 148:620–631. https://doi.org//10.1016/j.ecoenv.2017.11.008

Ghorbanpour M, Bhargava P, Varma A, Choudhary DK (2020) Biogenic nano-particles and their use in agro-ecosystems. Springer Singapore. https://doi.org/10.1007/978-981-15-2985-6

Gogos A (2015) Engineered nanomaterials in the agricultural environment: current state of applications and development of analytical methods. ETH Zurich. https://doi.org/10.3929/ethz-a-010473555

Gonçalves MF, Gomes SI, Scott-Fordsmand JJ, Amorim MJ (2017) Shorter lifetime of a soil invertebrate species when exposed to copper oxide nanoparticles in a full lifespan exposure test. Sci Rep 7:1–8. https://doi.org//10.1038/s41598-017-01507-8

Goswami L, Kim KH, Deep A, Das P, Bhattacharya SS, Kumar S, Adelodun AA (2017) Engineered nano particles: nature, behavior, and effect on the environment. J Environ Manage 196:297–315. https://doi.org/10.1016/j.jenvman.2017.01.011

Grillo R, Rosa AH, Fraceto LF (2015) Engineered nanoparticles and organic matter a review of the state-of-the-art. Chemosphere 119:608–619. https://doi.org/10.1016/j.chemosphere.2014.07.049

Gul HT, Saeed S, Khan FZA, Manzoor SA (2014) Potential of nanotechnology in agriculture and crop protection. Appl Sci Bus Econ 1:23–28

Guo B, Jiang J, Serem W, Sharma VK, Ma X (2019) Attachment of cerium oxide nanoparticles of different surface charges to kaolinite

molecular and atomic mechanisms. Environ Res Lett 177:108645. https://doi.org//10.1016/j.envres.2019.108645

He J, Wang D, Zhou D (2019) Transport and retention of silver nanoparticles in soil: effects of input concentration, particle size and surface coating . Sci Total Environ 648:102–108. https://doi.org/10.1016/j.scitotenv.2018.08.136

Hoppe M, Mikutta R, Utermann J, Duijnisveld W, Guggenberger G (2014) Retention of sterically and electrosterically stabilized silver nanoparticles in soils. Environ Sci Technol 48:12628–12635. https://doi.org/10.1021/es5026189

Hsieh SF (2010) Developing a high throughput screening approach to predict the potential toxicity of engineered nanomaterials. University of Massachusetts Lowell

Ishtiaq M, Al-Rashida M, Alharthy RD, Hameed A (2020) Ionic liquid–based colloidal nanoparticles: applications in organic synthesis. In: Metal nanoparticles for drug delivery and diagnostic applications. Elsevier, pp 279–299. https://doi.org/10.1016/B978-0-12-816960-5.00015-X

Ivask A, Kurvet I, Kasemets K, Blinova I, Aruoja V, Suppi S, Visnapuu M (2014) Size-dependent toxicity of silver nanoparticles to bacteria, yeast, algae, crustaceans and mammalian cells *in vitro*. PLoS ONE 9:e102108. https://doi.org//journal.pone.0102108

Jampílek J, Kráľová K (2017) Nanomaterials for delivery of nutrients and growth-promoting compounds to plants. In Prasad R, Kumar M, Kumar V (eds) Nanotechnology. Springer Singapore, pp 177–226. https://doi.org/10.1007/978-981-10-4573-8_9

Javed Z, Dashora K, Mishra M, Fasake VD, Srivastva (2019) A effect of accumulation of nanoparticles in soil health-a concern on future. https://doi.org/10.15761/FNN.1000181

Jiang X, Huang K, Deng D, Xia H, Hou X, Zheng C (2012) Nanomaterials in analytical atomic spectrometry. TrAC Trends Anal Chem 39:38–59. https://doi.org/10.1016/j.trac.2012.06.002

Joo SH, Feitz AJ, Sedlak DL, Waite TD (2005) Quantification of the oxidizing capacity of nanoparticulate zero-valent iron. Environ Sci Technol 39:1263–1268. https://doi.org/10.1021/es048983d

Jorge JR, AM Prokofiev, E De Lima, GF Rauch, E Veron, M Botta WJ, Langdon TG (2013) An investigation of hydrogen storage in a magnesium-based alloy processed by equal-channel angular pressing. Int J Hydrog 38:8306–8312. https://doi.org/10.1016/j.ijhydene.03.158

Kanel SR, Flory J, Meyerhoefer A, Fraley JL, Sizemore IE, Goltz MN (2015) Influence of natural organic matter on fate and transport of silver nanoparticles in saturated porous media laboratory experiments and modeling. J Nanopart Res 17:154. https://doi.org//10.1007/s11051-015-2956-y

Karnib M, Kabbani A, Holail H, Olama Z (2014) Heavy metals removal using activated carbon, silica and silica activated carbon composite. Energy Procedia 50:113–120. https://doi.org//10.1016/j.egypro.2014.06.014

Keller AA, Vosti W, Wang H, Lazareva A (2014) Release of engineered nanomaterials from personal care products throughout their life cycle. J Nanopart Res 16:248. https://doi.org//10.1007/s11051-014-2489-9

Khan I, Saeed K, Khan I, (2019) Nanoparticles properties applications and toxicities. Arab J Chem 12:908–931. https://doi.org/10.1016/j.arabjc.2017.05.011

Khin, MM Nair, AS Babu, VJ Murugan R, Ramakrishna S (2012) A review on nanomaterials for environmental remediation. Energy Environ Sci 5:8075–8109. https://doi.org/10.1039/C2EE 21818F

Kool PL, Ortiz MD, van Gestel CA (2011) Chronic toxicity of ZnO nanoparticles, non-nano ZnO and ZnCl2 to *Folsomia candida* Collembola in relation to bioavailability in soil. Environ Pollut 159:2713–2719. https://doi.org/10.1016/j.envpol.2011.05.021

Kumar R, Rawat KS, Mishra AK (2012) Nanoparticles in the soil environment and their behaviour an overview. J Appl Sci 4:310–324. https://doi.org/10.31018/jans.v4i2.270

Kumar, S, Kashyap P L, Rai P, Kumar R, Sharma S, Jasrotia P, Srivastava AK (2018) Microbial nanotechnology for climate resilient agriculture In: Microbes for climate resilient agriculture, Wiley-Blackwell, 279

Li F, Liang X, Li H, Jin Y, Jin J, He M, Klumpp E, Bol R (2020) Enhanced soil aggregate stability limits colloidal phosphorus loss potentials in agricultural systems. Environ Sci Eur 32:17. https://doi.org/10.1186/s12302-020-0299-5

Li M, Zhu L, Lin D (2011) Toxicity of ZnO nanoparticles to *Escherichia coli*: mechanism and the influence of medium components. Environ Sci Technol 45:1977–1983. https://doi.org//10.1021/es102624t

Li Q, Mahendra S, Lyon DY, Brunet L, Liga MV, Li D, Alvarez PJ (2008) Antimicrobial nanomaterials for water disinfection and microbial control: potential applications and implications. Water Res 42:4591–4602. https://doi.org/10.1016/j.watres.2008.08.015

Li W, Zhu X, Chen H, He Y, Xu J (2013) Enhancement of extraction amount and dispersibility of soil nanoparticles by natural organic matter in soils. In: Functions of natural organic matter in changing environment. Springer Dordrecht, pp 769–772. https://doi.org/10.1007/978-94-007-5634-2_139

Li YH, Ding J, Di LZ, Zhu Z, Y, Xu C, Wei B, (2003) Competitive adsorption of Pb^{2+}, Cu^{2+} and Cd^{2+} ions from aqueous solutions by multiwalled carbon nanotubes. Carbon 41:2787–2792. https://doi.org//10.1016/S0008-622303)00392-0

Lian J, Zhao L, Wu J, Xiong H, Bao Y, Zeb A, Liu W (2020) Foliar spray of TiO_2 nanoparticles prevails over root application in reducing Cd accumulation and mitigating Cd-induced phytotoxicity in maize *Zea mays L*. Chemosphere 239:124794. https://doi.org//10.1016/j.chemosphere.2019.124794

Liang J, Xia X, Zhang W, Zaman WQ, Lin K, Hu S, Lin Z (2017) The biochemical and toxicological responses of earthworm (*Eisenia fetida*) following exposure to nanoscale zerovalent iron in a soil system. Environ Sci Pollut Res 24:2507–2514. https://doi.org/10.1007/s11356-016-8001-6

Lin D, Tian X, Wu F, Xing B (2010) Fate and transport of engineered nanomaterials in the environment. J Environ Qual 39:1896–1908. https://doi.org//10.2134/jeq2009.0423

Lin D, Xing B (2007) Phytotoxicity of nanoparticles inhibition of seed germination and root growth. Environ Pollut 150:243–250. https://doi.org/10.1016/j.envpol.2007.01.016

Liu Y, Wang J (2019) Reduction of nitrate by zero valent iron ZVI-based materials a review. Sci Total Environ 671:388–403. https://doi.org/10.1016/j.scitotenv.2019.03.317

Luo ZX, Wang ZH, Xu B Sarakiotis IL, Du Laing, G, Yan CZ (2014) Measurement and characterization of engineered titanium dioxide nanoparticles in the environment. J Zhejiang Univ Sci 15:593–605. https://doi.org/10.1631/jzus.A1400111

Ma C, He XH, Ma Q, Huang R (2018) Deposition of engineered nanoparticles ENPs on surfaces in aquatic systems a review of interaction forces experimental approaches, and influencing factors. Environ Sci Pollut Res 25:33056–33081. https://doi.org/10.1007/s11356-018-3225-2s

Ma H, Williams PL, Stephen (2013) A diamond ecotoxicity of manufactured ZnO nanoparticles a review. Environ Pollut 172:76-85. https://doi.org/10.1016/j.envpol.2012.08.011

Mahapatra O, Bhagat M, Gopalakrishnan C, Arunachalam KD (2008) Ultrafine dispersed CuO nanoparticles and their antibacterial activity. J Exp Nanosci 3:185–193. https://doi.org//10.1080/17458080802395460

Makselon J, Schäffer A, Klumpp E (2018) Transport and retention behavior of silver nanoparticles in soil. Doctoral dissertation RWTH Aachen University. https://doi.org/10.18154/RWTH-2018-226795

Manjunatha SB, Biradar DP, Aladakatti YR (2016) Nanotechnology and its applications in agriculture a review J Farm Sci 29:1–13

Navratilova J, Praetorius A, Gondikas A, Fabienke W, Von der Kammer F, Hofmann T (2015) Detection of engineered copper nanoparticles in soil using single particle ICP-MS. Int J Env Res Pub 12:15756–15768. https://doi.org/10.3390/ijerph121215020

Nie X, Zhu K, Zhao S, Dai Y, Tian H Sharma VK, Jia H (2020) Interaction of Ag^+ with soil organic matter: elucidating the formation of silver nanoparticles. Chemosphere 243:125413 https://doi.org/10.1016/j.chemosphere.2019.12543

Noori A, White JC, Newman LA (2017) Mycorrhizal fungi influence on silver uptake and membrane protein gene expression following silver nanoparticle exposure. J Nanopart Res 19:66. https://doi.org/10.1007/s11051-016-3650-4

Novo M, Lahive E, Díez-Ortiz, M Matzke, M Morgan, AJ Spurgeon, DJ Svendsen C, Kille P (2015) Different routes same pathway molecular mechanisms under silver ion and nanoparticle exposures in the soil sentinel *Eisenia fetida*. Environ Pollut 205: 385–393. 10.1016 /j.envpol.2015.07.010

Nowack B, Ranville JF, Diamond S, Gallego-Urrea JA, Rose MC, Horne J, Koelmans N, Klaine AA (2012) Potential scenarios for nanomaterial release and subsequent alteration in the environment. Environ Toxicol Chem 31:50–59. https://doi.org//10.1002/etc.726

Nurmi JT, Tratnyek PG, Sarathy V, Baer DR, Amonette JE, Pecher K, Wang C, Linehan JC, Matson DW, Penn RL, Driessen MD (2005) Characterization and properties of metallic iron nanoparticles: spectroscopy, electrochemistry, and kinetics. Environ Sci Technol 39:1221–1230. https://doi.org//10.1021/es049190u

Olson MR, Blotevogel J, Borch T, Petersen MA, Royer RA, Sale TC (2014) Long-term potential of in situ chemical reduction for treatment of polychlorinated biphenyls in soils. Chemosphere 114:144–149. https://doi.org/10.1016/j.chemosphere.2014.03.109

Omar FM, Aziz HA, Stoll S (2014) Aggregation and disaggregation of ZnO nanoparticles influence of pH and adsorption of Suwannee River humic acid. Sci Total Environ 468:195–201. https://doi.org/10.1016/j.scitotenv.2013.08.044

Pan B, Xing B (2012) Applications and implications of manufactured nanoparticles in soils a review. Eur J Soil Sci 63:437–456. https://doi.org/10.1111/j.1365-2389.2012.01475

Patra AK, Adhikari T, Bhardwaj AK (2016) Enhancing crop productivity in salt-affected environments by stimulating soil biological processes and remediation using nanotechnology. Innovative saline agriculture. Springer, pp 83–103. https://doi.org/10.1111/j.1365-2389.2012 .01475.x

Peijnenburg WJ, Baalousha M, Chen J, Chaudry Q, Von der kammer, F Kuhlbusch, TA Lead J Nickel C, Quik, JT, Renker M, Wang Z (2015) A review of the properties and processes determining the fate of engineered nanomaterials in the aquatic environment. Crit Rev Environ Sci Technol 45: 2084–2134. https://doi.org/10.1080/10643389.2015.1010430

Pérez-Hernández H, Fernández-Luqueño F, Huerta-Lwanga E, Mendoza-Vega J, Álvarez-Solís José D (2020) Effect of engineered nanoparticles on soil biota do they improve the soil quality and crop production or jeopardize them. Land Degrad Dev. https://doi.org/10.1002/ldr.3595

Polte J, Erler R, Thunemann AF, Sokolov S, Ahner TT, Rademann K, Emmerling FK (2010) Nucleation and growth of gold nanoparticles studied via in situ small angle X-ray scattering at millisecond time resolution. ACS Nano 4:1076–1082. https://doi.org//10.1021/nn901499c

Pornwilard MM, Siripinyanond A (2014) Field-flow fractionation with inductively coupled plasma mass spectrometry past present and future. J Anal Spectrom 29:1739–1752. https://doi.org//10.1039/C4JA00207E

Pradhan S, Mailapalli DR (2017) Interaction of engineered nanoparticles with the agri-environment. J Agric Food Chem 65:8279–8294. https://doi.org/10.1021/acs.jafc.7b02528

Prasad R, Bhattacharyya A, Nguyen QD (2017) Nanotechnology in sustainable agriculture: recent developments, challenges, and perspectives. Front Microbiol 8:1014. https://doi.org//10.3389/fmicb.2017.01014

Pullagurala VLR, Adisa IO, Rawat S, White JC, Zuverza-Mena N, Hernandez-Viezcas JA, Peralta-Videa JR Gardea-Torresdey JL (2019) Fate of engineered nanomaterials in agroenvironments and impacts on agroecosystems. In: Exposure to engineered nanomaterials in the environment. Elsevier, pp 105–142. https://doi.org/10.1016/B978-0-12-814835-8.00004-2

Qiu X, Fang Z, Liang B, Gu F, Xu Z (2011) Degradation of decabromodiphenyl ether by nano zero-valent iron immobilized in mesoporous silica microspheres. J Hazard Mater 193:70–81. https://doi.org//10.1016/j.jhazmat.2011.07.024

Rajput VD (2018) Metal oxide nanoparticles: applications and effects on soil ecosystems. Soil contamination sources assessment and remediation. Nova Science Publishers Hauppauge, pp 81–106. https://doi.org/10.1007/s10653-019-00317-3

Raliya R Ed (2019) Nanoscale engineering in agricultural management. CRC Press. https://doi.org/10.1038/s41598-018-24871-5

Reginatto C, Cecchin I Heineck KS, Thomé A, Reddy KR (2020) Influence of nanoscale zero-valent iron on hydraulic conductivity of a residual clayey soil and modeling of the filtration parameter. Environ Sci Pollut Res 1-9. https://doi.org/10.1007/s11356-019-07197-1

Rico CM, Lee SC, Rubenecia R, Mukherjee A, Hon J Peralta-Videa JR, Gardea-Torresdey JL (2014) Cerium oxide nanoparticles impact yield and modify nutritional parameters in wheat *Triticum aestivum L.* J Agric Food Chem 62: 9669–9675. https://doi.org/10.1021/jf503526r

Rodrigues SM, Trindade T Duarte, AC Pereira, E Koopmans GF, Römkens PFAM (2016) A framework to measure the availability of engineered nanoparticles in soils: Trends in soil tests and analytical tools. TrAC Trends Analyt Chem 75:129–140. https://doi.org/10.1016/j.trac.2015.07.003

Sabir S, Arshad M, Chaudhari SK (2014). Zinc oxide nanoparticles for revolutionizing agriculture synthesis and applications Sci World J . https://doi.org/10.1155/2014/925494

Sajid M, Ilyas M, Basheer C, Tariq M, Daud M, Baig N, Shehzad F (2015) Impact of nanoparticles on human and environment review of toxicity factors exposures control strategies and future prospects. Environ Sci Pollut Res 22:4122–4143. https://doi.org//10.1007/s11356-014-3994-1

Saliba AM, De Assis, MC Nishi, R Raymond B, Marques EDA, Lopes UG Touqui, L Plotkowski MC (2006) Implications of oxidative stress in the cytotoxicity of *Pseudomonas aeruginosa* ExoU. Microbes Infect 8:450–459. https://doi.org/10.1016/j.micinf.2005.07.011

Santiago-Martín Ad, Constantin B, Guesdon G, Kagambega N, Raymond S, Cloutier RG(2016) Bioavailability of engineered nanoparticles in soil systems. J Hazard Toxic Radioact Waste 20: B4015001. https://doi.org/10.1061/(ASCE)HZ.2153-5515.0000263

Saxena R, Kumar M, Tomar RS (2020) Implementation of nanotechnology in agriculture system: a current perspective. In: Tomar RS, Jyoti A, Kaushik S (eds) Nanobiotechnology concepts and applications in health agriculture and environment, 121

Schlich K, Hund-Rinke K (2015) Influence of soil properties on the effect of silver nanomaterials on microbial activity in five soils. Environ Pollut 196:321–330. https://doi.org//10.1016/j.envpol.2014.10.021

Schultz C, Powell K, Crossley A, Jurkschat K, Kille P, Morgan AJ, Read D, Tyne W, Lahive E, Svendsen C, Spurgeon DJ (2015) Analytical approaches to support current understanding of exposure, uptake and distributions of engineered nanoparticles by aquatic and terrestrial organisms. Ecotoxicology 242:239–261. https://doi.org//10.1007/s10646-014-1387-3

Shrivastava M, Srivastav A, Gandhi S Rao, S Roychoudhury A, Kumar A, Singhal RK Jha, SK Singh SD (2019) Monitoring of engineered nanoparticles in soil-plant system. a review. Environ Nanotechnol Monit Manag 11:100218. https://doi.org/10.1016/j.enmm.2019.100218

Siani NG, Fallah S, Pokhrel LR, Rostamnejadi A (2017) Natural amelioration of zinc oxide nanoparticle toxicity in fenugreek (*Trigonella foenum-gracum*) by arbuscular mycorrhizal (*Glomus intraradices*) secretion of glomalin. Plant Physiol Biochem 112:227–238. https://doi.org/j.plaphy.2017.01.001

Tan W (2017) Exposure of commercial titanium dioxide and copper hydroxide nanomaterials on basil *Ocimum basilicum* a life cycle and transgenerational study. https://digitalcommons.utep.edu/open_etd/564

Tang W, W Zeng, GM Gong, JL Liang, J Xu P, Zhang C, Huang BB (2014) Impact of humic/fulvic acid on the removal of heavy metals from aqueous solutions using nanomaterials a review. Sci Total Environ 468:1014–1027. https://doi.org/10.1016/j.scitotenv.2013.09.044

Tarafdar J, Adhikari T (2015) Nanotechnology in soil science. In: Soil science: an introduction, pp 775–807

Tian Y, Gao B, Silvera-Batista C, Ziegler KJ (2010) Transport of engineered nanoparticles in saturated porous media. J Nanopart Res 12:2371–2380. https://doi.org//10.1007/s11051-010-9912-7

Tiede K, Boxall AB, Tear SP, Lewis J, David H, Hassellöv M (2008) Detection and characterization of engineered nanoparticles in food and the environment. Food Addit Contam 25:795–821. https://doi.org//10.1080/02652030802007553

Topuz E, Talinli I, Aydin E (2011) Integration of environmental and human health risk assessment for industries using hazardous materials a quantitative multi criteria approach for environmental decision makers. Environ Int 372:393–403. https://doi.org/10.1016/j.envint.2010.10.013

Torkzaban S, Bradford SA, Walker SL (2007) Resolving the coupled effects of hydrodynamics and DLVO forces on colloid attachment in porous media. Langmuir 23:9652–9660. https://doi.org//10.1021/la700995e

Tripathi Singh DK, Singh S, Pandey S, Singh R, Sharma VP, Prasad NC, Dubey SM, Chauhan NK (2017) An overview on manufactured nanoparticles in plants uptake translocation accumulation and phytotoxicity. Plant Physiol Biochem 110:2–12. https://doi.org//10.1016/j.plaphy.2016.07.030

Tsai CH, Chang WC, Saikia D Wu CE, Kao HM (2016) Functionalization of cubic mesoporous silica SBA-16 with carboxylic acid via one-pot synthesis route for effective removal of cationic dyes. J Hazard Mater 309: 236–248. https://doi.org/10.1016/j.jhazmat.2015.08.051

Tungittiplakorn, W Cohen C, Lion LW (2005) Engineered polymeric nanoparticles for bioremediation of hydrophobic contaminants. Environ Sci Technol 39: 1354–1358. https://doi.org/10.1021/es049031as

Ullah H, Li X Peng, L Cai Y, Mielke HW (2020) In vivo phytotoxicity, uptake, and translocation of PbS nanoparticles in maize *Zea mays* L. plants. Sci Total Environ 139558. https://doi.org/10.1016/j.scitotenv.2020.139558

Van Der Ploeg, MJ Handy, RD Heckmann, LH Van Der Hout A, Van Den Brink NW (2013) C60 exposure induced tissue damage and gene expression alterations in the earthworm *Lumbricus rubellus*. Nanotoxicology 7: 432–440. https://doi.org/10.3109/17435390.2012.668569

Velicogna JR, Schwertfeger DM, Jesmer AH, Scroggins RP, Princz JI (2017) The bioaccumulation of silver in *Eisenia andrei* exposed to silver nanoparticles and silver nitrate in soil. NanoImpact 6:11–18. https://doi.org//10.1016/j.impact.2017.03.001

Verma SK, Das AK, Patel MK, Shah A, Kumar V, Gantait S (2018) Engineered nanomaterials for plant growth and development a perspective analysis. Sci Total Environ 630:1413–1435. https://doi.org//10.1016/j.scitotenv.2018.02.313

Walden C, Zhang W (2016) Biofilms versus activated sludge: considerations in metal and metal oxide nanoparticle removal from wastewater. Environ Sci Technol 50(16):8417–8431. https://doi.org//10.1021/acs.est.6b01282

Wang H, Dong Y-N, Zhu M, Li X, Keller AA, Wang T, Li F (2015) Heteroaggregation of engineered nanoparticles and kaolin clays in aqueous environments. Water Res 80:130–138. https://doi.org//10.1016/j.watres.2015.05.023

Wang R, Dang F, Liu C, Wang DJ, Cui PX, Yan HJ, Zhou DM (2019) Heteroaggregation and dissolution of silver nanoparticles by iron oxide colloids under environmentally relevant conditions. Environ Sci Nano 6:195–206. https://doi.org/10.1039/C8EN00543E

Wu H, Fang H, Xu C, Ye J, Cai Q, Shi J (2020) Transport and retention of copper oxide nanoparticles under unfavorable deposition conditions caused by repulsive van der Waals force in saturated porous media. Environ Pollut 256:113400. https://doi.org/10.1016/j.envpol.2019.113400

Xiaoli F, Qiyue C, Weihong G, Yaqing Z, Chen H, Junrong W, Longquan S (2020) Toxicology data of graphene-family nanomaterials. Arch Toxicol, An update. https://doi.org/10.1007/s00204-020-02717-2

Xie Y, Cheng W, Tsang PE, Fang Z (2016) Remediation and phytotoxicity of decabromodiphenyl ether contaminated soil by zero valent iron nanoparticles immobilized in mesoporous silica microspheres. J Environ Manage 166:478–483. https://doi.org/10.1007/s00204-020-02717-2

Xu Y, Zhao D (2005) Removal of copper from contaminated soil by use of poly (amidoamine) dendrimers. Environ Sci 39:2369–2375. https://doi.org/10.1021/es040380e

Yang J, Hou B, Wang J, Bi TB, J, Wang N Li X, Huang X, (2019) Nanomaterials for the removal of heavy metals from wastewater. Nanomaterials. 9:424. https://doi.org/10.3390/nano9030424

You T, Liu D Chen J, Yang Z, Dou R, Gao X, Wang L (2018) Effects of metal oxide nanoparticles on soil enzyme activities and bacterial communities in two different soil types. J Soils Sedim 18:211–221. https://doi.org/10.1007/s11368-017-1716-2

Zahra Z, Arshad M, Rafique R, Mahmood A, Habib A, Qazi IA, Khan SA (2015) Metallic nanoparticle TiO_2 and Fe_3O_4 application modifies rhizosphere phosphorus availability and uptake by *Lactuca sativa*. J Agric Food Chem 63:6876–6882. https://doi.org/10.1021/acs.jafc. 5b01611

Zhang C, Hu Z, Deng B (2016) Silver nanoparticles in aquatic environments: physiochemical behavior and antimicrobial mechanisms. Water Res 88:403–427. https://doi.org/10.1016/j.watres.2015.10.025

Zhang J, Guo W Li Q, Wang Liu S (2018) The effects and the potential mechanism of environmental transformation of metal nanoparticles on their toxicity in organisms. Environ Sci Nano 5:2482–2499. https://doi.org/10.1039/C8EN00688A

Zhang Q, Zhang B, Wang C (2014) Ecotoxicological effects on the earthworm *Eisenia fetida* following exposure to soil contaminated with Imidacloprid. Environ Sci Pollut Res 21:12345–12353. https://doi.org/10.1007/s11356-014-3178-z

Zhang S, Li X Yang, Y Li Y, Chen J, Ding F (2019) Adsorption, transformation, and colloid-facilitated transport of nano-zero-valent iron in soils. Environ Pollut Bioavail 31: 208–218. https://doi.org/10.1080/26395940.2019.1608865

Zhang W, Long J, Geng J, Li J, Wei Z (2020) Impact of titanium dioxide nanoparticles on Cd phytotoxicity and bioaccumulation in rice *Oryza sativa* L. Int J Env Res Pub Health 17:2979. https://doi.org/10.3390/ijerph17092979

Zhang W, Musante C, White JC, Schwab P, Wang Q, Ebbs SD, Ma X (2017) Bioavailability of cerium oxide nanoparticles to *Raphanus sativus L.* in two soils. Plant Physiol Biochem 110:185–193. https://doi.org/10.1016/j.plaphy.2015.12.013

Zhang Y, Tang ZR, Fu X, Xu YJ (2010) TiO_2– graphene nanocomposites for gas-phase photocatalytic degradation of volatile aromatic pollutant: is TiO_2– graphene truly different from other TiO_2–carbon composite materials. ACS Nano 4:7303–7314. https://doi.org//10.1021/nn1024219

Zhao L, Peralta-Videa, JR Ren, M Varela-Ramirez A Li C, Hernandez-Viezcas, JA Aguilera RJ, Gardea-Torresdey JL (2012) Transport of Zn in a sandy loam soil treated with ZnO NPs and uptake by corn plants Electron microprobe and confocal microscopy studies. Chem Eng J 184:1–8. https://doi.org/10.1016/j.cej.2012.01.041

Zou Y, Wang X, Khan A, Wang P, Liu Y, Alsaedi A, Hayat T, Wang X (2016) Environmental remediation and application of nanoscale zero-valent iron and its composites for the removal of heavy metal ions: a review. Environ Sci Technol 50:7290–7304. https://doi.org//10.1021/acs.est.6b01897

Effect of Engineered Nanoparticles on Soil Attributes and Potential in Reclamation of Degraded Lands

Vipin Kumar Singh, Rishikesh Singh, Ajay Kumar, and Rahul Bhadouria

Abstract

Rapid upsurge in the discipline of nanoscience and technology has led to emergence of myriads of nanoparticles. Apart from substantial application in medicine, textile, food science, and environmental technology, nanoparticles have received considerable application and immense opportunities in agricultural practices. Given the inherent potential, nanoparticle based on zinc, iron, manganese, copper, titanium, and mixtures thereof has been successfully employed in agricultural lands. Although negative consequences of nanoparticle application are well recognized, the judicious application of various nanoparticles in agriculture could improve the soil productivity in a better way in contrast to currently used strategies. Therefore, assessment of soil attributes may provide important insight on possible threats of nanoparticles in agro-ecosystem. The modulation in characteristics like pH, moisture content, soil organic matter, nutrient and mineral composition, microbial attributes, fauna and enzymatic activities to a great extent after the introduction of nanoparticle in agroecosystem is documented. Unprecedented rise in agricultural technologies and accelerated application of agrochemicals are the important phenomena responsible for massive degradation of agricultural lands worldwide causing decline in crop productivity. Nanotechnology could provide important platform for efficient restoration of degraded land areas. This chapter has reviewed on application of engineered nanoparticles in (a) improving agricultural productivity, (b) important techniques for nanoparticle quantification, (c) impact on soil characteristics, and (d) potential in management of degraded lands.

Keywords

Agriculture • Land degradation • Nutrient cycling • Restoration • Soil enzymes • Soil organic matter

1 Introduction

Globally very large areas of lands are described to be ecologically degraded putting great risks to goal of food security (Xie et al. 2020; Morales and Zuleta 2020). Land degradation is observed as a detrimental phenomenon influencing the productivity of soils. The important factors influencing land degradation include introduction of innovative agricultural technologies and intensive application of agrochemicals (Ouyang et al. 2018). Although research and development in farm machineries have improved crop productivity multifolds, negative consequences on soil productivity could not be denied (Shah et al. 2017). For instance, soil compaction, changes in soil properties and perturbation in soil organism, especially annelids and arthropods by modern plowing instruments may substantially deteriorate the soil health, eventually overall crop productivity (Beylich et al. 2010). Alteration in soil micropores, macropores, moisture content, and soil aggregate structure by agri-instruments could be important contributors for rapid decline in soil natural productivity. Excessive application of agrochemicals like fertilizers, herbicides,

V. K. Singh (✉)
Department of Botany, Centre of Advanced Study, Institute of Science, Banaras Hindu University, Varanasi, 221005, India
e-mail: vipinks85@gmail.com

R. Singh
Institute of Environment and Sustainable Development, Banaras Hindu University, Varanasi, Uttar Pradesh 221005, India
e-mail: rishikesh.iesd@gmail.com

A. Kumar
Agriculture Research Organization (ARO), Volcani Center, P.O. Box 15159, 7528809 Rishon LeZion, Israel
e-mail: ajaykumar_bhu@yahoo.com

R. Bhadouria
Department of Botany, University of Delhi, New Delhi, 110021, India
e-mail: rahulbhadouriya2@gmail.com

© Springer Nature Switzerland AG 2021
P. Singh et al. (eds.), *Plant-Microbes-Engineered Nano-particles (PM-ENPs) Nexus in Agro-Ecosystems*, Advances in Science, Technology & Innovation, https://doi.org/10.1007/978-3-030-66956-0_8

weedicides, pesticides, and insecticides is well recognized to interfere with the soil physical, chemical, and biological characteristics (Belay et al. 2002; Zhang et al. 2008; Afsar et al. 2017; Daam et al. 2020). The introduction of agrochemicals alone or in combination with organic amendments may considerably modify pH, moisture content, aggregate structure, porosity, bulk density, metal enrichment, water holding attributes, ion exchange characteristics of soil (Hati et al. 2006; Carbonell et al. 2011; Yargholi and Azarneshan 2014) and activity of organisms including microbes (Rahman et al. 2020), arthropods, annelids (Frampton et al. 2006), etc., leading to loss in productivity potential of agro-ecosystems (Förster et al. 2006).

Land degradation exerting degenerating impacts on natural environment (Wang et al. 2020) is widely reported across the globe influencing the crop productivity, therefore economic status of both developing and developed countries. Restoration of such ecologically disturbed soil could be helpful in meeting the exponentially rising demand of food. Restoration of degraded lands is chiefly based on physico-chemical and biological strategies with each method having advantages as well as disadvantages (Silva et al. 2015; Mohammed and Denboba 2020; Singh et al. 2020). The application of nanotechnology producing enormous quantity of nanomaterials possessing potential in management of degraded soil is quite attractive and promising. The nanoparticles comprising of both metals and non-metals could be exploited to facilitate the restoration of degraded land areas (Fajardo et al. 2019; Latif et al. 2020). Some of the worth mentioning nanoparticles having significance in the management of ecologically unhealthy soil, contaminated water, and wastewater include carbon, manganese, iron, and titanium (An and Zhao 2012; Ghasemi et al. 2017; Gong et al. 2018; Yang et al. 2020).

Metal-based nanoparticles after entry into agro-ecosystem may get access to different environmental components. Incorporation of metals released, apart from nanoparticles itself in food chain, ultimately threatens human health (Tombuloglu et al. 2020; Rajput et al., 2020b). Precise determination of nanoparticles, therefore, is necessary to assess the impact to natural ecosystem. Development of rapid assessment techniques would not only help mitigate the toxicity but also transfer and accumulate in other environmental matrices.

Nanoparticles of different metals have received considerable attention in agricultural practices with an objective to improve the functionality and thereby productivity of degraded lands. Land management practices deploying nanoparticles have the potential to help resurrect the productivity of ecologically disturbed soils. For instance, nanostructured formulations of nitrogen- and phosphorus-based fertilizers could help improve the crop productivity (Sekhon 2014) by substantially modifying the

soil properties to a greater extent. The introduction of engineered nanoparticles (ENPs) into agro-ecosystems may directly and indirectly modify the soil characteristics. Alterations in humic substances and bacterial community characteristics upon the application of metal oxide nanoparticles in soil (Ben-Moshe et al. 2013; Rajput et al. 2018) are presented. In addition, minor changes in soil macroscopic attributes had also been observed. Although most of the investigations have demonstrated the negative consequences of nanoparticles application to soil environment (Rajput et al. 2020c), the beneficial impacts on soil are also documented. The contribution of iron oxide nanoparticles in sequestration of environmentally hazardous metals includes arsenic, manganese, chromium, cadmium, and lead (Shipley et al. 2011), therefore reduction in toxicity leads to improvement in soil productivity and is of immense ecological significance. Therefore, nanoparticles are helpful in soil amelioration leading to creation of additional land (Liu and Lal 2012) for agricultural activities. Extensive investigations on ENPs exhibiting compatibility with soil components may be helpful in improving the productivity of degraded lands. Exploration of the mechanism of soil productivity improvement caused by certain ENPs may provide important boulevard for the management of less productive soils in order to feed the continuously rising human population. Fate and transport in soil environment as well as detailed understandings of ENPs uptake would facilitate in escaping the toxicity to soil microbes and invertebrates.

The present chapter offers recent information concerned with ENPs application in agro-ecosystems, quantification techniques, impacts on soil physical, chemical and biological characteristics, and potential opportunities in reclamation responsible for improved productivity of less fertile soil.

2 Engineered Nanoparticle Application in Agriculture

Because of unique physico-chemical characteristics, so far, myriads of nanoparticles have been used in agriculture in order to improve the crop productivity. Nanoparticles comprising of single metal as well as complexes of metals have been employed in agriculture to meet the rising demand of global food. Additionally, the wide applications of non-metal-based nanomaterials like carbon are also reported. A systematic review dealing with contribution of considerably less explored silicon nanoparticles in agriculture is presented by Rastogi et al. (2019). Study on role of nanoformulated zinc and silicon in enhancement of mango productivity by mitigation of salt stress as achieved by foliar spray is recently demonstrated by Elsheery et al. (2020). The concentrations of nanozinc and nanosilicon used either singly or in combination were in the range of 50–150 and 150–300 mg/L, respectively. The simultaneous combination treatment with 100 mg/L nanozinc and 150 mg/L nanosilicon was found to improve not only the resistance and fruit quality but also the productivity of mango trees under salt stressed conditions. The involvement of biologically synthesized zinc oxide nanoparticles in management of seed-borne plant pathogen is recently documented (Lakshmeesha et al. 2020). With increase in concentration, zinc oxide nanoparticle having size 30–40 nm with hexagonal structure led to growth suppression of fungal phytopathogen *Cladosporium cladosporioides* and *Fusarium oxysporum*. Treatment with nanoparticles caused alterations in level of fungal ergosterol, peroxidation of lipid molecules and modulations in membrane functionality, implying the utilization of nanoparticles as economical strategy in minimizing fungal pathogen-induced losses in crop productivity. Zinc oxide nanoparticles serving as antifungal agent against *Colletotrichum* species responsible for anthracnose disease in coffee are reported by Mosquera-Sánchez et al. (2020), suggesting implications in sustainable crop protection. At 15 mM concentration, the nanoparticle treatment was observed to significantly inhibit the fungal growth within six days. Apart from inhibition of phytopathogens, the nanoparticulate forms of fertilizers referred as nanofertilizers may be used in agriculture to enhance the productivity of important crops and the efficiency of fertilizers (Ramírez-Rodríguez et al. 2020; Yusefi-Tanha et al. 2020). Since the biological activity of ENPs is affected much by type, concentration, size (Yusefi-Tanha et al. 2020), metals and complexes, pathogen selected, and most importantly the characteristics of environmental matrices like soil and water, the selection of apposite nanoparticle is a pre-requisite for experiencing optimum beneficial effect. Furthermore, small-scale field investigations should also be conducted prior to large application in agro-ecosystems to avoid the environmental toxicity of metal and non-metal derived nanoparticles.

3 Techniques for Quantification of Nanoparticles

Extensive utilization of nanopesticides and nanofertilizers in agriculture has introduced unexpectedly large quantities of different nanoparticles in soil environment, posing undesirable effects (Carley et al. 2020). The concentrations of nanoparticles beyond certain limits are reported to exert toxicity to soil microbes and invertebrates. Surprising, to date, no regulatory limits have been set for different nanoparticles in water and soil environment. The precise identification, characterization, and determination using advanced instrumentation techniques, therefore, are

inevitable to mitigate the toxicity of nanoparticles to agricultural soils.

Quantification of engineered nanoparticles consisting of gold, silver, and cerium based on inductively coupled plasma mass spectrometry (ICP-MS) is reported by Gschwind et al. (2013) and results were comparable to other quantifying methods. The microdrop generator integrated with ICP-TOF-MS has been described for the determination of silver and gold nanoparticle mixture (Borovinskaya et al. 2014). Recently, simultaneous identification and quantification of titanium nanoparticles employing single particle ICP-MS equipped with TEM-EDS are presented by Wu et al. (2020). The developed method was able to determine the nanoparticle concentrations within the limits of 10^2 particles/ml.

Development of field-based techniques for rapid assessment of even minute concentrations of various nanoparticles from different soil components would facilitate the employment of appropriate strategies for evaluating the ecological risks (Wu et al. 2020) and maintenance of continuously deteriorating soil health. Further, the improvement in limit of detection (LOD) and limit of quantification (LOQ) could be helpful in measuring the traces of nanoparticles. In addition, the precise determination of nanoparticles is affected considerably by extraction methods, substances used for dispersion (Bland and Lowry 2020), types of soil, and instrumental sensitivity.

4 Impact of Nanoparticle Application on Soil Characteristics

Considerable rise in fabrication of varied metal and non-metal nanoparticles followed by application for multiple agricultural purposes has caused enhanced exposure and entry into soil environment (Ben-Moshe et al. 2013; Sun et al. 2020), consequently causing food chain contamination (Rajput et al. 2020a; b). The interaction of ENPs with soil is complex because of substantial variations in soil composition as well as prevailing environmental conditions. After introduction into terrestrial environment, nanoparticles may characteristically modulate the physical, chemical, and biological characteristics of soil (Samanta and Mandal 2017).

4.1 Soil pH

Soil pH is important parameter governing the growth and development of plant as well as soil microbial community structures and functions. The interaction of ENPs with soil may modulate the pH and varies significantly for different soil types (Conway and Keller 2016). The introduction of nanoparticles comprising of titanium, copper, and cerium in the range of 100 mg/kg was exhibited to raise the pH of loam soil, nevertheless, reduction in pH was observed for sandy loam soil. The modification in pH was not influenced by varying concentrations of nanoparticles. Displacement of soil associated ions by addition of nanoparticles was considered as an important factor responsible for pH modification. Modifications in soil pH are also attributed to the interaction of ENPs with plant roots. Alteration in secretory products of plant root possibly induced by nanoparticle addition, resulting into changes in soil pH is documented by Rossi et al. (2018). Nevertheless, direct evidences regarding modification in soil pH rendered by enhancement in root exudates under the influence of nanoparticles are not reported.

4.2 Cation Exchange Capacity

Cation exchange capacity is increasingly associated with potential to hold nutrients as well as environmental contaminants, hence acting like an important parameter pointing toward soil chemical attributes. The mobilization of nanoparticles in terrestrial environment is also modulated substantially by soil cation exchange capacity. The binding of engineered nanoparticles to minerals present in soil (Zhao et al. 2012) could greatly influence the inherent cation exchange capacity. However, to date, limited studies have been conducted pertaining to impact of nanoparticle addition to soil onto cation exchange characteristics. Controlled greenhouse condition-based experiment performed by De Souza et al. (2019) demonstrated rise in rhizospheric ion exchange capacity due to presence of ENPs consisting of iron oxides. In a similar manner, increase in ion exchange potential induced by silver nanoparticles biofabricated via the action of leaf extract is described recently by Das et al. (2019). Another investigation showing modulation in cation exchange capacity galvanized by interaction of cerium oxide nanoparticle with the soil mineral kaolinite leading to surface charge density variation has been indicated by Guo et al. (2019).

4.3 Nutrient and Mineral Characteristics

Nutrients and minerals present in the soil are important constituents governing the growth and development of various agricultural crops. Terrestrial incorporation of nanoparticles is considered to modify the availability of important mineral elements present in soil, thereby nutritional quality of cultivated crops. Intergenerational impact of cerium oxide nanoparticles treatment to wheat responsible for alterations in minerals and nutrients content of root and grain as evident through synchrotron X-ray fluorescence

spectroscopy, elemental analysis and X-ray absorption near edge spectroscopy is presented by Rico et al. (2017). Both generation treatments with ENPs had considerable effect on nutritional attributes as compared to only second generation treatment. Study on the impact of ENPs including cerium dioxide and titanium dioxide nanoparticle influencing the phytoavailability of beneficial nutrient elements nitrogen, phosphorus, and zinc as well as hazardous metal ion varying with soil characteristics is demonstrated recently by Duncan and Owens (2019). The observed effects on metal and nutrient phytoavailability were ascribed to competition and antimicrobial action of investigated nanoparticles.

Addition of titanium dioxide nanoparticle into biosolids generally applied for soil fertilization is documented to reduce the bioavailability of important elements including manganese, zinc, iron, and phosphorus by 65%, 20%, and 27%, respectively (Bellani et al. 2020), and to some extent was influenced by the amount and size of nanoparticles applied. The reduced availability may be attributed to interaction between minerals and highly reactive nanoparticle surface. In addition, the soil amendment with biosolids spiked with titanium dioxide modulated the nutrient composition of grown pea plants causing decline in level of manganese, zinc, potassium, and phosphorus in root and shoot. Therefore, to avoid the non-target impact of ENPs on soil ecosystem, extensive greenhouse condition should be conducted prior to recommendation for field application.

4.4 Soil Organic Matter

The heterogeneous soil organic matter resulting from living matter both by biological and non-biological processes greatly regulates multitude of ecological functions in terrestrial environment (Wiesmeier et al. 2019). The characteristics of pores present in organic matter considerably determine acquired air volume, reaction ability, water holding potential, and environmental fate of externally sorbed substances (de Jonge et al. 1996; Pignatello 1998). Being sink of numerous environmental contaminants, different nanoparticles of anthropogenic origin from different sources are expected to accumulate in soil ecosystem. The addition of metal-based nanoparticles consisting of copper oxide and iron oxide into soil with no obvious alterations in organic materials has been registered by Ben-Moshe et al. (2013). Nevertheless, modifications in content of humic substances as deciphered through fluorescence spectroscopy were recorded. Investigation on influence of platinum nanoparticles on features of soil organic matter (SOM) has been represented recently (Komendová et al. 2019). Nanoparticles with 3 nm size diminished the evaporation enthalpy of water molecules present in SOM and facilitated loss of water from soil. Further, the addition of nanoparticle enhanced the morphological firmness. The increased concentration of platinum nanoparticles reduced the amount of water in SOM and catalyzed the crystallization of aliphatic fractions.

4.5 Soil Microbial Characteristics

The incorporation of environmentally hazardous nanoparticles in agricultural soils may significantly hamper the normal ecological functioning of existing microbial communities (Navarro et al. 2008). Addition of nanoparticles into soil leading to variations in microbial characteristics in terms of bacterial community constitution based on denaturing gradient gel electrophoresis (DGGE) is described by Ben-Moshe et al. (2013). Diminished soil microbial performance and biomass upon challenged with multiwalled carbon nanotube (MWCNT) are narrated by different workers (Chung et al. 2011; Chen et al. 2018). Reduction in microbial biomass carbon and nitrogen, together with the upsurge in metabolic quotient by MWCNT, silver nanoparticles and titanium dioxide nanoparticle is reported by Xin et al. (2020). The observed effects were dose-dependent and much apparent at higher concentrations of MWCNT, nanosilver, and titanium dioxide. Modifications in metabolic profiles of bacterial communities surviving in three different soil types under the presence of ENPs consisting of silver and zinc oxide are recently demonstrated by Chavan and Nadanathangam (2020). However, titanium dioxide nanoparticle did not exert observable differences. The supplementation with silver and zinc oxide nanoparticle also led to substantial changes in selected diversity indices.

4.6 Soil Enzymes

Soil enzymes are important contributor of agro-ecosystem regulating cycling of different nutrients and the introduction of nanoparticles may likely hinder the natural phenomena (Shin et al. 2012). Inhibitory action of silver nanoformulation on enzymatic activities of calcareous soil is well documented (Rahmatpour et al. 2017). The silver nanoparticles exerted greater inhibition over enzymatic activities in comparison with bulk silver ions. Experimental investigation indicating restrictions in soil enzymatic activities including dehydrogenase, urease, and phosphatase by the action of MWCNT, nanosilver, and nanotitanium oxide is registered currently by Xin et al. (2020). The increased deployment of nanoformulated pesticides is considered to hinder the natural soil biogeochemical cycling of beneficial elements. The inhibitory effects of copper oxide-based nanoparticles at the concentrations 10, 100, and 500 mg/kg during 60 h

exposure affecting the enzymes involved in denitrification and electron transfer phenomenon are demonstrated (Zhao et al. 2020). The introduction of nanoparticles reduced 10–42% activities of nitrate reductase, nitric oxide reductase, and retarded denitrification process resulting into diminished emission of N_2O. The observed impacts were attributed to the inhibitory action of copper ions released from nanoparticles. The observed negative effects of nanoparticle addition on soil enzymes are considered to be the resultant of: (a) interaction of released metals with sulfhydryl group of active site of enzyme (Liau et al. 1997), and (b) direct interaction of nanoparticles with soil enzymes (Wigginton et al. 2010).

4.7 Soil Annelids and Arthropods

Soil invertebrates like annelids and arthropods are important fauna affecting the characteristics of different agricultural soils. Numbers of studies have indicated the impact of ENPs on normal cellular functioning, reproductive processes, and behavioral responses in given environmental conditions (Shoults-Wilson et al. 2011; Schlich et al. 2013; Kwak et al. 2014). The toxicity of silver nanoparticles to model soil annelid *Eisenia fetida* through similar pathways causing perturbation in ribosomal activity, metabolic processes associated with sugar and protein, and interferences in energy generation mechanisms has been established by Novo et al. (2015). Impact assessment of powdered zinc oxide nanoparticles to annelid *Enchytraeus crypticus* in gel-based media suggesting toxicity to soil organism is illustrated (Hrdá et al. 2016). The toxicity in terms of mortality was affected by size of agglomerated nanoparticle and method of media preparation for treatment. The annelid mortality upon exposure differed in the range of 28.9–34.4% and 0–66.6% for the two different treatment methods using the nanoparticle concentrations 50, 100, 200, 500, and 1000 mg/kg. Recently, silver nanoparticle-induced toxicological effects in terms of reproduction and mortality on soil arthropod *Folsomia candida* after four week exposure is represented by Hlavkova et al. (2020). Silver nanoparticles had higher EC_{50} value as compared to bulk silver ions implying lesser toxicity to tested invertebrate. Further, silver nanoparticles in the concentration range 166–300 mg kg^{-1} dry weight did not exert toxicity.

Apart from type and amount of nanoparticles, the observed effect in agro-environment is influenced by properties of given soil (Xin et al. 2020). The impact of nanoparticle on soil characteristics is essential to investigate the ecotoxicity in order to safeguard the agricultural ecosystem. The extensive investigation would help regulate the quantity of nanoparticle to be used for agricultural purposes. In addition, the predecided dosage of different nanoparticles, therefore, would help to improve the productivity of soil in an economical manner.

5 Application of Nanoparticles in Reclamation of Degraded Agricultural Lands

Myriads of nanoparticles with diverse application in environmental decontamination, medicine, and enhancement in agricultural productivity based on different physico-chemical methods are synthesized to date. Varying degree of influences of ENPs on crops, soil properties, and ecological functioning is reported by various authors. Apart from considerable toxicological impacts on agro-ecosystem, engineered nanomaterials could be exploited successfully for the reclamation of ecologically degraded lands. Successful reclamation would provide additional cultivable land areas to meet the rising demand of food crops. A schematic representation of ENPs impacts on different soil properties and their potential for degraded soil reclamation has been depicted in Fig. 1.

Reclamation involves sequestration of hazardous contaminants and improvement in soil characteristics leading to enhanced soil productivity. A detailed account pertaining to contribution of engineered nanomaterials in sustainable management of mine areas and other ecologically disturbed soil is elaborated by Liu and Lal (2012). The review explained the application of zeolites and nanoparticles of iron oxide, phosphorus, iron sulfide, zero-valent iron, and carbon nanotubes for efficient decontamination of land areas affected by mining activities. The combined action of synthesized nanomaterials and conventional treatment methods was also suggested to help minimize the cost required for improving the characteristics of degraded land areas.

The porous zeolites may serve as important materials for the remediation of contaminated lands (Li et al. 2018) and are described to be available in the soil, but the content typically present is very low. The most dominating zeolite existing in soil is clinoptilolite. Zeolite-based nanomaterials hold promising potential in improving the characteristics of soil due to rise in water retention potential, enhancement in clay shift proportions, augmentation of nutritional features, and efficient sequestration of toxic substances (Ming and Allen 2001). In addition, both natural and synthesized zeolites are able to potentially adsorb the noxious heavy metals occurring in contaminated soils, thereby minimizing the threats to human health and environment. Treatment of mine soil with synthetic zeolites at the rate of 0.5–5% weight basis, culminating into substantial decline of labile and readily accessible heavy metals like zinc, lead, copper, and cadmium by 42–72% is illustrated by Edwards et al. (1999). Apart from surface binding, increase in soil pH rendered by zeolite introduction into soil was also ascribed to elimination of heavy metals. Similar investigations pointing toward the decreased availability of heavy metals after soil application of zeolites at 0.5 to 16 weight % are also documented (Shanableh and Kharabsheh 1996; Lin et al. 1998; Moirou et al. 2001). In addition to extraction of heavy metals from contaminated soil, zeolites have the tendency to efficiently adsorb the radionuclides like cesium and strontium, hence potential to reduce the availability in cultivated plants (Ming and Allen 2001). Githinji et al. (2011) have presented the considerable contribution of zeolites, having size 0.55–0.60 mm, in reducing the soil bulk density and twofold enhancements in water availability. Role of zeolites in remediation of vanadium contaminated soil facilitated by stabilization process is recently demonstrated by Yang et al. (2020). The study concluded modulation in soil pH as an important factor controlling the stabilization of vanadium. The application of zeolites in a given agro-ecosystem should

Fig. 1 Effect of engineered nanoparticles (NPs) on soil attributes and potential in land reclamation/restoration

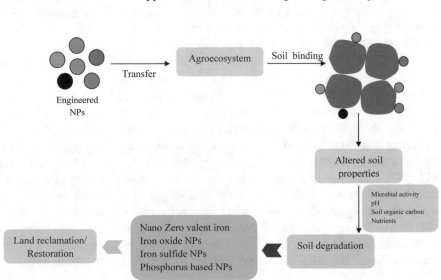

be optimized as the particle size and amount used greatly modulates the soil physical attributes.

The naturally existing soil iron oxide nanoparticles having average size ranging from 5 to 100 nm, possess reactive sites with inherent ability to adsorb varieties of organic and inorganic contaminants through the process like surface binding (Bigham et al. 2002). The efficiency in rapid adsorption, minimal chances of secondary contamination, and ecofriendly nature has fostered the engineered iron oxide nanoparticles with multiple applications including remediation of contaminated water as well as soil. Some of the widely applied iron oxide nanomaterials described for extraction of heavy metals such as copper, chromium, nickel, lead, arsenic, and zinc are goethite, magnetite, hematite, and maghemite. Column-based investigation indicating arsenic immobilization for more than four months in soil amended with 15% nanomagnetite and 100 $\mu g \ l^{-1}$ arsenic spiked at a rate of 0.3 ml h^{-1}, in contrast to soil without amendment, has been presented by Shipley et al. (2011). However, after the elapse of 208 days, a total of 20% arsenic was noticed to be leached from column. The study further suggested the simultaneous removal of 12 other metals.

The strong reductant nanozero-valent iron (nZVI) had been synthesized with an objective to degrade the organic contaminants including pesticides and petrochemical products (Zhang 2003). Intriguingly, nZVI may also serve as an important material for sequestration of various heavy metals from terrestrial system. Because of reducing action of nZVI, metal ions with higher oxidation states like chromium and uranium are transformed to corresponding lower oxidation states and reduce the toxicity, as well as solubility and mobilization in soil environment by the process referred as reductive immobilization. Numerous studies have shown the efficiency of nZVI in immobilization of uranium in comparison with reductants including iron fillings, lead sulfide, and iron sulfide, because of large size conferred by small size, increased reactivity, and release of reactive iron produced. Reduction of approximately 98% hexavalent chromium to trivalent form assisted by the catalytic activity of nZVI, leading to reduction in toxicity to soil is demonstrated by Franco et al. (2009). Similar observation on reductive immobilization of higher oxidation state chromium (hexavalent) in soil, with the resultant decline in ecotoxicity to soil, is also documented (Ponder et al. 2000; Xu and Zhao 2007). The application of engineered graphene oxide nanoparticles as a promising tool in management of heavy metal contaminated soil responsible for immobilization of copper, lead, and cadmium, in contrast to mobilization of arsenic and phosphorus has been reported by Baragaño et al. (2020). In addition, phosphate and iron sulfide-based nanomaterials (Liu et al. 2020; Rodríguez-Seijo et al. 2020) and carbon nanotubes (Liu et al. 2018; Egbosiuba et al. 2020) are also remarkably annotated for possessing promising potential in removal of heavy metals and organic contaminants.

The employment of ENPs, although, for reclamation of contaminated agro-ecosystem is quite attractive, the fate and toxicity to environmental components must be extensively investigated for safe application. The behavior of ENPs incorporated into soil environment is significantly modified by soil attributes, prevailing environmental conditions as well as its own size and morphology. The optimization of dose for different soil types, and different organic and inorganic contaminants are crucial steps toward application of ENPs in agro-ecosystems.

6 Conclusion and Future Perspectives

Engineered nanomaterials are continuously gaining importance in varied disciplines like medicine, electronics, environment, and agriculture. The increased applications in agriculture as nanofertilizers and nanopesticides have improved the productivity of agro-ecosystem multifolds. However, the nanoparticle incorporation in food crops, toxicity to human health, and negative consequences on soil properties including enzymes, microbial diversity, soil nutrient cycling, and ecotoxicity to soil dwelling annelids and arthropods have questioned their application for enhancing the crop productivity. The employment of ENPs, therefore, must be based on extensive ecotoxicity appraisal to beneficial non-target organisms as well as humans exposed via agronomic crops. In addition to augmentation of soil productivity, ENPs could be applied for restoration of ecologically disturbed sites like mining affected cultivable sites. The soil reclamation using zeolites and iron oxide nanoparticles, however, is in infant stage, implying further research work in this direction.

Since the dose of applied ENPs varies according to the nature of nanoparticles and soil characteristics, deciding optimum dose so as to minimize the residues left over in agro-environment is a crucial step and need much experimental work. The techniques for identification and quantification of ENPs should be improved in order to minimize the impact on natural environment and associated health hazards. Investigation on sources of nanoparticles, fate, and transport in soil environment is another area of research for protection of soil health. Further, there is urgent need to set the regulatory limits for different nanoparticles currently being applied in agro-ecosystem to prevent excessive accumulation in soil as well as crop products.

References

Afsar MA, Khalil SK, Wahab S, Khalil IH, Khan AZ, Khattak MK (2017) Impact of various ratios of nitrogen and sulfur on maize and soil pH in semiarid region. Commun Soil Sci Plant Anal 48(8):825–834

An B, Zhao D (2012) Immobilization of As (III) in soil and groundwater using a new class of polysaccharide stabilized Fe–Mn oxide nanoparticles. J Hazard Mat 211:332–341

Baragaño D, Forján R, Welte L, Gallego JLR (2020) Nanoremediation of As and metals polluted soils by means of graphene oxide nanoparticles. Sci Rep 10(1):1–10

Belay A, Claassens A, Wehner FC (2002) Effect of direct nitrogen and potassium and residual phosphorus fertilizers on soil chemical properties, microbial components and maize yield under long-term crop rotation. Biol Fert Soils 35(6):420–427

Bellani L, Siracusa G, Giorgetti L, Di Gregorio S, Castiglione MR, Spanò C, Muccifora S, Bottega S, Pini R, Tassi E (2020) TiO_2 nanoparticles in a biosolid-amended soil and their implication in soil nutrients, microorganisms and *Pisum sativum* nutrition. Ecotox Environ Safe 190:110095

Ben-Moshe T, Frenk S, Dror I, Minz D, Berkowitz B (2013) Effects of metal oxide nanoparticles on soil properties. Chemosphere 90 (2):640–646

Beylich A, Oberholzer HR, Schrader S, Höper H, Wilke BM (2010) Evaluation of soil compaction effects on soil biota and soil biological processes in soils. Soil Till Res 109(2):133–143

Bigham JM, Fitzpatrick RW, Schulze DG (2002) Iron oxides. Soil Mineral Environ Appl 7:323–366

Bland GD, Lowry GV (2020) Multi-step method to extract moderately soluble copper oxide nanoparticles from soil for quantification and characterization. Anal Chem 92(14):9620–9628

Borovinskaya O, Gschwind S, Hattendorf B, Tanner M, Günther D (2014) Simultaneous mass quantification of nanoparticles of different composition in a mixture by microdroplet generator-ICPTOFMS. Anal Chem 86(16):8142–8148

Carbonell G, de Imperial RM, Torrijos M, Delgado M, Rodriguez JA (2011) Effects of municipal solid waste compost and mineral fertilizer amendments on soil properties and heavy metals distribution in maize plants (*Zea mays* L.). Chemosphere 85(10):1614–1623

Carley L, Panchagavi R, Song X, Davenport S, Bergemann CM, McCumber AW, Gunsch CK, Simonin M (2020) Long-term effects of copper nanopesticides on soil and sediment community diversity in two outdoor mesocosm experiments. Environ Sci Technol 54 (14):8878–8889

Chavan S, Nadanathangam V (2020) Shifts in metabolic patterns of soil bacterial communities on exposure to metal engineered nanomaterials. Ecotox Environ Safe 189:110012

Chen M, Zhou S, Zhu Y, Sun Y, Zeng G, Yang C, Xu P, Yan M, Liu Z, Zhang W (2018) Toxicity of carbon nanomaterials to plants, animals and microbes: recent progress from 2015-present. Chemosphere 206:255–264

Chung H, Son Y, Yoon TK, Kim S, Kim W (2011) The effect of multi-walled carbon nanotubes on soil microbial activity. Ecotox Environ Safe 74(4):569–575

Conway JR, Keller AA (2016) Gravity-driven transport of three engineered nanomaterials in unsaturated soils and their effects on soil pH and nutrient release. Water Res 98:250–260

Daam MA, Garcia MV, Scheffczyk A, Römbke J (2020) Acute and chronic toxicity of the fungicide carbendazim to the earthworm *Eisenia fetida* under tropical versus temperate laboratory conditions. Chemosphere 255:126871

Das G, Patra JK, Debnath T, Ansari A, Shin HS (2019) Investigation of antioxidant, antibacterial, antidiabetic, and cytotoxicity potential of silver nanoparticles synthesized using the outer peel extract of *Ananas comosus* (L.). PloS one 14(8):e0220950

de Jonge H, Mittelmeijer-Hazeleger MC (1996) Adsorption of CO_2 and N_2 on soil organic matter: nature of porosity, surface area, and diffusion mechanisms. Environ Sci Technol 30(2):408–413

De Souza A, Govea-Alcaide E, Masunaga SH, Fajardo-Rosabal L, Effenberger F, Rossi LM, Jardim RF (2019) Impact of Fe_3O_4 nanoparticle on nutrient accumulation in common bean plants grown in soil. SN Appl Sci 1(4):308

Duncan E, Owens G (2019) Metal oxide nanomaterials used to remediate heavy metal contaminated soils have strong effects on nutrient and trace element phytoavailability. Sci Tot Environ 678:430–437

Edwards R, Rebedea I, Lepp NW, Lovell AJ (1999) An investigation into the mechanism by which synthetic zeolites reduce labile metal concentrations in soils. Environ Geochem Health 21(2):157–173

Egbosiuba TC, Abdulkareem AS, Kovo AS, Afolabi EA, Tijani JO, Roos WD (2020) Enhanced adsorption of As (V) and Mn (VII) from industrial wastewater using multi-walled carbon nanotubes and carboxylated multi-walled carbon nanotubes. Chemosphere 254:126780

Elsheery NI, Helaly MN, El-Hoseiny HM, Alam-Eldein SM (2020) Zinc oxide and silicone nanoparticles to improve the resistance mechanism and annual productivity of salt-stressed mango trees. Agronomy 10(4):558

Fajardo C, Costa G, Nande M, Martín C, Martín M, Sánchez-Fortún S (2019) Heavy metals immobilization capability of two iron-based nanoparticles (nZVI and Fe3O4): soil and freshwater bioassays to assess ecotoxicological impact. Sci Tot Environ 656:421–432

Förster B, Garcia M, Francimari O, Römbke J (2006) Effects of carbendazim and lambda-cyhalothrin on soil invertebrates and leaf litter decomposition in semi-field and field tests under tropical conditions (Amazonia, Brazil). Eur J Soil Biol 42:S171–S179

Frampton GK, Jänsch S, Römbke S-F, J, Van den Brink PJ, (2006) Effects of pesticides on soil invertebrates in laboratory studies: a review and analysis using species sensitivity distributions. Environ Toxicol Chem 25(9):2480–2489

Franco DV, Da Silva LM, Jardim WF (2009) Reduction of hexavalent chromium in soil and ground water using zero-valent iron under batch and semi-batch conditions. Water Air Soil Pollut 197(1–4):49–60

Ghasemi E, Heydari A, Sillanpää M (2017) Superparamagnetic Fe3O4@ EDTA nanoparticles as an efficient adsorbent for simultaneous removal of Ag (I), Hg (II), Mn (II), Zn (II), Pb (II) and Cd (II) from water and soil environmental samples. Microchem J 131:51–56

Githinji LJ, Dane JH, Walker RH (2011) Physical and hydraulic properties of inorganic amendments and modeling their effects on water movement in sand-based root zones. Irrig Sci 29(1):65–77

Gong X, Huang D, Liu Y, Peng Z, Zeng G, Xu P, Cheng M, Wang R, Wan J (2018) Remediation of contaminated soils by biotechnology with nanomaterials: bio-behavior, applications, and perspectives. Crit Rev Biotechnol 38(3):455–468

Gschwind S, Hagendorfer H, Frick DA, Günther D (2013) Mass quantification of nanoparticles by single droplet calibration using inductively coupled plasma mass spectrometry. Anal Chem 85 (12):5875–5883

Guo B, Jiang J, Serem W, Sharma VK, Ma X (2019) Attachment of cerium oxide nanoparticles of different surface charges to kaolinite: molecular and atomic mechanisms. Environ Res 177:108645

Hati KM, Mandal KG, Misra AK, Ghosh PK, Bandyopadhyay KK (2006) Effect of inorganic fertilizer and farmyard manure on soil physical properties, root distribution, and water-use efficiency of soybean in Vertisols of central India. Biores Technol 97(16):2182–2188

Hlavkova D, Beklova M, Kopel P, Havelkova B (2020) Effects of silver nanoparticles and ions exposure on the soil invertebrates *Folsomia candida* and *Enchytraeus crypticus*. Bull Environ Cont Toxicol 105:244–249

Hrdá K, Opršal J, Knotek P, Pouzar M, Vlček M (2016) Toxicity of zinc oxide nanoparticles to the annelid *Enchytraeus crypticus* in agar-based exposure media. Chem Pap 70(11):1512–1520

Komendová R, Žídek J, Berka M, Jemelková M, Řezáčová V, Conte P, Kučerík J (2019) Small-sized platinum nanoparticles in soil organic matter: influence on water holding capacity, evaporation and structural rigidity. Sci Tot Environ 694:133822

Kwak JI, Lee WM, Kim SW, An YJ (2014) Interaction of citrate-coated silver nanoparticles with earthworm coelomic fluid and related cytotoxicity in *Eisenia andrei*. J Appl Toxicol 34(11):1145–1154

Lakshmeesha TR, Murali M, Ansari MA, Udayashankar AC, Alzohairy MA, Almatroudi A, Alomary MN, Asiri SMM, Ashwini BS, Kalagatur NK, Nayak CS (2020) Biofabrication of zinc oxide nanoparticles from *Melia azedarach* and its potential in controlling soybean seed-borne phytopathogenic fungi. Saudi J Biol Sci 27 (8):1923

Latif A, Sheng D, Sun K, Si Y, Azeem M, Abbas A, Bilal M (2020) Remediation of heavy metals polluted environment using Fe-based nanoparticles: mechanisms, influencing factors, and environmental implications. Environ Pollut 264:114728

Li Z, Wang L, Meng J, Liu X, Xu J, Wang F, Brookes P (2018) Zeolite-supported nanoscale zero-valent iron: New findings on simultaneous adsorption of Cd (II), Pb (II), and As (III) in aqueous solution and soil. J Hazard Mat 344:1–11

Liau SY, Read DC, Pugh WJ, Furr JR, Russell AD (1997) Interaction of silver nitrate with readily identifiable groups: relationship to the antibacterialaction of silver ions. Lett Appl Microbiol 25(4):279–283

Lin CF, Lo SS, Lin HY, Lee Y (1998) Stabilization of cadmium contaminated soils using synthesized zeolite. J Hazard Mat 60 (3):217–226

Liu R, Lal R (2012) Nanoenhanced materials for reclamation of mine lands and other degraded soils: a review. J Nanotechnol. https://doi. org/10.1155/2012/461468

Liu Y, Huang Y, Zhang C, Li W, Chen C, Zhang Z, Chen H, Wang J, Li Y, Zhang Y (2020) Nano-FeS incorporated into stable lignin hydrogel: a novel strategy for cadmium removal from soil. Environ Pollut 264:114739

Ming DW, Allen ER (2001) Use of natural zeolites in agronomy, horticulture and environmental soil remediation. Rev Mineral Geochem 45(1):619–654

Mohammed SA, Denboba MA (2020) Study of soil seed banks in ex-closures for restoration of degraded lands in the central Rift Valley of Ethiopia. Scientific Reports 10(1):1–9

Moirou A, Xenidis A, Paspaliaris I (2001) Stabilization Pb, Zn, and Cd-contaminated soil by means of natural zeolite. Soil Sediment Cont 10(3):251–267

Morales NS, Zuleta GA (2020) Comparison of different land degradation indicators: do the world regions really matter? Land Degrad Dev 31(6):721–733

Mosquera-Sánchez LP, Arciniegas-Grijalba PA, Patiño-Portela MC, Guerra–Arias BE, Muñoz-Florez JE, Rodríguez-Páez JE (2020) Antifungal effect of zinc oxide nanoparticles (ZnO-NPs) on *Colletotrichum* sp., causal agent of anthracnose in coffee crops. Biocat Agr Biotechnol 25:101579

Navarro E, Baun A, Behra R, Hartmann NB, Filser J, Miao AJ, Quigg A, Santschi PH, Sigg L (2008) Environmental behavior and ecotoxicity of engineered nanoparticles to algae, plants, and fungi. Ecotoxicol 17(5):372–386

Novo M, Lahive E, Díez-Ortiz M, Matzke M, Morgan AJ, Spurgeon DJ, Svendsen C, Kille P (2015) Different routes, same pathways:

molecular mechanisms under silver ion and nanoparticle exposures in the soil sentinel *Eisenia fetida*. Environ Pollut 205:385–393

Ouyang W, Lian Z, Hao X, Gu X, Hao F, Lin C, Zhou F (2018) Increased ammonia emissions from synthetic fertilizers and land degradation associated with reduction in arable land area in China. Land Degrad Dev 29(11):3928–3939

Pignatello JJ (1998) Soil organic matter as a nanoporous sorbent of organic pollutants. Adv Colloid Interface Science 76:445–467

Ponder SM, Darab JG, Mallouk TE (2000) Remediation of Cr (VI) and Ph (II) aqueous solutions using supported, nanoscale zero valent iron. Environ Sci Technol 34(12):2564–2569

Rahman MM, Nahar K, Ali MM, Sultana N, Karim MM, Adhikari UK, Rauf M, Azad MAK (2020) Effect of long-term pesticides and chemical fertilizers application on the microbial community specifically anammox and denitrifying bacteria in rice field soil of Jhenaidah and Kushtia District. Bangladesh. Bull Environ Cont Toxicol 104(6):828–833

Rahmatpour S, Shirvani M, Mosaddeghi MR, Nourbakhsh F, Bazarganipour M (2017) Dose–response effects of silver nanoparticles and silver nitrate on microbial and enzyme activities in calcareous soils. Geoderma 285:313–322

Rajput V, Minkina T, Ahmed B, Sushkova S, Singh R, Soldatov M, Laratte B, Fedorenko A, Mandzhieva S, Blicharska E, Musarrat J (2020) Interaction of copper-based nanoparticles to soil, terrestrial, and aquatic systems: critical review of the state of the science and future perspectives. Rev Environ Cont Toxicol 252:51–96

Rajput V, Minkina T, Mazarji M, Shende S, Sushkova S, Mandzhieva S, Burachevskaya M, Chaplygin V, Singh A, Jatav H (2020) Accumulation of nanoparticles in the soil-plant systems and their effects on human health. Ann Agr Sci 65(2):137–143

Rajput V, Minkina T, Sushkova S, Behal A, Maksimov A, Blicharska E, Ghazaryan K, Movsesyan H, Barsova N (2020) ZnO and CuO nanoparticles: a threat to soil organisms, plants, and human health. Environ Geochem Health 42(1):147–158

Rajput VD, Minkina T, Sushkova S, Tsitsuashvili V, Mandzhieva S, Gorovtsov A, Nevidomskyaya D, Gromakova N (2018) Effect of nanoparticles on crops and soil microbial communities. J Soil Sediments 18(6):2179–2187

Ramírez-Rodríguez GB, Dal Sasso G, Carmona FJ, Miguel-Rojas C, Pérez-de-Luque A, Masciocchi N, Guagliardi A, Delgado-López JM (2020) Engineering biomimetic calcium phosphate nanoparticles: a green synthesis of slow-release multinutrient (NPK) nanofertilizers. ACS Appl Bio Materials 3(3):1344–1353

Rastogi A, Tripathi DK, Yadav S, Chauhan DK, Živčák M, Ghorbanpour M, El-Sheery NI, Brestic M (2019) Application of silicon nanoparticles in agriculture. 3 Biotech 9(3):90

Rico CM, Johnson MG, Marcus MA, Andersen CP (2017) Intergenerational responses of wheat (*Triticum aestivum* L.) to cerium oxide nanoparticles exposure. Environ Sci: Nano 4(3):700–711

Rodríguez-Seijo A, Vega FA, Arenas-Lago D (2020) Assessment of iron-based and calcium-phosphate nanomaterials for immobilisation of potentially toxic elements in soils from a shooting range berm. J Environ Manage 267:110640

Rossi L, Sharifan H, Zhang W, Schwab AP, Ma X (2018) Mutual effects and in planta accumulation of co-existing cerium oxide nanoparticles and cadmium in hydroponically grown soybean (*Glycine max* (L.) Merr.). Environ Sci: Nano 5(1):150–157

Samanta PK, Mandal AK (2017) Effect of nanoparticles on biodiversity of soil and water microorganism community. J Tissue Sci Eng 8:196

Schlich K, Klawonn T, Terytze K, Hund-Rinke K (2013) Effects of silver nanoparticles and silver nitrate in the earthworm reproduction test. Environ Toxicol Chem 32(1):181–188

Sekhon BS (2014) Nanotechnology in agri-food production: an overview. Nanotechnol Sci Appl 7:31. https://doi.org/10.2147/NSA.S39406

Shah AN, Tanveer M, Shahzad B, Yang G, Fahad S, Ali S, Bukhari MA, Tung SA, Hafeez A, Souliyanonh B (2017) Soil compaction effects on soil health and cropproductivity: an overview. Environ Sci Pollut Res 24(11):10056–10067

Shanableh A, Kharabsheh A (1996) Stabilization of Cd, Ni and Pb in soil using natural zeolite. J Haz Mat 45(2–3):207–217

Shin YJ, Kwak JI, An YJ (2012) Evidence for the inhibitory effects of silver nanoparticles on the activities of soil exoenzymes. Chemosphere 88(4):524–529

Shipley HJ, Engates KE, Guettner AM (2011) Study of iron oxide nanoparticles in soil for remediation of arsenic. J Nanoparticle Res 13(6):2387–2397

Shoults-Wilson WA, Zhurbich OI, McNear DH, Tsyusko OV, Bertsch PM, Unrine JM (2011) Evidence for avoidance of Ag nanoparticles by earthworms (Eisenia fetida). Ecotoxicol 20 (2):385–396

Silva LC, Doane TA, Corrêa RS, Valverde V, Pereira EI, Horwath WR (2015) Iron-mediated stabilization of soil carbon amplifies the benefits of ecological restoration in degraded lands. Ecolo App 25 (5):1226–1234

Singh S, Jaiswal DK, Krishna R, Mukherjee A, Verma JP (2020) Restoration of degraded lands through bioenergy plantations. Restor Ecol 28(2):263–266

Sun W, Dou F, Li C, Ma X, Ma LQ (2020) Impacts of metallic nanoparticles and transformed products on soil health. Crit Rev Environ Sci Technol 1–30. https://doi.org/10.1080/10643389.2020.1740546

Tombuloglu H, Ercan I, Alshammari T, Tombuloglu G, Slimani Y, Almessiere M, Baykal A (2020) Incorporation of micro-nutrients (nickel, copper, zinc, and iron) into plant body through nanoparticles. J Soil Sci Plant Nutr. https://doi.org/10.1007/s42729-020-00258-2

Wang J, Wei H, Cheng K, Ochir A, Davaasuren D, Li P, Chan FKS, Nasanbat E (2020) Spatio-temporal pattern of land degradation from 1990 to 2015 in Mongolia. Environ Dev 34:100497

Wiesmeier M, Urbanski L, Hobley E, Lang B, von Lützow M, Marin-Spiotta E, van Wesemael B, Rabot E, Ließ M, Garcia-Franco N, Wollschläger U (2019) Soil organic carbon storage as a key function of soils-a review of drivers and indicators at various scales. Geoderma 333:149–162

Wigginton NS, Titta AD, Piccapietra F, Dobias JAN, Nesatyy VJ, Suter MJ, Bernier-Latmani R (2010) Binding of silver nanoparticles to bacterial proteins depends on surface modifications and inhibits enzymatic activity. Environ Science Technol 44(6):2163–2168

Wu S, Zhang S, Gong Y, Shi L, Zhou B (2020) Identification and quantification of titanium nanoparticles in surface water: a case study in Lake Taihu. China. J Hazard Mat 382:121045

Xie H, Zhang Y, Wu Z, Lv T (2020) A bibliometric analysis on land degradation: current Status, development, and future directions. Land 9(1):28

Xin X, Zhao F, Zhao H, Goodrich SL, Hill MR, Sumerlin BS, Stoffella PJ, Wright AL, He Z (2020) Comparative assessment of polymeric and other nanoparticles impacts on soil microbial and biochemical properties. Geoderma 367:114278

Xu Y, Zhao D (2007) Reductive immobilization of chromate in water and soil using stabilized iron nanoparticles. Water Res 41 (10):2101–2108

Yang J, Gao X, Li J, Zuo R, Wang J, Song L, Wang G (2020) The stabilization process in the remediation of vanadium-contaminated soil by attapulgite, zeolite and hydroxyapatite. Ecol Eng 156:105975

Yang JW, Fang W, Williams PN, McGrath JW, Eismann CE, Menegário AA, Elias LP, Luo J, Xu Y (2020) Functionalized mesoporous silicon nanomaterials in inorganic soil pollution research: opportunities for soil protection and advanced chemical imaging. Curr Pollut Rep 6:264–280

Yargholi B, Azarneshan S (2014) Long-term effects of pesticides and chemical fertilizers usage on some soil properties and accumulation of heavy metals in the soil (case study of Moghan plain's (Iran) irrigation and drainage network). Int J Agr Crop Sci 7(8):518

Yusefi-Tanha E, Fallah S, Rostamnejadi A, Pokhrel LR (2020) Zinc oxide nanoparticles (ZnONPs) as novel nanofertilizer: Influence on seed yield and antioxidant defense system in soil grown soybean (Glycine max cv. Kowsar). Sci Tot Environ 738:140240

Zhang H, Wang B, Xu M (2008) Effects of inorganic fertilizer inputs on grain yields and soil properties in a long-term wheat–corn cropping system in South China. Comm Soil Sci Plant Anal 39(11–12):1583–1599

Zhang WX (2003) Nanoscale iron particles for environmental remediation: an overview. J Nanoparticle Res 5(3–4):323–332

Zhao LJ, Peralta-Videa JR, Hernandez-Viezcas JA, Hong J, Gardea-Torresdey JL (2012) Transport and retention behavior of ZnO nanoparticles in two natural soils: effect of surface coating and soil composition. J Nano Res 17:229–242

Zhao S, Su X, Wang Y, Yang X, Bi M, He Q, Chen Y (2020) Copper oxide nanoparticles inhibited denitrifying enzymes and electron transport system activities to influence soil denitrification and N_2O emission. Chemosphere 245:125394

Engineered Nanoparticles as Nanofertilizers and Biosensors

Advances of Engineered Nanofertilizers for Modern Agriculture

Theivasanthi Thirugnanasambandan

Abstract

This chapter focuses on the applications of nanotechnology in fertilizers and utilizations of engineered nanomaterials as fertilizers for plants. Nanotechnology is a promising technology that has vast applications. Its utilization in agriculture and related fields are myriad. Recently, many products are developed in parallel with the growing of agri-nanotechnology. Fertilizers are the material that contains nutrients that are essential for the plants. Nanofertilizer is the new generation fertilizer. During the applications in soils or in plant tissues, it delivers the nutrients better than the conventional fertilizers. Nanotechnology is applied in the preparation of nanofertilizer. This technology is highly interdisciplinary in nature and it is closely connected with various technologies of biology such as biotechnology, nano-biotechnology, bio-nanotechnology, and agri-nanotechnology. These properties of nanotechnology lead to the development of the plants benevolent nanofertilizers and related engineered nanomaterials. These materials are in the nanometer size range particularly size less than the plants cells. Primary nutrients, metal oxides nanoparticles (zinc oxide and iron oxide), zeolite nanofertilizers, and carbon nanomaterial-based fertilizers are discussed in this chapter. Nanotechnology applications in fertilizers like slow release or controlled release fertilizers (hydroxyapatite nanoparticles coated urea, polymer coated fertilizers) are explored. Apart from these, applications of coating technology in fertilizers using bio-polymers (such as chitosan and thermoplastic starch) and sulfur are explained. The reviewed literatures reveal that the nanofertilizers will dominate the modern agriculture.

Keywords

Chitosan fertilizers • Hydroxyapatite urea • Metal oxides fertilizers • Nano-biotechnology • Polymer fertilizers • Starch fertilizers • Sulfur fertilizers • Zeolites fertilizers

Abbreviations

Ca	Calcium
$CaCO_3$	Calcium carbonate
CEC	Cation exchange capacity
CeO_2	Cerium oxide
CS	Chitosan
CuO	Copper oxide
EC	Ethyl cellulose
EC	Electrical conductivity
Fe	Ferrous
Fe_2O_3	Iron oxide
HANPs	Hydroxyapatite nanoparticles
KH_2PO_4	Potassium dihydrogen phosphate
KNO_3	Potassium nitrate
MAP	Monoammonium phosphate
Mg	Magnesium
Mn	Manganese
$MnSO_4$	Manganese sulfate
NPs	Nanoparticles
pH	Power of hydrogen or potential for hydrogen
PVA	Polyvinyl alcohol
PVP	Polyvinylpyrrolidone
SAP	Superabsorbent polymer
SPAD	Nitrogen concentration and chlorophyll content
TiO_2	Titanium oxide
WNLCF	Water and nutrient loss control fertilizer
Zn	Zinc
ZnO	Zinc oxide

T. Thirugnanasambandan (✉)
International Research Centre, Kalasalingam Academy
of Research and Education (Deemed University), Krishnankoil,
Tamil Nadu 626126, India
e-mail: ttheivasanthi@gmail.com

© Springer Nature Switzerland AG 2021
P. Singh et al. (eds.), *Plant-Microbes-Engineered Nano-particles (PM-ENPs) Nexus in Agro-Ecosystems*,
Advances in Science, Technology & Innovation,
https://doi.org/10.1007/978-3-030-66956-0_9

1 Introduction

Fertilizer is a material which is applied in the soil to supply nutrients for the growth of plants. Considerable amount of the applied fertilizer is wasted by water or wind before it is used by plants. It can be utilized in a better way with the help of modern advanced technologies. Nanotechnology is the emerging technology that can support for the various fertilization practices to meet increasing demands of food. Heavy usage of fertilizers results in accumulation of fertilizers in water bodies thus causing eutrophication problems.

Chemical fertilizers affect the soil mineral balance which in turn decreases the soil fertility. Fertilizer formulations made using engineered nanoparticles improve the uptake in plant cells and minimize the nutrient loss. In addition, they increase the rate of seed germination, seedling growth, photosynthetic activity, nitrogen metabolism, synthesis of carbohydrate and protein as well (Solanki et al. 2015). Nanomaterials produced by applying nanotechnology have properties different from their bulk materials. Particle size of these materials is less than 100 nm (at least in any one dimension). The large surface area and more active sites of these materials lead them to function efficiently. Their property like compatibility with flexible substrates is useful in several agricultural applications.

Recently, nanotechnology-based products are developed for the utilizations in agriculture. Agri-nanotechnology products such as nanofertilizers, nano-biofertilizers, biofertilizers, nano-pesticides, nano-nutrients, agricultural nanosensors, storage materials for food grains or agricultural harvested products protection, and food packaging/protection materials are modernizing the agriculture and allied fields. They are developed in parallel with the development of the emerging agri-nanotechnology.

Fertilizers are the materials that applied in agricultural activities, i.e., to supply macro or micronutrients or both to the plants. Applications of fertilizers are considerably focused on primary macronutrients. Plants mostly utilize the macronutrients in more quantity (according to the name, the demand for macronutrients is in macrolevel). It leads to lack of macronutrients availability in the agricultural land. Hence, it is essential to complete the demand. In micronutrients case, plants consume less quantity, i.e., microlevel. It leads to the availability of micronutrients in the agricultural land. Contrary to macronutrients, the demand for micronutrients is in microlevel or negligible. In slow or controlled release fertilizers, nutrients are coated by nano-coatings that help to release the nutrients in slow or controlled method.

For the growth of the plants, microelements such as iron, cobalt, copper, selenium, zinc, molybdenum, and other metals are essential. They constitute biologically active compounds like proteins, enzymes, hormones, vitamins, and pigments in plants. Nanopowders of the said microelements can be developed as innovative fertilizers. The advantages of these fertilizers over traditional fertilizers are: increasing the level of resistance to pests and diseases; their consumption is only one gram per ton of processed seeds; reduction of procedures involved; finally decreasing the costs of labor and operating agricultural equipment (NUST MISIS 2017).

The micronutrient calcium can be supplied to plants with the nanoparticles such as Ca-NPs, $CaCO_3$ NPs, and hydroxyapatite NPs. Ca-NPs can improve the seedling growth in plants. Mg-NPs are superior to regular Mg salt by improving the uptake of Mg in plant stems and leaves. Fe NPs are able to increase the chlorophyll contents in leaves. Mn-NPs are applied to replace the conventional $MnSO_4$ salt (Liu et al. 2015). Novel fertilizers can be made using various nanoparticles to increase the crop production.

ZnO nanoparticles stimulate the lateral roots that modify the root architecture and increase the overall uptake of nutrients in wheat plant. Shoot growth is stimulated in bean, chickpea, and green pea with a low dose (1 kg per 100 mg) of ZnO nanoparticles. Stimulation of chlorophyll production increases the rate of photosynthesis and reduction of the severity of chlorosis in plants. Such stimulation is achieved by iron oxide and manganese nanoparticles. TiO_2 nanoparticles are able to increase the RuBisCO activase enzyme activity and chlorophyll production in spinach. Stimulation of the root growth in soybean and cilantro is performed by CeO_2 nanoparticles. These nanoparticles also prevent membrane peroxidation and leakage in maize by inducing the activity of antioxidative enzymes. CuO nanoparticles allow for high uptake of cognate element into the plant. It improves the level of the essential nutrient elements. Slow releasing of fertilizers helps to avoid leaching and fixation of nutrients and makes the nutrients available in proper time. This method also rectifies the overuse of fertilizer (Dimkpa et al. 2014).

Fertilizer leaching is the loss of water-soluble plant nutrients. It leads to the natural environment concern like groundwater contamination. It is caused by the dissolution of fertilizers and different biocides (such as pesticides, herbicides, insecticides, and fungicides) due to rain and irrigation (Wikipedia 2014). Excess NO_3 ions of the nitrogen fertilizers applied are not absorbed by plants or soils which are leached into groundwater (Lin et al. 2001). Phosphorus loss is a major threat to manage the surface water quality. It plays a major role in the eutrophication of surface waters (Carpenter et al. 1998). It does not interact with soil particles through adsorption and desorption. However, soils rich in iron (ferrihydrite) and aluminum oxides or hydroxides (gibbsite) retain phosphorus (Borling 2003; Schoumans

2015). They will release the P into the soil solution. Also, changing the chemical conditions of the soil leads to the P leaching (Shenker et al. 2004; Zak and Gelbrecht 2007).

Considerable quantity of fertilizers is lost while applying that leads to environmental problems. Localized applications of fertilizers (such as salts of ammonia, nitrate, urea, and phosphate compounds) in large quantity produce harmful effects (Trenkel 1997; Ombodi and Saigusa 2000). Soils do not retain nitrates for future utilizations. Also, plants absorb different nutrients at various times (Smart fertilizer 2020). Nanomaterials have applications in slow and controlled release fertilizers that reduce the fertilizer consumption and environmental pollution as well (Wu and Liu 2008). Slow and controlled release fertilizers prevent leaching by releasing the nutrients in a controlled manner.

Engineered nanomaterials can be used in minimum concentration and thereby minimize environmental pollution. The advantages of nanotechnology lie in crop growth, enhance the fertilizer use, reduce nutrient losses, and minimize the adverse environmental impacts. The size of the nanoparticles is in the nanometer level. Hence, they can enter easily into plant cells since the plant cells are in micrometer range.

Sulfur nano-coatings applied on fertilizers are beneficial to the sulfur deficient soils (Santosa et al. 1995; Brady and Weil 2017). Nanocoated urea and phosphate are prepared to release the fertilizers in slow or controlled manner. They will release the nutrients slowly in accordance with the demands of the soil and crops. Biodegradable and biocompatible materials such as chitosan nanoparticles (bio-polymer) are useful in the preparation of controlled release NPK fertilizer materials, i.e., urea, calcium phosphate, and potassium chloride (Corradini et al. 2010). Kaolin and polymeric biocompatible nanoparticles are used to prepare slow release fertilizers (Wilson et al. 2008).

Sabir et al. (2014) have demonstrated that applying nanocalcite ($CaCO_3$-40%) with nano-SiO_2 (4%), MgO (1%), and Fe_2O_3 (1%) enhances the uptake of Ca, Mg, and Fe. It also enhances the intake of the P along with the micronutrients like Zn and Mn (Sabir et al. 2014). Figure 1 exhibits the different applications related to nanotechnology and engineered nanomaterials in agriculture such as slow and controlled released nanofertilizers, nano-based target delivery (nano-carriers), nano-pesticides, and nano-sensors. These applications enhance the plant growth, productivity, and yield ultimately (Yilen et al. 2019).

Utilization of engineered nanomaterials (like nanofertilizers, nano-pesticides, and nano-sensors) in agriculture can increase crop yield by influencing availability of nutrient in soil and uptake by crops. Engineered nanomaterials can control the crop diseases by minimizing the pathogens

activities directly (through several mechanisms that includes releasing of reactive oxygen species). Also, they control disease indirectly by enhancing crop nutrition and plant defense mechanisms as well. Efficient use of these materials may replace conventional fertilizers and pesticides that ultimately minimize the environmental impact (Adisa et al. 2019).

2 Fertilizers

2.1 Fertilizers in Agriculture

Fertilizers improve the agricultural productivity. However, the disproportionate utilization of chemical fertilizers causes damages to soil. Also, it decreases the available area (with good condition soil) which is necessary for crop production. Sustainable agriculture suggests reducing the utilization of agrochemicals. Advancements in nanotechnology (like enhanced crop productivity) are applied to overcome the agricultural crisis that leads to sustainability (Priyom Bose 2020).

Some complex fertilizers are harmful to the crops. For example, to supply potassium to the crop instead of using potassium chloride as a fertilizer, potassium nitrate (KNO_3) can be used. Potassium chloride contains chloride which is harmful to the crops. On the other hand, KNO_3 contains more nitrate than ammonium. The uptake of essential nutrition elements like K, Ca, and Mg is impaired by ammonium. Hence, KNO_3 is a better option than using potassium chloride and ammonium (Israelagri.com 2016).

2.2 Classification of Fertilizers

Fertilizers are the chemicals or natural substances to supply essential nutrients for the plant growth to maintain the soil fertility. Benton (2012) has reported that fertilizers can be classified in many approaches: Firstly, fertilizers can be classified depending upon the contents, i.e., single nutrient fertilizers or straight fertilizers (e.g., nitrogen—N, phosphorus—P, or potassium—K) and multi-nutrient fertilizers or complex fertilizers (e.g., two or more nutrients—N and P); secondly, based on the inorganic and organic content, i.e., inorganic fertilizers and organic fertilizers. Inorganic fertilizers do not contain carbon materials. They are prepared using several chemical treatments. Hence, they are also called as synthetic fertilizers. All organic fertilizers should have carbon content and they can be derived from plant and/or animal sources or recycled materials of plant or animal source or both (Benton 2012).

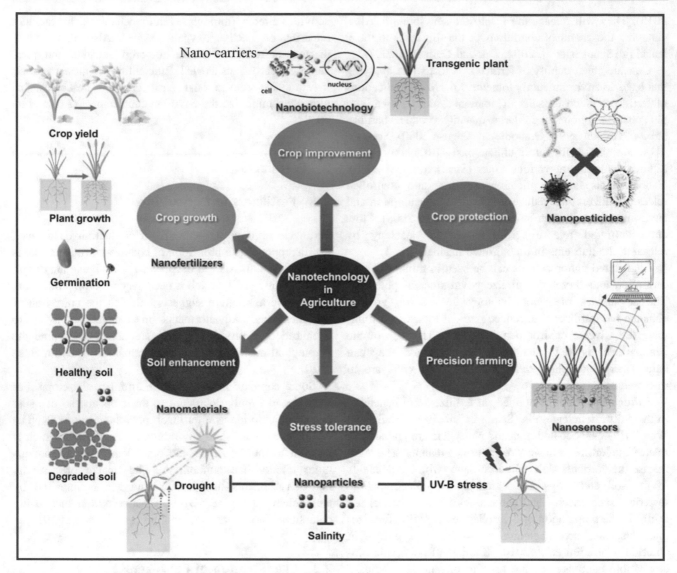

Fig. 1 Applications related to nanotechnology and nanomaterials in agriculture. Slow and controlled released nanofertilizers enhance plant growth, productivity, and yield. Nano-based target delivery (gene transfer) is useful in plants development. Nano-pesticides control pathogens and useful to protect the plants efficiently. Nano-sensors (with computerized controls) are utilized in precision farming. Engineered nanomaterials are applied in plant stress and soil enhancement. *Source* Yilen et al. (2019), with permission

2.3 Macro and Micronutrients

Nitrogen, phosphorus, and potassium are called as main or primary macronutrients. Among them: Nitrogen (N) is vital for the development of leaves; phosphorus (P) is necessary for the development of roots, flowers, fruits, and seeds. Activities such as stem growth, water transportation inside the plants, and promotion of flowering and fruiting need potassium (K) macronutrient. Calcium (Ca), magnesium (Mg), and sulfur (S) are known as secondary macronutrients. Copper (Cu), iron (Fe), manganese (Mn), molybdenum (Mo), zinc (Zn), chloride (Cl), and boron (B) are the essential micronutrients.

Silicon (Si), cobalt (Co), and vanadium (V) are some of the micronutrients that have less importance.

Micronutrients like zinc, copper, and molybdenum can be supplied to plants as water-soluble salts. Iron is transformed into insoluble compounds at moderate soil pH and phosphate concentrations. This transformation causes bio-unavailability. Hence, it is supplied in the form of chelated complex like EDTA derivative. The requirements of micronutrients are depending upon the plants, for example, sugar beets need boron and legumes need cobalt (Scherer 2009). Likewise, nanofertilizers also have macronutrient and micronutrient contents.

2.4 Major Elements of Plants

Plants are created by the composition of four main elements, i.e., carbon, hydrogen, oxygen, and nitrogen. Apart from the life of plants, they have major roles in the creation of entire biological system (including human, animals, and microorganisms) and maintenance of life. These elements present in carbohydrate, protein, and fat that are utilized as food/feed by human and animals. Phosphate is essential for the DNA, ATP (energy carrier of cells), some lipids, and bones of human and animals.

Plants can get hydrogen and oxygen from water and carbon from carbon dioxide. Nitrogen is present in the atmosphere. However, plants are unable to use it. Plants require it in a fixed form because of its major role in the development of proteins, DNA, and other important components like chlorophyll. Some bacteria and legumes fix atmospheric nitrogen (N_2) by ammonia conversion. Phosphate plays a major role in the DNA, ATP, and some lipids production of the plants (Wiki/Fertilizer 2018). Figure 2 depicts that plants utilize carbon dioxide and water (getting carbon, hydrogen, and oxygen) during photosynthesis to produce starch and sugar or glucose.

The deficiency of micronutrients reduces the crops productivity that ultimately affects the human health (while consuming the low nutrient foods). For example, iron deficiency causes anemia and affecting growth, reproductive health, and cognitive performance in humans (Swaminathan et al. 2013; Monreal et al. 2016). Hence, imbalance (surplus supply or deficiency) of the main elements, macronutrients, and micronutrients in plants will affect the living things of entire biological system including plants. It emphasizes the significance of the fertilizers.

3 Engineered Nanofertilizers

3.1 Technology of Nanofertilizers

Nanotechnology develops agricultural products such as nanofertilizers, nano-herbicides, nano-pesticides, nano-fungicides, and nano-sensors (Duhan et al. 2017). Also, nanotechnology supports for agriculture by making nanoscale carriers, bio-remediation of pesticides, wastewater treatment, enzymatic sensors, nano-lingocellulose, and clay nanotubes (Dasgupta et al. 2015). Many countries are applying nanotechnology in agriculture and food sectors. It will support to meet the demands and to feed of the increased population (Ali et al. 2014).

Nanofertilizers are defined as the synthesized or modified form of traditional fertilizers or fertilizers bulk materials or extracted from different vegetative or reproductive parts of the plant by different chemical, physical, mechanical, or biological methods with the help of nanotechnology used to improve soil fertility, productivity, and quality of agricultural produces. Nanoparticles can be made from fully bulk materials (Brunnert et al. 2006).

3.2 Classification of Nanofertilizers

Different classifications and types of nanofertilizers are shown in Fig. 3. Classifications of nanofertilizers, i.e., nutrient-based, action-based, and based on the quantity applied are shown in Fig. 3a–c, respectively. Like conventional fertilizers (as explained earlier), nanofertilizers also have macro and micronutrients. However, in the case of nanofertilizers, the size of the nutrients (it may be macronutrients or micronutrients) is in nanoscale range. Ruiqiang and Rattan (2016) reported about the nutrient-based classification of nanofertilizers. Figure 3a shows this classification of nanofertilizers based on nutrients. Nanofertilizers supply nutrients to the plants that improve the plant growth and yields. Also, they are applied to enhance the performance of conventional fertilizers. They are divided into four classes: macronutrient nanofertilizers (e.g., apatite nanoparticles), micronutrient nanofertilizers (e.g., iron oxide NPs and zinc oxide NPs), nutrient-loaded nanofertilizers (e.g., zeolites), and plant growth stimulating nanomaterials (e.g., carbon nanomaterials). Developing the macronutrient nanofertilizers (nitrogen and phosphorus) is necessary to improve agricultural activities and to reduce environmental problems (Ruiqiang and Rattan 2016).

Based on the actions, nanofertilizers are categorized as control or slow release fertilizers, water, and nutrient loss control fertilizers (WNLCF), magnetic, or nanocomposite fertilizers combined nanodevices (Lateef et al. 2016; Panpatte et al. 2016). This classification is shown in Fig. 3b. Priyom Bose (2020) has divided the nanofertilizers (based on the quantity applied) into three types: (i) nanoscale fertilizers: these are nano-sized particles that have nutrients, (ii) nanoscale coating fertilizers: nanoparticles coated or loaded on traditional fertilizers, and (iii) nanoscale additive fertilizers: these fertilizers are the traditional fertilizers mixed with the nano-sized additives (Priyom Bose 2020). Figure 3c shows the classification based on the quantity applied.

Nanofertilizers combined nanodevices are designed with a nanonetwork to monitor the plants. The monitoring system comprises of nano and microscale network devices. These devices and the data collected by them are managed by control units. The transmitters (nano-sensors) collect data and transmit to receivers (micro-devices). Finally, the data are relayed to the Internet through gateways (Dufresne 2000; Luca 2018). Attapulgite or Palygorskite clay is one of the varieties of fuller's earth clay material. Magnesium aluminum phyllosilicate is the chemical content and

Fig. 2 Sources of carbon, hydrogen, and oxygen to the plants. Plants utilize them and produce starch and sugar by photosynthesis process

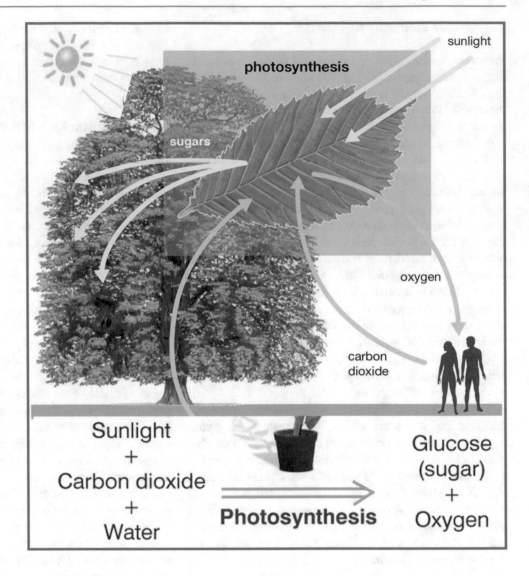

(Mg, Al)$_2$Si$_4$O$_{10}$(OH)·4(H$_2$O) is the chemical formula of this clay material. Several researchers have reported about mixing of attapulgite-polymer complex with conventional fertilizer to utilize as WNLCF.

While applying nitrogen fertilizer, a portion of this fertilizer causes environmental pollution due to various activities (like runoff, leaching, and volatilization) which can be avoided by high-performance WNLCF. Addition of high-energy electron beam dispersed attapulgite–sodium polyacrylate–polyacrylamide complex into traditional fertilizer is a method to prepare WNLCF. The attapulgite-polymer complex serves as the water and nutrient loss control agent in WNLCF. It retains the water and nutrient effectively that prevents the water and nutrient loss. Ultimately, it leads to efficient utilization of water and nutrient reduces the pollution risk caused by the fertilizer (Zhou et al. 2015).

A biomass-based, multifunctional controlled release fertilizer (BMCF) is a cost-effective water and nutrient loss control fertilizer. It enhances the utilization of nutrient and crop production as well. Also, it reduces the adverse effects like environment pollution. BMCF is designed by using co-granulated ammonium zinc phosphate and urea in attapulgite matrix. This fertilizer core is coated by cellulose acetate butyrate coating initially. Again it is coated by carboxymethyl chitosan-g-poly(acrylic acid)/attapulgite superabsorbent composite as an outer coating. BMCF decreases nitrogen leaching loss and runoff. Also, it enhances the moisture retention of soil and restructures the acidity and alkalinity of soil (Wang et al. 2014).

3.3 Benefits of Nanofertilizers

Nanofertilizers are produced with intention to regulate the nutrients supply and consumption to meet demands of the crops with minimum loss. Conventional nitrogenous fertilizers reduce the efficiency of fertilizing activities (due to

Fig. 3 Different classification and types of nanofertilizers. **a** Nutrient-based nanofertilizers, **b** nanofertilizers based on the actions, and **c** nanofertilizers based on the quantity applied

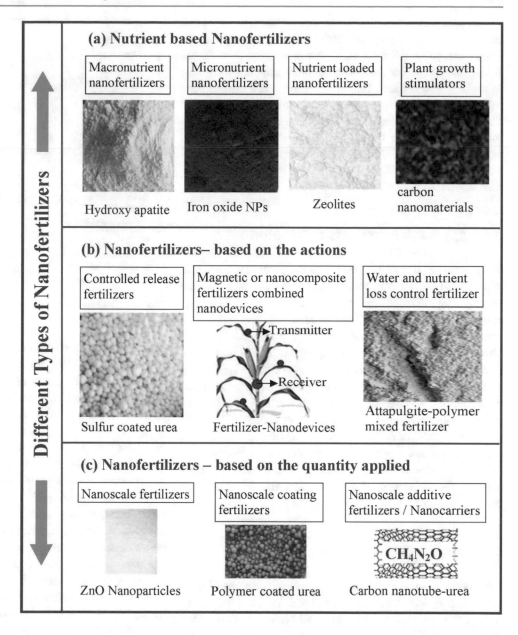

Different Types of Nanofertilizers

(a) Nutrient based Nanofertilizers

| Macronutrient nanofertilizers | Micronutrient nanofertilizers | Nutrient loaded nanofertilizers | Plant growth stimulators |

Hydroxy apatite   Iron oxide NPs   Zeolites   carbon nanomaterials

(b) Nanofertilizers– based on the actions

| Controlled release fertilizers | Magnetic or nanocomposite fertilizers combined nanodevices | Water and nutrient loss control fertilizer |

Transmitter

Receiver

Sulfur coated urea   Fertilizer-Nanodevices   Attapulgite-polymer mixed fertilizer

(c) Nanofertilizers – based on the quantity applied

| Nanoscale fertilizers | Nanoscale coating fertilizers | Nanoscale additive fertilizers / Nanocarriers |

CH_4N_2O

ZnO Nanoparticles   Polymer coated urea   Carbon nanotube-urea

leaching, evaporation, and degradation) that ultimately makes loss (Mia et al. 2015; Wang et al. 2015; Yang et al. 2016). In the case of nitrogenous nanofertilizers, the release of fertilizer-N is regulated by nanoformulations. It reduces the nutrient loss by improving the interaction of nutrients with crops and avoiding the interaction of nutrients with soils, microorganisms, water, and air (Dwivedi et al. 2016; Panpatte et al. 2016).

Nanofertilizers improve crop growth/yield, nutrient utilization; reduce expenditures of fertilizer and cultivation. Optimum quantity supply of nanofertilizers enhances crop growth but beyond that level decreases the crop growth (due to the toxicity of nutrient). ZnO nanofertilizer enhances the seed germination and plant growth considerably (Singh et al. 2017). Applying the nano-formulated or nano-entrapped

micronutrients as slow or controlled release fertilizers will improve the soil health and uptake by plants. It ultimately enhances the growth and productivity of crops (Peteu et al. 2010). Nanomaterials (like magnesium hydroxide) are utilized for the early germination of seeds. They break the seed dormancy and they may increase the chlorophyll content of the plants. Overall, nanomaterials treatment of the seeds enhances the germination percentage and plant growth as well. Hence, these nanoparticles can be utilized as the efficient nano-nutrients for the plant growth promotion (Shinde et al. 2020). Carbon nanomaterials, ZnO nanoparticles, and iron oxide nanomaterials are applied as pre-soaking and seed germination technology. For these kinds of applications, bio-synthesized nanomaterials are the suitable material instead of chemically synthesized nanomaterials.

Encapsulating the microorganisms (bacteria or fungi) improves the nitrogen, phosphorus, and potassium availability that support for the plant growth (Priyom Bose 2020). Encapsulation of nutrients with nanomaterials is a technique to make nanofertilizers. Initially, nanomaterials are prepared in physical (top-down) or chemical (bottom-up) method. In the next step, nutrients are encapsulated by nanoporous materials or polymer thin film coating or nano-emulsions of cationic (NH_4^+, K^+, Ca^{2+}, Mg^{2+}) or nutrients surface modified with anionic (NO_3^-, PO_4^-, SO_4^-) nutrients (Subramanian et al. 2015; Panpatte et al. 2016;).

3.4 Application Methods of Nanofertilizers

When nanofertilizers are applied to the plants in soil application method, the soil mixed nanoparticles enter into the plants using the routes such as root hairs, lenticles, mucilage, and exoates. In addition, microorganisms are utilized in these activities. Xylem transport plays a major role in the absorption of the soil mixed nanoparticles. Direct interaction between fertilizers and soil systems of this application method leads to some undesirable consequences such as soil acidification, wastage of fertilizers as well. In the case of foliar application method, aerosol nanoparticles penetrate into the plants directly. Stomata, trichomes, hydathodes, lenticles, and cuticle wounds are the possible entry routes. In this case, phloem transport plays a major role. Figure 4 exhibits the difference between the soil and foliar applications of fertilizers.

4 Bio-synthesis of Nanomaterials

Utilization of toxic or biologically incompatible materials in plants and agriculture as fertilizers, pesticides, and herbicides or in any form will cause harm to all biological organisms. Applying the bio-synthesized nanomaterials (as nanofertilizers or other applications) instead of chemically synthesized nanomaterials will reduce the bio-incompatibility. For biological applications, bio-compatibility is essential. Possibilities of bio-compatibility are more in bio-synthesis method comparing to other synthesis methods. During the preparation of nanomaterials, various chemicals are utilized. Hence, applying the resultant materials and the residues of these chemically synthesized nanomaterials in agriculture may cause untoward effects. In the case of bio-synthesis or green synthesis of nanomaterials, chemicals are avoided except the pre-cursor materials.

Dhillon et al. (2012) reported that bio-synthesis of nanoparticles using biological materials such as plant extracts (leaves, flowers, stem, fruit peels, and seeds), bacteria, fungi, and algae results in several benefits, i.e.,

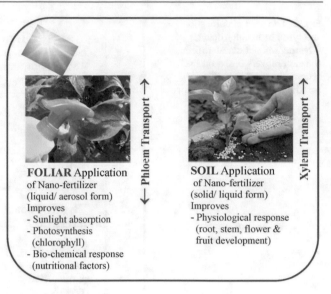

Fig. 4 Foliar and soil application methods of nanofertilizers in plants

eco-friendliness and bio-compatibility. In this synthesis, toxic chemicals are not utilized (Dhillon et al., 2012). Plant-mediated synthesis of metal nanoparticles (like gold, silver, copper, and iron) and metal oxide nanoparticles (like titanium oxide and zinc oxide) are more reliable, inexpensive, and eco-friendly approach. Figure 5a, b exhibit the bio-reduction process and bio-reduction mechanism related to the plant-mediated bio-synthesis of metallic nanoparticles, respectively (Khandel et al. 2018).

Nano-ecotoxicology is the toxic effects of the nanomaterials released into the environment and biological systems (like humans, animals, plants, fungi, and microbes). Humans and animals are exposed to nanomaterials in several ways via air, water, and consuming food accumulated with nanomaterials (Pachapur et al. 2015). While utilizing nanomaterials, high priority should be given for safety. Toxicity mainly focuses on human beings and protecting them, but the ecotoxicity intends to protect the various levels of trophic organism and ecosystems. Ecotoxicity includes natural mechanisms and the environmental factors related to the bioavailability (Rana and Kalaichelvan 2013).

5 Nanofertilizers of Macro and Micronutrients

5.1 Hydroxyapatite Nanoparticles (HA NPs)

Hydroxyapatite nanoparticles (HA NPs) are the major source of the fertilizers. It is able to supply both macronutrient and micronutrient. It is used either alone or mixed with other fertilizers. It is applied in the coating of fertilizers to control the release of the fertilizers. As per the report of Kottegoda

Fig. 5 Schematic diagram showing the plant-mediated synthesis of metallic nanoparticles. **a** Bio-reduction process metal salt solution by plant extract, **b** bio-reduction mechanism involved in the bio-synthesis of metallic nanoparticles. *Source* Khandel et al. (2018), with permission

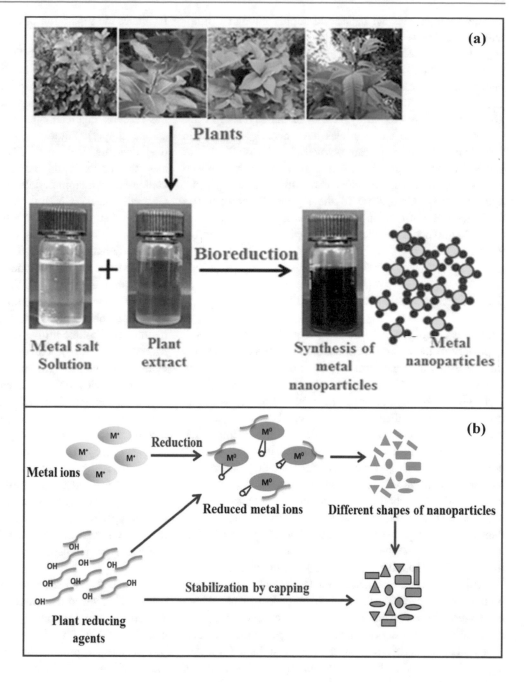

et al. (2017), HA NPs can be applied as phosphorous fertilizers. They can also supply calcium in addition to phosphorous. In wet soil, urea breaks down quickly and the formation of ammonia occurs. The ammonia enters the atmosphere as nitrogen dioxide which is the main greenhouse gas associated with agriculture. This decomposition limits the application of more urea as fertilizer. To avoid the breakdown of urea, slow releasing of urea is done. For this purpose, urea is coated with HA NPs. The HA NPs coated urea release the nitrogen slowly, i.e., 12 times slower than urea without HA NPs coating (Kottegoda et al. 2017). The slow release of phosphorous helps plants to take up the nutrient continuously as they grow. Slow release of phosphorous can be achieved with the help of HA NPs. Application of chemical fertilizers results in soil acidification. Hence, the cost of reversing soil pH to optimal is also extremely high. The advantage of HA NPs is that it does not change soil pH when phosphorous is released. When plants grow, different types of organic acids like oxalic acid and citric acid are released. They dissolve the HA NPs which makes the phosphorous availability to the plants (Phys.org. 2015).

Soluble phosphate salts cause surface water eutrophication. Solid phosphates supply low level nutrient P. Synthetic apatite nanoparticles act better and supply enough P nutrients to the plants compared to the soluble and solid counterparts. The greenhouse experiment conducted on soybean (*Glycine max* (L.) Merrill) shows that applying of synthetic apatite nanoparticles increases the growth rate (32.6%), seed yield (20.4%), and biomass productions (above ground by 18.2% and below-ground by 41.2%) of the soybean. In this experiment, apatite nanoparticles have been synthesized in wet chemical route with carboxymethyl cellulose (CMC) as stabilizing agent. The shape of the synthesized hydroxyapatite nanoparticles is spherical and particle size is 15 nm approximately. Comparing to the regular P fertilizer (Ca $(H_2PO_4)_2$), the utilization of apatite nanoparticles improves the yield and decreases the water eutrophication (Liu and Lal 2015a).

5.2 Carbon Nanomaterials

Carbon is one of the main elements required for the plants. It is present in the all organic materials. Plants get the carbon mainly from air in the form of carbon dioxide. Verma et al. (2019) reported that the impacts of carbon nanomaterials on plant growth (from enhanced crop yield to acute cytotoxicity) have been studied by many researchers. The concentration of the carbon nanomaterial is more important. Vegetative growth and yield of fruit/seed increase at lower concentration of carbon nanomaterials but they decrease at higher concentrations of carbon nanomaterials. At lower concentrations, carbon nanomaterials are able to increase water uptake and transport, seed germination, and antioxidant activities (Verma et al. 2019). The supportive factors available at lower concentrations of carbon nanomaterials improve the vegetative growth and yield of fruit/seed.

Multi-walled carbon nanotubes (MWCNTs) are applied with urea fertilizer for the growth of paddy plants. Functionalized carbon nanotubes are unique since they are attached with a variety of functional groups on their surface. This makes the carbon nanotubes material suitable for lots of applications. MWCNTs are functionalized with 4wt% of carboxyl (–COOH) functional groups. The functionalization enhances the efficacy of urea fertilizer as plant nutrition for (local MR219) paddy. About 0.6wt% of functionalized MWCNTs is grafted onto urea fertilizer. The experiment is performed using a pot under exposure to natural light. After 14, 35, and 55 days, the crop growth of plants significantly increased. The homogeneous grafting of functionalized MWCNTs onto the urea leads to such beneficial result (Yatim et al. 2018). Functionalization of MWCNTs assists in attaching urea fertilizer onto MWCNTs. The bonding between urea and MWCNTs can be confirmed using spectroscopy and chemical characterization techniques such as FT-IR and total N analysis. The functionalization process facilitates the separation of nanotube bundles into individual tubes (Yatim et al. 2015).

Carbon nanotubes (CNTs) are synthesized in chemical vapor deposition (CVD) method. Zaytseva and Neumann (2016) have explained the synthesis and applications of carbon nanomaterials. Figure 6a shows the CVD reactor that has reaction chamber and tubes (for inert gas and hydrocarbon supply). Figure 6b, c exhibits the base-growth and tip-growth mechanism of CNT growth. Figure 6d enumerates the agricultural and environmental applications of carbon-based nanomaterials (Zaytseva and Neumann 2016). Generally, in SWCNTs production, methane gas is utilized and the substrate is heated up to 850–1000 °C. In the case of MWCNTs production, ethylene or acetylene gas is utilized and the substrate is heated up to 550–700 °C. Carbon is produced due to thermal decomposition of hydrocarbons. After producing a certain concentration of carbon, semi-fullerene cap is formed. In the next stage, the growth of cylindrical nanotube is formed by carbon flow from the hydrocarbon source on the catalyst (Matsuzawa et al. 2014; Morsy et al. 2014). Formation of semi-fullerene cap and cylindrical nanotube growth can be seen in Fig. 6b, c.

Nanoparticles can be used as potential plant growth regulator. Preparation of the slow releasing Cu–Zn micronutrient carrying carbon nanofibers (CNFs) is an easy method. It can be done through dispersing the micronutrient (Cu–Zn/CNFs) in a polymeric formulation of PVA–starch. Applying the prepared micronutrient increases the plant height significantly. The translocation of the Cu–Zn/CNFs from roots to shoots is analyzed. Scavenging of reactive oxygen species by the micronutrient nanoparticles is confirmed by measuring the quantity of superoxide anion radicals and hydrogen peroxide present in the plant (Kumar et al. 2018).

Banana peel pieces have been blended with tap water using a high-speed mechanical blender which is then mixed with potassium hydroxide. The prepared slurry has been heated at 100 °C for 30 min (Fig. 7a). This thermo-chemical process leads to produce the nanofertilizer. Figure 7b shows the TEM image of the nanofertilizer. Figure 7c shows the histogram analysis of the particles. The average particle size of nanofertilizer is found to be 40 nm. Elemental analysis reveals that chelated potassium, chelated iron, urea, citric acid, amino acids, protein, and tryptophan are the some materials present in the nanofertilizer. This nanofertilizer can be applied to increase the germination of seeds in crops such as tomato and fenugreek (Hussein et al. 2019).

Banana peel consists of Na^+, K^+, P, Ca^{++}, Fe^{+++}, and Mg^{++}. Mixing of potassium hydroxide with banana peel helps to break lignin and cellulose. Presence of urea, citric acid, amino acids, tryptophan, and protein liberate minerals. It leads to the plant germination efficiently (Aboul-Enein et al. 2016). Graphene oxide helps to release the potassium

Fig. 6 Schematic diagram showing the synthesis of carbon nanotubes (CNTs) in chemical vapor deposition method. **a** Simple chemical vapor deposition (CVD) reactor, **b** and **c** base-growth and tip-growth method of CNT growth mechanism, respectively, and **d** various applications of carbon-based nanomaterials in agricultural and environmental sectors. *Source* Zaytseva and Neumann (2016), with permission

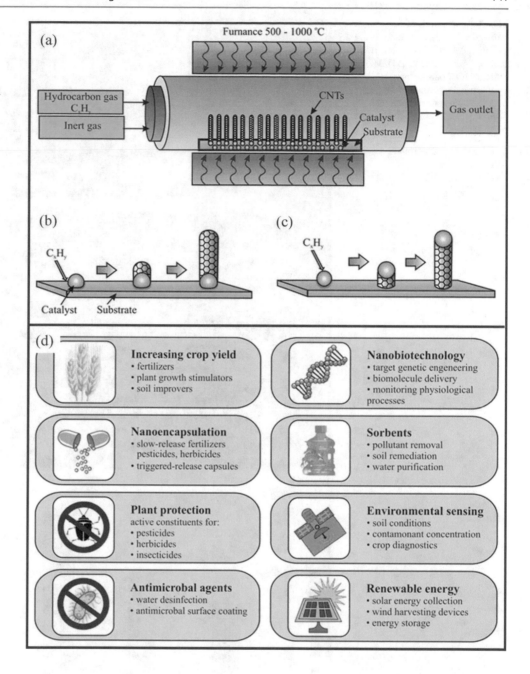

nitrate slowly. It prolongs the time of action and reduces loss by leaching (Shalaby et al. 2016).

5.3 Zinc Oxide Nanoparticles

Zinc is a micronutrient that removes the zinc deficiency of the soil. It enhances the various parts and activities of the plants (like shoot, root, biomass, activities of chlorophyll, protein, antioxidant, and enzyme related). ZnO nanoparticles have better solubility than bulk ZnO particles. Tarafdar et al. (2014) have reported that bio-synthesized zinc nanoparticles are used as nanofertilizer application in the plant pearl millet

(*Pennisetum americanum* L.) cv. HHB 67. Considerable improvement in the various contents and activities of the plant such as shoot length (15%), root area (24%), root length (4%), plant dry biomass (13%), chlorophyll (24%), total soluble leaf protein (39%), and enzyme activities (acid phosphatase: 77%, alkaline phosphatase: 62%, phytase: 322%, and dehydrogenase: 21%) are found. After the application of zinc nanofertilizer, grain yield at the maturity of crop up to 38% has improved (Tarafdar et al. 2014).

Zinc oxide and titanium dioxide nanoparticles are incorporated on the leaves of the tomato plants using novel aerosolization techniques. As a result, light and minerals are absorbed more effectively by the plants and the fruit had

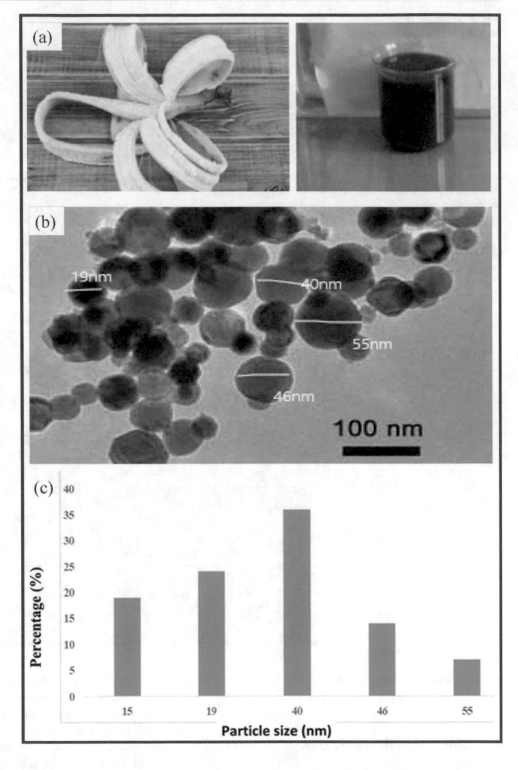

higher antioxidant content. These nanoparticles can act like a biofertilizer by secreting enzymes. These enzymes trigger bacterial microbes in the soil to turn the nutrients into plants usable form. By this method, the formation of stable complexes is prevented in the soil. The tomatoes also possess more lycopene (antioxidant) which is useful to reduce the risk of various diseases like cancer, heart disease, and age-related eye disorders. Also, these tomatoes can reduce malnutrition and child mortality by supplying more nutrients (Raliya et al. 2015). Bulk ZnO is less soluble in water. This drawback can be avoided ZnO nanoparticles. Milani et al. (2012) have reported that ZnO nanoparticles can be used instead of bulk ZnO particles to remove the zinc deficiency in the soil. The solubility of these nanoparticles is high when compared to its bulk counterpart. These nanoparticles can be coated on macronutrient fertilizers such as urea and monoammonium phosphate (MAP). The results demonstrated that MAP granules released more Zn than urea granules because of the more acidity produced by MAP granules (Milani et al. 2012).

Green synthesized (using a soil fungus) ZnO nanoparticles are sprayed on the leaves of mung bean plants. Zinc oxide nanoparticles are used to mobilize native phosphorus in the soil. They help in utilizing phosphorus in a sustainable way. ZnO nanoparticle interacts with the enzymes such as phosphatases and phytase that mobilize the complex form of phosphorus in the soil into a form that plants can absorb. This can increase the phosphorous uptake by 11%. These nanoparticles increase the root volume, stem height, and phosphorous-mobilizing soil microbial population. Also, toxicity studies have been performed to ensure the safety in the plant. The nanoparticles did not accumulate in the mung bean seeds beyond the safe limit. Green synthesis makes the nanoparticles coated with fungal proteins which prevent the direct contact between soil and the nanoparticles (Raliya et al. 2016).

Excess nitrogen and phosphorus fertilizers are fixed in the soil while applying fertilizers in the conventional method. In this method, they form chemical bonds with other elements and become unavailable for plants. The nitrogen and phosphorus are sent into rivers, lakes, and bays which results in environmental problems. Nanotechnology allows the usage of small quantities of fertilizers. Nanomaterials can be applied in the soil or sprayed onto their leaves. Foliar application is good for the environment because they do not come in contact with the soil. Since the particles are nanometer in size, plants can absorb more efficiently than via soil application.

5.4 Iron Oxide Nanoparticles

Iron is a micronutrient. Iron oxide nanoparticles can be utilized as fertilizers (to remove iron deficiency) and as seed pre-soaking solutions. Rui et al. (2016) have reported that nanotechnology can give solution to solve the shortcoming present in the traditional fertilizers. Plants like peanut (*Arachis hypogaea* L.) are highly sensitive to Fe deficiency. Fe participates in physiological processes such as chlorophyll bio-synthesis, respiration, and redox reactions. Fe_2O_3 NPs are able to increase root length, plant height, biomass, and SPAD values in peanut plants (Rui et al. 2016). In this case, Fe_2O_3 NPs support for the plants health by overcoming the weakness of the traditional fertilizers.

Iron oxide nanoparticles (Fe NPs) can be applied as next-generation iron deficiency fertilizers. Iron oxide is used as seed pre-soaking solutions. This technique is an environment friendly because it uses less fertilizer. Effects of iron oxide nanoparticles at low and high concentrations, at varied pH and the effect on embryonic root growth in legumes have been analyzed. The results show that iron oxide nanoparticles improve root growth by 88–366% at low concentrations (Palchoudhury et al. 2018).

The effects of iron nanoparticles (Fe NPs) on the anatomical and ultrastructural responses of *Capsicum annuum* L. have been studied. Iron nanoparticles show positive effects only at low concentrations which is confirmed by light and electron microscope analyses. SEM and TEM analysis results are shown in Fig. 8a–c. Iron nanoparticles are able to support for the plant growth by altering the leaf organization, increasing the chloroplast number, and regulating the development of vascular bundles. Fe NPs are absorbed in the roots and transported to the central cylinder in bio-available forms. However, in high concentrations, Fe NPs are found to be aggregated into cell walls and transported via the apoplastic pathway in the roots, which may potentially block the transfer of iron nutrients. Figure 8d, e show the effects of Fe NPs at different concentrations. Low concentration yields better plant growth (Yuan et al. 2018).

6 Slow/Controlled Release Fertilizers

6.1 Properties of Slow/controlled Release Fertilizers

Coating on the fertilizers improves many properties of the fertilizers. Ultimately, it leads to the enhancement of plants health, growth of the plants, and yield at the maturity. Fertilizers can be released in a slow or controlled manner using polymers. For the preparation of slow or controlled release fertilizers, the nutrient is the main active material. It is kept in the central portion. It is covered externally using natural or synthetic polymer coating. This polymer coating controls the release of active nutrient material present in the central portion. The nutrients will be released in a slow or controlled manner after the damaging of external polymer coating. Generally, the damage of the polymer is caused by water, microbes, and physical forces.

Slow release fertilizers are less soluble in water. They are slowly broken down by microbial action. Controlled release fertilizers are soluble fertilizers coated with materials like sulfur and polymers. In foliar application of fertilizers, the nanoparticles are transported through phloem tissues. Hence, the direct interaction of fertilizers with soil systems is avoided. As per the report of Haifa Group, both slow and controlled release fertilizers release the nutrients slowly. However, there are some differences in between them such as releasing mechanism, releasing factors, and longevity.

Fig. 8 Effect of the iron nanoparticles (Fe NPs) on *Capsicum annuum* L. plant. Electron microscope images of Fe NPs: **a** SEM image, **b**, **c** TEM image, **d** photograph of *C. annuum* L., **e** Fe NPs concentration vs plant growth (plant height). Fe NPs at low concentration promotes plant growth better than the Fe NPs at high concentration. *Source* Yuan et al. (2018), with permission)

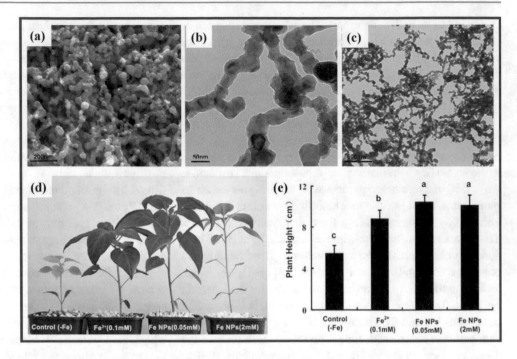

Slow release fertilizers have less control in releasing of nutrient. Factors such as soil moisture, temperature, and pH affect the releasing ability. In the case of controlled release fertilizers, soil temperature only affects the nutrients release (Haifa Group 2020).

6.2 Synthetic Polymer Coating

Polymer coated fertilizers are suitable for high-value applications because it reduces the nutrient loss. In Japan, for rice plant more than 70% of polymer coated fertilizers are used (out of total fertilizers utilizations). These fertilizers offer more sophisticated nitrogen release pattern. The nutrient release is controlled by diffusion which is constant over time. Also, it depends on the coating thickness, chemical constituents, temperature, and moisture. Controlled release fertilizers release nutrients at a rate driven by temperature and moisture of the root zone. Nutricot, osmocot, and polyon are some of the marketed products. The coatings of these commercial products are tough, resist to damage, and thin (Naik et al. 2017). Subbarao et al. (2013) have mentioned in their report about the preparation of slow release fertilizer. In this method, potash and wet clay are mixed together. Then, the mixture is made as pellets by casting in cylindrical molds. Finally, these pellets are coated with polyacrylamide to achieve slow release of potash fertilizer (Subbarao et al. 2013). Polyacrylamide is a water-soluble polymer. Damaging of this polymer plays a major role in the releasing of potash. The hydrophilic nature of this polymer leads to the damage.

Slow release nanocomposite nitrogen fertilizers are prepared with polyacrylamide hydrogel or polycaprolactone (less than 4% by weight) with a high nutrient load (75% by weight). For this preparation, plastic mixture extrusion method is adopted. This preparation can be scaled up for large-scale granule production without additional or increasing costs (Pereira et al. 2015). For the controlled release of urea fertilizers, urea is coated with sulfur. In the next step, the coated granule is sealed by polymer coating. The coating can be degraded by microbial, chemical, and physical processes. The releasing time of the fertilizer is decided by the thickness of coating and permeability. These factors can be affected by temperature and moisture (Trenkel 2010). This technology is applicable in high-value crops, environmentally sensitive areas, and fields highly susceptible to N losses (Pioneer.com. 2020).

Sulfur can be sprayed in molten form over urea granules. Then, sealant wax is applied over this to close any cracks or imperfections present in the coating. Other polymers used in sulfur coating include resin-based polymers, polyesters, and low-permeability polyethylene polymers for controlled release of fertilizers. Figure 9a shows the sulfur sprayed urea granules and chemical structure of urea. Figure 9b explains the urea release from sulfur and polymer coated urea. The nutrient releasing mechanism (caused by the damaging of the outer coating) is also explained. Pioneer (2020) reported that addition of aldehydes with urea reduces the solubility of urea (Pioneer.com. 2020). Aldehydes are mixed with urea to prepare the products such as urea–formaldehyde and methylene urea. Clapp (2001) has reported that the reaction of aldehyde and ammonia or primary amine of excess urea

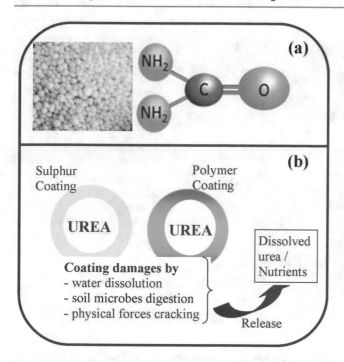

Fig. 9 Coating on urea: **a** sulfur sprayed on urea granules and chemical structure of urea, **b** the slow/controlled release of urea from the sulfur/polymer coated urea. Initially, the coating is damaged by water, microbes, and physical forces. Water moves into the coating and dissolves the nutrients. Then, the dissolved nutrients are released

(present in an aqueous medium) leads to the formation of a product called urea-triazone (Clapp 2001). The end product (urea-triazone) obtained in this process can be utilized in controlled releasing fertilizers applications.

6.3 Coating by Biological Products

Natural polymers are hydrophilic, eco-friendly, cost effective, easily available, and biodegradable. They can be prepared in various forms such as micro-particles, nanoparticles, beads, and hydrogels. The advantages of slow release are to increase the water holding capacity, aeration, soil permeability, and microbial activity. Polysaccharides such as starch, cellulose, dextrans, chitosan, pectin, guar gum, cyclodextrins, and alginate are utilized in the preparation of carriers. These carriers act as nano-carrier of bioactive compounds in agricultural applications (Campos 2013). Figure 10 explains the challenges in the utilization of conventional fertilizers (active material). Polymer-based nutrient delivery system is an alternative to solve the challenges.

Natural and biodegradable polymers such as tamarind powder, guar gum, and xanthan gum are used for the coating of urea. In this case, diatomite with epichlorohydrin is utilized as a crosslinker. *Diatomaceous* earth is used as a medium to grow plants because it is able to hold *fertilizers* and release to the roots. This fertilizer possesses high

nutrients, slow release property, and good water retention capacity (Mukerabigwi et al. 2015). Guar gum or guran is obtained from the guar plant seed, i.e., *Cyamopsis tetragonoloba* (L.) TAUB. Xanthan gum is prepared by gram-negative bacteria *Xanthomonas campestris*.

A composite made of poly(vinyl alcohol), horn meal, rapeseed cake, glycerol, and phosphogypsum is utilized for the coating of fertilizers. The composition of the fillers decides the mechanical, sorption properties, water vapor permeability, solubility in water, and the dimensional stability of the composite films. This kind of encapsulation leads to increase the releasing time of the fertilizers which is useful in the cultivation of tomato sprouts. This fertilizer is working well on the development of the roots of the plants (Treinyte et al. 2017).

Biochar is a carbon material made from biomass. This charcoal is utilized as a soil amendment material that enhances plant growth and crop yields. Particularly, it is useful in fields with depleted soils and lower level organic resources, nutrients, and water. Chen et al. (2018) reported about the biochar-polymers complex coating. This complex contains copolymer of PVA and polyvinylpyrrolidone (PVP) and biochar. It is applied as coating material in slow release fertilizers. Biochar helps to: decrease water absorbency of copolymer; increase degradability; improve the slow releasing property of urea. Particularly, the biochar made from rice plant exhibits an excellent release behavior, i.e., 65.28% nutrient leaching (Chen et al. 2018).

6.4 Starch in Slow/controlled Releasing

Demand and utilizations of sulfur-coated fertilizers decrease because of its high cost, process complexity, and inconsistent results. Instead of sulfur, bio-polymer coating on fertilizers is applied. Azeem et al. (2016) have reported that coating of fertilizers with synthetic polymers (such as polyethylene, polystyrene, polyacrylamide, and polysulfone) is also not economical and non-biodegradable. Utilization of biodegradable and low-cost material such as starch as coating material is an alternate way. A coating material is prepared using starch and polyvinyl alcohol (binder). Starch-based coating of fertilizers in proper thickness allows for promising controlled release characteristics (Azeem et al. 2016). Double-coated slow release fertilizer is prepared using ethyl cellulose as inner coating and starch-based superabsorbent polymer (starch-SAP) as outer coating. The fertilizer particles coated with starch-SAP shows superior slow release properties. The starch-SAP coated fertilizer offers reduced nitrogen release rate and steady release behavior for a period longer than 96 h to potato plants (Qiao et al. 2016).

Efficient fertilization practices can be developed with the help of nanoparticles and polymers. Urea is coated with

Fig. 10 Schematic diagram showing the plant growth factor delivery system. Utilization of conventional fertilizers (active material) in agriculture creates some problems including environmental and health issues. Polymer (polysaccharides)-based nutrient (bioactive compounds) delivery system for targeting applications is an alternative to solve these issues. It has benefits like slow release of nutrients and extending the duration of action

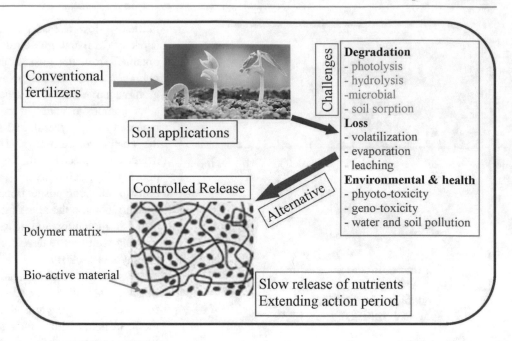

thermoplastic starch. Hydroxyapatite nanoparticles are dispersed in this urea matrix. These coating and dispersion lead to produce urea-hydroxyapatite nanocomposite. This nanocomposite controls the N-release and increases the P-availability in soil. The interaction between hydroxyapatite and urea reduces the phosphorus immobilization that increases the P-availability (Giroto et al. 2017).

6.5 Chitosan in Slow/controlled Releasing

Chitosan (CS) nanoparticle possesses polymeric cationic, biodegradable, bioabsorbable, and bactericidal characteristics. Hence, it can be a best candidate in controlled release of fertilizers. Chitosan nanoparticles are prepared by polymerizing methacrylic acid for the incorporation with NPK fertilizers. The slow release concept in fertilizers is able to save fertilizer consumption and minimize the environmental pollution (Corradini et al. 2010). In this way, chitosan nanoparticles loaded with nitrogen, phosphorous, and potassium are applied for wheat plants. Chitosan supports for the growth of roots, shoots, and leaves of plants. Nano-chitosan-NPK fertilizer increase harvest index, crop index, and mobilization index of the wheat plants (Abdel-Aziz et al. 2016). Crosslinking of chitosan with suberoyl chloride enhances the controlled release properties and mechanical strength of chitosan by forming a three-dimensional network structure. Chen et al. (2013) reported about the permeabilities of plant nutrients such as N, P, K, Zn^{2+}, and Cu^{2+} and plant growth regulator (naphthylacetic acid). N/P/K permeability is the important parameters to evaluate the controlled release fertilizer.

Utilization of crosslinking agent (suberoyl chloride) in crosslinking of N-phthaloyl acylated chitosan improves the properties (such as film-forming ability, mechanical property, and hydrophobicity) of chitosan membranes. Adding small amount of suberoyl chloride improves the properties but the excessive crosslinking leads to poor permeability. The macroelements (N, P, K), microelements (Zn^{2+} and Cu^{2+}), and plant growth regulator (naphthylacetic acid) releasing amount with different crosslinking densities (from 0 to 7.4%) is shown in Fig. 11 (Chen et al. 2013).

Crosslinking decreases the permeability of macroelements, microelements, and NAA which confirms that the crosslinked materials are suitable to use as controlled release microelement fertilizers. The permeability is low when crosslinking increases. Penetration of materials through crosslinked N-phthaloyl acylated chitosan membrane is confirmed from Fig. 11. Material releasing amount is less in crosslinked membrane and it is high in membrane without crosslinking. All curves exhibit the time-dependent release pattern. Material releasing amount is high when time increases (Chen et al. 2013).

6.6 Polyurethane in Slow/controlled Releasing

Controlled release fertilizers are made by coating fertilizers using polymers like polyurethane. They can be synthesized from low-cost, biodegradable, and renewable cottonseed oil. The specialty of this coating over conventional methods is increased surface roughness, reduced surface energy, and superhydrophobic nature. The superhydrophobic nature offers the non-wetting contact of water in gas state instead of

Fig. 11 Nutrients release vs time curves. Permeability of macro and micronutrients through crosslinked N-phthaloyl acylated chitosan membrane (with different crosslinking densities). **a** Urea; **b** phosphorus; **c** potassium; **d** zinc; **e** copper; **f** NAA. crosslinking densities ■ 0%; ● 2.9%; ▲ 4.4%; ▼ 5.9%; ◆ 7.4%. *Source* Chen et al. (2013), with permission

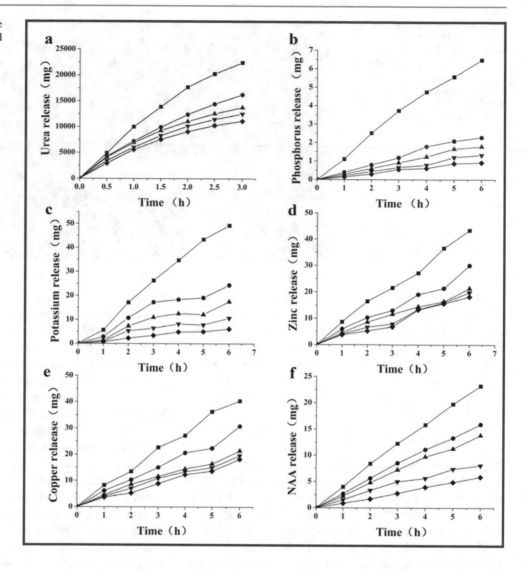

in liquid state (Xie et al. 2017). The oxidation and brittleness of sulfur are preventing sulfur coating in slow release fertilizers. Also, it is not economical to use synthetic polymers for coating fertilizers. Hence, the best alternate available coating is using natural polymer and it is a good water absorber. However, the nutrient release longevity is available for less than 30 days. This duration will not meet nitrogen supply requirements for crops. To overcome the drawback of the bio-resources, a new technology is adopted. In this method, natural polymers are converted into bio-polyols that can work better. The fertilizer is coated with polyurethane made from wheat straw. Solvents like ethylene glycol/ethylene carbonate are used to liquefy the wheat straw. The liquid is added with polymethylene polyphenyl isocyanate and castor oil to get bio-based polyurethane (Lu et al. 2015).

6.7 Zeolites in Slow/controlled Releasing

Zeolites are aluminosilicates of sodium, potassium, calcium, and barium. Their applications lie in cation exchanges and molecular sieves. SEM images of different zeolite particles are shown in Fig. 12. These zeolite particles have properties like adsorption of urease enzyme. Hence, they (except zeolite L) are useful in the preparation of urea-sensitive biosensors (Kucherenko et al. 2015). Compared to the conventional fertilizers, releasing of the fertilizer contents is slow and more while applying the zeolites mixed/coated macro or micronutrients. Zeolites reduce the nutrient loss by controlling the release and improving uptake. Yuvaraj et al. (2018) reported about the modification of zeolites for slow release fertilizer application. Zeolites modified by a surfactant hexadecyltrimethylammonium bromide are treated

Fig. 12 SEM images of zeolite particles. **a** Nanozeolite beta, **b** nanozeolite L, **c** 80 nm silicalite-1, **d** 160 nm silicalite-1, **e** 450 nm silicalite-1, **f** mesoporous silica spheres, **g** zeolite L (*Source* Kucherenko et al. 2015, with permission)

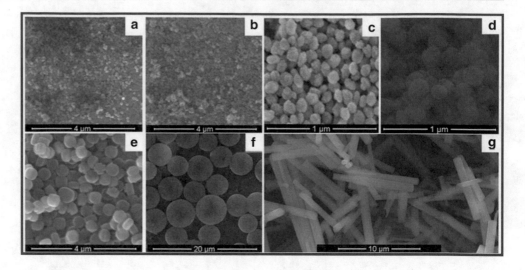

with KH_2PO_4 for preparation of slow release of phosphorus fertilizer. From this fertilizer, phosphorus is released even after 1080 h (Bansiwal et al. 2006). Zinc is utilized by plants as a micronutrient up to 2–3%. To avoid zinc fixation in the soil, nanozeolites are synthesized by ball milling and fortified with zinc by loading zinc sulfate. Zinc fertilizer coated with nanozeolites releases zinc for a period of 1176 h (Yuvaraj et al. 2018).

Phosphorus and potassium are incorporated in zeolite to form a nanofertilizer. Release of phosphorous and potassium from this nanofertilizer is higher than the conventional fertilizer. Also, the accumulation of phosphorous and potassium are more in plants while applying this zeolite-based nanofertilizer. After the application of this nanofertilizer, the soil possesses better pH, moisture, EC, CEC, and availability of P and K (Rajonee et al. 2017). Utilization of porous nanomaterials (zeolites, clay, or chitosan) in fertilizer applications considerably reduces nitrogen loss by controlling the release and improving uptake (Millan et al. 2008; Abdel-Aziz et al. 2016; Panpatte et al. 2016). Ammonium mixed zeolites improve the solubility of phosphate that leads to improving the availability of phosphorus (Dwivedi et al. 2016).

7 Influences of Nanofertilizers on the Soil and Crop Plants

It is analyzed and observed from the various literatures while applying the nanofertilizer, it is essential to consider some factors like concentration of nanofertilizers, biocompatibility, solubility, nutrient releasing period, control over nutrient release, encapsulation/coating of nanofertilizer, and size of nanomaterial. Some of the factors have influences on the agricultural output and yield of the plants. They are enumerated below.

- Application of chemical fertilizers has drawback like soil acidification. Slow releasing fertilizers avoid it.
- Utilization of toxic or biologically incompatible materials causes harm to all biological organisms. Biocompatibility is an essential one. It can be achieved by utilizing bio-synthesized nanomaterials instead of chemically synthesized nanomaterials.
- The concentration of the nanomaterial (nanofertilizer) is more important. Applying the nanofertilizer (like carbon nanomaterials and iron nanoparticles) up to optimum quantity or at lower concentration yields better results. Improvement in yields is due to the increased water uptake and transport, seed germination, and antioxidant activities.
- Applying the nanofertilizer at higher concentration decreases the yields. Due to aggregation at cell walls that blocks the nutrients transfer and antioxidant activities.
- Encapsulation of microorganisms enhances the N, P, and K availability that stimulate the plant growth.
- Sulfur-coated urea releases N slowly due to gradual microbial, chemical, and physical degradation process.
- Urea reacted with aldehydes compounds release their N slowly.
- Hydroxyapatite nanoparticles coating on urea reduces the conversion of urea into ammonia and release the nitrogen slowly.
- Hydroxyapatite nanoparticles release phosphorous and improve the phosphorous availability to the plants without changing soil pH or soil acidification.
- Hydroxyapatite nanoparticles decrease the water eutrophication that improves the yield.
- Hydroxyapatite nanoparticles can release both phosphorous and calcium.
- ZnO nanoparticles have solubility higher than the bulk ZnO.

- Porous nanomaterials (ammonium charged zeolites, graphene oxide, and nanocalcite) reduce nitrogen loss in considerable quantity by controlling the nutrient release that enhances the plant uptake process.
- In controlled release fertilizers, the thickness of coating and permeability determines the releasing time of the fertilizer.
- Ammonium mixed zeolites enhance the solubility of phosphate.

8 Conclusion

Nanofertilizer is the engineered nanomaterial prepared by applying nanotechnology. Particles of this material are in the nanometer size range. Due to the presence of nano-sized particles (particularly size less than the plants cells), this new generation fertilizer has an ability to reduce the fertilizer consumption and to deliver the nutrients better than the conventional fertilizers. It is observed from the literature that the nanofertilizers will play a major role in the modern agriculture. Engineered nanomaterials are applied in plant stress, pathogens controlling, target delivery (gene transfer), plants development, and soil enhancement. Nanomaterials like magnesium hydroxide, carbon nanomaterials, ZnO nanoparticles, and iron oxide nanomaterials are utilized in pre-soaking of seeds and seed germination technology. They improve the solubility of the nutrients. Some nanofertilizers have good water retention capacity. Comparing to the conventional fertilizers, they release the nutrients and reduce the nutrient loss in a better way. Slow and controlled released nanofertilizers augment the plant growth that leads to higher level plant productivity and crop yield.

Fertilizers are expected to do more functions like stimulation of soil microorganisms and mobilization of phosphorus and potassium to make them easily available to the plants. To cope up with the agriculture market, a fertilizer should be fast acting and long-lasting. The fertilizer should provide support to better yield (with a uniform, dense, or full growth) and eliminate entanglement. It should promote humus formation and soil health. Advanced technologies (like nanotechnology, nanofertilizers, and biofertilizers) support to fertilizers for the supply of nutrients properly.

Acknowledgements The author expresses immense thanks to her husband Mr. G. Sankar for his assistances in this work. Also, she acknowledges the assistances of International Research Center, Kalasalingam Academy of Research and Education (Deemed University), Krishnankoil-626 126, (India) for providing necessary supports and facilities.

References

Abdel-Aziz HMM, Hasaneen MNA, Omer AM (2016) Nano chitosan-NPK fertilizer enhances the growth and productivity of wheat plants grown in sandy soil. Span J Agric Res 14:17. https://doi.org/10.5424/sjar/2016141-8205

Aboul-Enein AM, Salama ZA, Gaafar AA, Aly HF, A bou-Elella F, Habiba AA (2016) Identification of phenolic compounds from banana peel (Musa aradaisica L.) as antioxidant and antimicrobial agents. J Chem Pharm Res 8(4):46–55

Adisa IO, Reddy VL, Peralta-Videa JR et al (2019) Recent advances in nano-enabled fertilizers and pesticides: a critical review of mechanisms of action. Environ Sci Nano 6(7):2002–2030. https://doi.org/10.1039/C9EN00265K

Ali MA, Rehman I, Iqbal A, Din S, Rao AQ, Latif A, Samiullah TR, Azam S, Husnain T (2014) Nanotechnology, a new frontier in agriculture. Adv Life Sci 1(3):129–138

Babar A, KuShaari K, Man Z (2016) Effect of coating thickness on release characteristics of controlled release urea produced in fluidized bed using waterborne starch biopolymer as coating material. Procedia Eng 148:282–289

Bansiwal AK, Rayalu SS, Labhasetwar NK, Juwarkar AA, Devotta S (2006) Surfactant-modified zeolite as a slow release fertilizer for phosphorus. J Agric Food Chem 54(13):4773–4779

Borling K (2003) *Phosphorus sorption, accumulation and leaching.* Diss. Sveriges lantbruksuniv. Acta Universitatis Agriculturae Sueciae. Agraria, 1401–6249; 428

Brady NR, Weil RR (2017) The nature and properties of soils. Prentice Hall, New Jersey, pp 415–473. ISBN: 9780133254525

Brunner TJ, Wick P, Manserp S, Grass RN, Limbach LK, Bruinink A, Stark WJ (2006) In vitro cytotoxicity of oxide nanoparticles: comparison to asbestos, silica, and the effect of particle solubility. Environ Sci Technol 40:4374–4381

Carpenter SR, Caraco NF, Correll DL, Howarth RW, Sharpley AN, Smith VH (1998) Nonpoint pollution of surface waters with phosphorus and nitrogen. Ecol Appl 8:559–568. https://doi.org/10.1890/1051-0761(1998)008[0559:NPOSWW]2.0.CO;2

Chen C, GaoZ QX, Hu S (2013) Enhancement of the controlled-release properties of chitosan membranes by crosslinking with suberoyl chloride. Molecules 18(6):7239–7252. https://doi.org/10.3390/molecules18067239

Chen S, Yang M, Ba C, YuS JY, Zou H, Zhang Y (2018) Preparation and characterization of slow-release fertilizer encapsulated by biochar-based waterborne copolymers. Sci Total Environ 615:431–437

Clapp J (2001) Urea-triazone N characteristics and uses. In: Proceedings of the 2nd international nitrogen conference on science and policy. Scientific World 1(S2), pp 103–107. ISSN 1532-2246. https://doi.org/10.1100/tsw.2001.356

Corradini E, MouraMR De, Mattoso LHC (2010) A preliminary study of the incorporation of NPK fertilizer into chitosan nanoparticles. Express Polym Lett 4(8):509–515

Dasgupta N, Ranjan S, Mundekkad D, Ramalingam C, Shanker R, Kumar A (2015) Nanotechnology in agro-food: from field to plate. Food Res Internat 69:381–400

Dhillon GS, Brar SK, Kaur S, Verma M (2012) Green approach for nanoparticle biosynthesis by fungi: current trends and applications. Crit Rev Biotechnol 32:49–73

Dimkpa C (2014) Potential of nanotechnology in crop fertilization: current state and future perspectives. In: World fertilizer congress, vol 16, p 46. https://allanore.mit.edu/Global_Challenges_files/Proceedigns_16WFC_high_resolution_1.pdf

Dufresne A, Dupeyre D, Vignon MR (2000) Cellulose microfibrils from potato tuber cells: processing and characterization of starch-cellulose microfibril composites. J Appl Poly Sci 76:2080–2092. https://doi.org/10.1002/(SICI)1097-4628(20000628)76:143.0.CO;2-U

Duhan JS, Kumar R, Kumar N, Kaur P, Nehra K, Duhan S (2017) Nanotechnology: the new perspective in precision agriculture. Biotechnol Rep 1:11–23

Dwivedi S, Saquib Q, Al-Khedhairy AA (2016) Musarrat J (2016) Understanding the role of nanomaterials in agriculture. In: Singh DP, Singh HB, Prabha R (eds) Microbial inoculants in sustainable agricultural productivity. Springer; New Delhi, India, pp 271–288

Giroto AS, Guimarães GGF, Foschini M, Ribeiro C (2017) Role of slow-release nanocomposite fertilizers on nitrogen and phosphate availability in soil. Sci Rep 7:46032. https://doi.org/10.1038/srep46032

Haifa Group (2020) Slow release fertilizer versus controlled release fertilizer (online) Available at: https://www.haifa-group.com/knowledge_center/articles/haifa_articles/slow_release_fertilizer.aspx. Accessed 8 June 2020. https://doi.org/10.1002/14356007.a10_323.pub3

Hussein HS, Shaarawy HH, Hussien NH, Hawash SI (2019) Preparation of nano-fertilizer blend from banana peels. Bull Nat Res Centre 43(1):26. https://doi.org/10.1186/s42269-019-0058-1

Israelagri.com (2016) A new generation of open field fertilizer (online). Available at: https://www.israelagri.com/?CategoryID=484&ArticleID=1208. Accessed 6 June 2020

Jones Jr B (2012) Inorganic chemical fertilisers and their properties. In: Plant nutrition and soil fertility manual, 2nd edn. CRC Press. ISBN 978-1-4398-1609-7. eBook ISBN 978-1-4398-1610-3

Khandel P, Yadaw RK, Soni DK et al (2018) Biogenesis of metal nanoparticles and their pharmacological applications: present status and application prospects. J Nanostruct Chem 8:217–254. https://doi.org/10.1007/s40097-018-0267-4

Kottegoda N, Sandaruwan C, Priyadarshana G, Siriwardhana A, Rathnayake UA, Arachchige DMB, Kumarasinghe AR, Dahanayake D, Karunaratne V, Amaratunga GAJ (2017) Urea-hydroxyapatite nanohybrids for slow release of nitrogen. ACS Nano 11(2):1214–1221. https://doi.org/10.1021/acsnano.6b07781

Kucherenko I, Soldatkin O, Kasap BO, Kirdeciler SK, Kurc BA, Jaffrezic-Renault N, Soldatkin A, Lagarde F, Dzyadevych S (2015) Nanosized zeolites as a perspective material for conductometric biosensors creation. Nanoscale Res Lett 10(1):209. https://doi.org/10.1186/s11671-015-0911-6

Kumar R, Ashfaq M, Verma N (2018) Synthesis of novel PVA–starch formulation-supported Cu–Zn nanoparticle carrying carbon nanofibers as a nanofertilizer: controlled release of micronutrients. J Mat Sci 53(10):7150–7164

Lateef A, Nazir R, Jamil N, Alam S, Shah R, Khan MN, Saleem M (2016) Synthesis and characterization of zeolite based nano–composite: an environment friendly slow release fertilizer. Micro-porous Microporous Mater 232:174–183. https://doi.org/10.1016/j.micromeso.2016.06.020

Lin BL, Sakoda A, Shibasaki R, Suzuki M (2001) A modelling approach to global nitrate leaching caused by anthropogenic fertilisation. Water Res 35(8):1961–1968. https://doi.org/10.1016/S0043-1354(00)00484-X

Liu R, Lal R (2015) Potentials of engineered nanoparticles as fertilizers for increasing agronomic productions. Sci Total Environ 514:131–139

Liu R, Lal R (2015a) Synthetic apatite nanoparticles as a phosphorus fertilizer for soybean (Glycine max). Sci Rep 4:5686. https://doi.org/10.1038/srep05686

Lu P, Zhang Y, Jia C, Wang C, Li X, Zhang M (2015) Polyurethane from liquefied wheat straw as coating material for controlled release fertilizers. BioResources 10(4):7877–7888

Luca M (2018) Nanotechnology in agriculture: new opportunities and perspectives. https://doi.org/10.5772/intechopen.74425. https://www.intechopen.com/books/new-visions-in-plant-science/nanotechnology-in-agriculture-new-opportunities-and-perspectives

Matsuzawa Y, Takada Y, Kodaira T, Kihara H, Kataura H, Yoshida M (2014) Effective nondestructive purification of single-walled carbon nanotubes based on high-speed centrifugation with a photochemically removable dispersant. J Phys Chem C 118:5013–5019

Miao YF, Wang ZH, Li SX (2015) Relation of nitrate N accumulation in dryland soil with wheat response to N fertilizer. Field Crops Res 170:119–130. https://doi.org/10.1016/j.fcr.2014.09.016

Milani N, McLaughlin MJ, Stacey SP, Kirby JK, Hettiarachchi GM, Beak DG, Cornelis G (2012) Dissolution kinetics of macronutrient fertilizers coated with manufactured zinc oxide nanoparticles. J Agric Food Chem 60(16):3991–3998

Millan G, Agosto F, Vazquez M, Botto L, Lombardi L, Juan L (2008) Use of clinoptilolite as a carrier for nitrogen fertilizers in soils of the Pampean regions of Argentina. Cienc Investig Agrar 35:293–302

Monreal CM, DeRosa M, Mallubhotla SC, Bindraban PS, Dimkpa C (2016) Nanotechnologies for increasing the crop use efficiency of fertilizer-micronutrients. Biol Fertil Soils 52:423–437. https://doi.org/10.1007/s00374-015-1073-5

Morsy M, Helal M, El-Okr M, Ibrahim M (2014) Preparation, purification and characterization of high purity multi-wall carbon nanotube. Spectrochim Acta a Mol Biomol Spectrosc 132:594–598

Mukerabigwi JF, Wang Q, Ma X, Liu M, Lei S, Wei H, Huang X, Cao Y (2015) Urea fertilizer coated with biodegradable polymers and diatomite for slow release and water retention. J Coatings Technol Res 12(6):1085–1094

Naik MR, Kumar BK, Manasa K (2017) Polymer coated fertilizers as advance technique in nutrient management. Asian J Soil Sci 12(1):228–232

NUST MISIS (2017) Nanovitamins for Macroharvests: new generation fertilizers based on metal nanoparticles increased crop productivity by 25% (online). En.misis.ru. Available at: https://en.misis.ru/university/news/science/2017-11/4987/. Accessed 5 June 2020

Ombodi A, Saigusa M (2000) Broadcast application versus band application of polyolefin-coated fertilizer on green peppers grown on andisol. J Plant Nutr 23(10):1485–1493

Pachapur V, Brar SK, Verma M, Surampalli RY (2015) Nano-ecotoxicology of natural and engineered nanomaterials for animals and humans. Nanomater Environ: 421–437. https://doi.org/10.1061/9780784414088.ch16

Palchoudhury S, Jungjohann KL, Weerasena L, Arabshahi A, Gharge U, Albattah A, Miller J, Patel K, Holler RA (2018) Enhanced legume root growth with pre-soaking in α-Fe$_2$O$_3$ nanoparticle fertilizer. RSC Adv 8(43):24075–24083

Panpatte DG, Jhala YK, Shelat HN, Vyas RV (2016) Microbial inoculants in sustainable agricultural productivity. Springer, New Delhi, India. Nanoparticles: The next generation technology for sustainable agriculture, pp 289–300

Pereira EI, Da Cruz CC, Solomon A, Le A, Cavigelli MA, Ribeiro C (2015) Novel slow-release nanocomposite nitrogen fertilizers: the impact of polymers on nanocomposite properties and function. Ind Eng Chem Res 54(14):3717–3725

Peteu SF, Oancea F, Sicuia OA, Constantinescu F, Dinu S (2010) Responsive polymers for crop protection. Polymers 2:229–251. https://doi.org/10.3390/polym2030229

Phys.org. (2015) Researchers identify behaviors of nanoparticle that shows promise as nanofertilizer (online) Available at: https://phys.org/news/2015-09-behaviors-nanoparticle-nanofertilizer.html#nRlv. Accessed 5 June 2020

Pioneer.com. (2020) Controlled-release nitrogen fertilizers (online) Available at: https://www.pioneer.com/us/agronomy/controlled_release_nitrogen.html Accessed 8 June 2020.

Priyom Bose P (2020) Uses of nanotechnology in fertilizers (online). AZoNano.com. Available at: https://www.azonano.com/article.aspx?ArticleID=5446. Accessed 8 June 2020

Qiao D, Liu H, Yu L, Bao X, Simon GP, Petinakis E, Chen L (2016) Preparation and characterization of slow-release fertilizer encapsulated by starch-based superabsorbent polymer. Carb Pol 147:146–154

Rajonee AA, Zaman S, Huq SMI (2017) Preparation, characterization and evaluation of efficacy of phosphorus and potassium incorporated nano fertilizer. Adv Nanoparticles 6(02):62

Raliya R, Nair R, Chavalmane S, Wang WN, Biswas P (2015) Mechanistic evaluation of translocation and physiological impact of titanium dioxide and zinc oxide nanoparticles on the tomato (Solanum lycopersicum L.) plant. Metallomics 7:1584–1594. https://doi.org/10.1039/C5MT00168D

Raliya R, Tarafdar JC, Biswas P (2016) Enhancing the mobilization of native phosphorus in mung bean rhizosphere using ZnO nanoparticle synthesized by soil fungi. J Agric Food Chem 7. https://doi.org/10.1021/acs.jafc.5b05224

Ramos EVC, De Oliveira JL, Fraceto LF, Singh B (2015) Polysaccharides as safer release systems for agrochemicals. Agron Sust Dev 35 (1):47–66. https://doi.org/10.1007/s13593-014-0263-0

Rana S, Kalaichelvan PT (2013) Ecotoxicity of nanoparticles. ISRN Toxicology 2013. Article ID 574648. https://doi.org/https://doi.org/10.1155/2013/574648

Rui M, Ma C, Hao Y, Guo J, Rui Y, Tang X, Zhao Q, et al. (2016) Iron oxide nanoparticles as a potential iron fertilizer for peanut (Arachis hypogaea). Front Plant Sci 7

Ruiqiang L, Lal R (2016) Nanofertilizers. In: Lal R (ed) Encyclopedia of soil science. CRC Press. ISBN 9781498738903. https://doi.org/10.1081/E-ESS3-120053199

Sabir A, Yazar K, Sabir F, Kara Z, Yazici MA, Goksu N (2014) Vine growth, yield, berry quality attributes and leaf nutrient content of grapevines as influenced by seaweed extract (Ascophyllum nodosum) and nanosize fertilizer pulverizations. Sci Hortic 175:1–8. https://doi.org/10.1016/j.scienta.2014.05.021

Santoso D, Lefroy RD, Blair GJ (1995) Sulfur and phosphorus dynamics in an acid soil/crop system. Soil Res 33(1):113–124

Scherer HW, Mengel K, Kluge G, Severin K (2009) Fertilizers. In: Ullmann's encyclopedia of industrial chemistry. Wiley-VCH, Weinheim. https://doi.org/10.1002/14356007.a10_323.pub3

Schoumans O (2015) Phosphorus leaching from soils: process description, risk assessment and mitigation. Diss. Wageningen University and Research Centre

Shalaby TA, Bayoumi Y, Abdalla N, Taha H, Alshaal T, Shehata S, Amer M, Domokos-Szabolcsy É (2016) El-Ramady H (2016) Nanoparticles, soils, plants and sustainable agriculture. In: Shivendu R, Nandita D, Eric L (eds) Nanoscience in food and agriculture 1. Springer; Cham, Switzerland, pp 283–312

Shang Y, Hasan MK, Ahammed GJ, Li M, Yin H, Zhou J (2019) Applications of nanotechnology in plant growth and crop protection: a review. Molecules24(14):2558. https://doi.org/https://doi.org/10.3390/molecules24142558

Shenker M, Seitelbach S, Brand S, Haim A, Litaor MI (2004) Redox reactions and phosphorus release in re-flooded soils of an altered wetland. Eur J Soil Sci 56:515–525. https://doi.org/10.1111/j.1365-2389.2004.00692.x

Shinde S, Paralikar P, Ingle AP, Rai M (2020) Promotion of seed germination and seedling growth of Zea mays by magnesium hydroxide nanoparticles synthesized by the filtrate from Aspergillus niger. Arab J Chem 13(1):3172–3182. https://doi.org/10.1016/j.arabjc.2018.10.001

Singh MD, Chirag G, Om Prakash P, Hari Mohan M, Prakasha G, Vishwajith, (2017) Nano-fertilizers is a new way to increase nutrients use efficiency in crop production. Int J Agri Sci 9(7):3831–3833

Smart Fertilizer (2020) Five ways to minimize nitrate leaching (Online). 1 Aug 2020. Available from: https://www.smart-fertilizer.com/articles/ways-to-minimize-nitrate-leaching/

Solanki P, Bhargava A, Chhipa H, Jain N, Panwar J (2015) Nano-fertilizers and their smart delivery system. In: Nanotechnologies in food and agriculture. Springer International Publishing, pp 81–101

Subbarao ChV, Kartheek G, Sirisha D (2013) Slow release of potash fertilizer through polymer coating. Int J Appl Sci Eng 11(1):25

Subramanian KS, Manikandan A, Thirunavukkarasu M, Rahale CS (2015) Nano-fertilizers for balanced crop nutrition. In: Rai M, Ribeiro C, Mattoso L, Duran N (eds) Nanotechnologies in food and agriculture. Springer; Cham, Switzerland, pp 69–80

Swaminathan S, Edward BS, Kurpad AV (2013) Micronutrient deficiency and cognitive and physical performance in Indian children. Eur J Clin Nutr 67:467–474. https://doi.org/10.1038/ejcn.2013.14

Tarafdar JC, Raliya R, MahawarH RI (2014) Development of zinc nanofertilizer to enhance crop production in pearl millet (Pennisetum americanum). Agricultural Research 3(3):257–262

Treinyte J, Grazuleviciene V, Paleckiene R, Ostrauskaite J, Cesoniene C (2017) Biodegradable polymer composites as coating materials for granular fertilizers. J Poly Environ: 1–12

Trenkel ME (1997) Controlled-release and stabilized fertilizers in agriculture, vol 11. International Fertilizer Industry Association, Paris

Trenkel ME (2010) Slow- and controlled-release and stabilized fertilizers: an option for enhancing nutrient use efficiency in agriculture. International Fertilizer Industry Association (IFA) Paris, France. ISBN 978-2-9523139-7-1

Verma SK, Das AK, Gantait S, Kumar V, Gurel E (2019) Applications of carbon nanomaterials in the plant system: a perspective view on the pros and cons. Sci Total Environ 667:485–499

Wang X, Lü S, Gao C, Xu X, Wei Y, Bai X, Feng C, Gao N, Liu M, Wu L (2014) Biomass-based multifunctional fertilizer system featuring controlled-release nutrient, water-retention and amelioration of soil. RSC Adv 4:18382–18390. https://doi.org/10.1039/C4RA00207E

Wang ZH, Miao YF, Li SX (2015) Effect of ammonium and nitrate nitrogen fertilizers on wheat yield in relation to accumulated nitrate at different depths of soil in drylands of China. Field Crops Res 183:211–224. https://doi.org/10.1016/j.fcr.2015.07.019

Wiki/fertilizer (2018) Fertilizer (online) Available at: https://en.wikipedia.org/wiki/Fertilizer. Accessed 6 Jun. 2018

Wikipedia (2014) Leaching (agriculture) (Online). 1 Aug 2020. Available from: https://en.wikipedia.org/wiki/Leaching_(agriculture)

Wilson MA, Tran NH, Milev AS, Kannangara GK, Volk H, Lu GM (2008) Nanomaterials in soils. Geoderma 146(1–2):291–302

Wu L, Liu M (2008) Preparation and properties of chitosan-coated NPK compound fertilizer with controlled-release and water-retention. Carbo Pol 72(2):240–247

Xie J, Yang Y, Gao B, Wan Y, Li YC, Xu J, Zhao Q (2017) Biomimetic superhydrophobic biobased polyurethane-coated fertilizer with atmosphere "Outerwear." ACS App Mat Interfac 9(18): 15868–15879

Yang H, Xu M, Koide RT, Liu Q, Dai Y, Liu L, Bian X (2016) Effects of ditch-buried straw return on water percolation, nitrogen leaching and crop yields in a rice-wheat rotation system. J Sci Food Agric 96:1141–1149. https://doi.org/10.1002/jsfa.7196

Yatim NM, Shaaban A, Dimin MF, Yusof F (2015) Statistical evaluation of the production of urea fertilizer-multiwalled carbon nanotubes using plackett burman experimental design. Proc-Social Behav Sci 195:315–323

Yatim NM, Shaaban A, Dimin MF, Yusof F, Razak JA (2018) Effect of functionalised and non-functionalised carbon nanotubes-urea fertilizer on the growth of paddy. Trop Life Sci Res 29(1):17

Yuan J, Chen Y, Li H, Lu J, Zhao H, Liu M, Nechitaylo GS, Glushchenko NN (2018) New insights into the cellular responses to iron nanoparticles in Capsicum annuum. . Sci Rep 8(1):1–9. https://doi.org/10.1038/s41598-017-18055-w

Yuvaraj M, Subramanian KS (2018) Development of slow release Zn fertilizer using nano-zeolite as carrier. . J Plant Nutr 41(3):311–320

Zak D, Gelbrecht J (2007) The mobilisation of phosphorus, organic carbon and ammonium in the initial stage of fen rewetting (a case study from NE Germany). Biogeochemistry 85:141–151. https://doi.org/10.1007/s10533-007-9122-2

Zaytseva O, Neumann G (2016) Carbon nanomaterials: production, impact on plant development, agricultural and environmental applications. Chem Biol Technol Agric:17. https://doi.org/https://doi.org/10.1186/s40538-016-0070-8

Zhou L, Cai D, He L, Zhong N, Yu M, Zhang X, Wu Z (2015) Fabrication of a high-performance fertilizer to control the loss of water and nutrient using micro/nano networks. ACS Sustain Chem Eng 3(4):645–653. https://doi.org/10.1021/acssuschemeng.5b00072

Nano-fertilizers and Nano-pesticides as Promoters of Plant Growth in Agriculture

Niloy Sarkar, Swati Chaudhary, and Mahima Kaushik

Abstract

A significant shift towards sustainable agriculture has been observed in the past decade in order to address the nutritional security of global population along with a focus on minimizing environmental impact as much as possible. This can be achieved by using nanotechnology in agriculture field mainly in the form of engineered nanoparticles (ENPs)-based nano-fertilizers and nano-pesticides. This is due to the fact that very small amount of conventional fertilizers actually reach the targeted site, which can be due to leaching of chemicals, microbial degradation, run-off, evaporation or hydrolysis. These fertilizers in excess amount severely affect the nutrient equilibrium of the soil. Unlike conventional agrochemicals, ENPs-based nano-fertilizers and nano-pesticides have increased selectivity, gradual release dynamics and resistance to physiochemical degradation. This reduces the environmental accumulation and thus, ill effects on the agro-ecosystem. The use of these nano-agrochemicals also improves the crop productivity by increasing the availability of nutrients in soil and their uptake by plants. These nano-materials can reduce the occurrence of crop diseases by acting upon pathogens directly, through several mechanisms. This chapter elaborates on the role of ENPs-based nano-fertilizers and nano-pesticides in the growth of plants. It also elucidates various types, mechanisms, benefits and potential applications of such novel nano-agrochemicals, which are necessary to attain more sustainable agriculture practices.

Keywords

Engineered nanoparticles • Herbicides • Nano-agrochemicals • Nano-toxicity • Sustainable agriculture

1 Introduction

The global food security and its protection faces challenges due to increased global population and changes in the dietary conditions. The significant factors that create hindrance in achieving the global food security include low soil nutrients, climatic conditions such as drought and flood, and agricultural crop pests. There are approximately 22,000 species of plant pathogens that attack the crops globally (Adisa et al. 2019). Plant pests include a wide variety of organisms like nematodes, molluscs, arthropods, weeds, etc. which reduce the crop yield. Pests are thought to destroy about 18% of crops globally. It is estimated that pesticides enhance the growth of approximately 70% of all crops globally, whereas without the use of the pesticides, production of fruits, vegetables and cereals may decrease up to 78, 54 and 32%, respectively (Zhang et al. 2018). The global average of pesticide consumption stands at 2 million tons annually, out of which, India's contribution is only 3.75% (Devi et al. 2017). The qualities of an ideal pesticide should include feasible degradability, high selectivity, potent ability and stability. Chemical engineers and agro-scientists are constantly working towards development of ideal pesticides, which fulfil all the above-mentioned criterion. However, overuse and misuse of pesticides cause harm not only to human beings but also to the non-target biota, as the pesticide residues are transported to adjoining natural ecosystems. Most at risk are those who come into direct contact

N. Sarkar · M. Kaushik (✉)
Nano-Bioconjugate Chemistry Lab, Cluster Innovation Centre, University of Delhi, Delhi, 110007, India
e-mail: mkaushik@cic.du.ac.in

N. Sarkar
Department of Environmental Studies, University of Delhi, Delhi, India

S. Chaudhary
Department of Applied Science, M.S.I.T., GGSIP University, New Delhi, India

© Springer Nature Switzerland AG 2021
P. Singh et al. (eds.), *Plant-Microbes-Engineered Nano-particles (PM-ENPs) Nexus in Agro-Ecosystems*,
Advances in Science, Technology & Innovation,
https://doi.org/10.1007/978-3-030-66956-0_10

with the pesticides, such as factory workers and agriculturalists, who are involved in the manufacture and application (Carvalho et al. 2017).

Pesticides are applied on the field either in the form of powder, solution or emulsion, from where approximately 2–25% of pesticides move away from the target, whereas about 80–90% volatilizes into the atmosphere in the period of few days (Aktar et al. 2009). Pesticides can be leached by surface run-off and end up in water bodies. As a general rule, the high volatility or instability of the pesticides lead to the farthest deposition of it from the application site. Synthetic pesticides, which are non-biodegradable, can persist in the soil for weeks to years and have been classified as persistent organic pollutants (POPs). First-generation organochloride pesticides were especially persistent, hence because of this reason and development of pest resistance urged their replacement by organophosphates, which are less persistent in the environment (Caravalho et al. 2017). However, the problem is that many animals across related phylogenies share similar physiologies and are also affected by the pesticides. Animals such as bees, shrimps and crabs are direct non-target casualties of pesticide application. Both target (pests) and non-target species are consumed within the food chain and concentrate within the biomass in higher trophic levels and have been termed as biological magnification. Overuse of pesticides cause population of beneficial soil bacteria to decline, which are responsible for long-term fixing of nutrients, e.g. nitrogen fixating bacteria (Aktar et al. 2009). Ultimately pesticides make their way to humans, where they have been related with diseases such as cancer, obesity and endocrine disruption (Caravalho et al. 2017). Phasing out of POPs was agreed upon in the Stockholm Convention, which includes several notorious pesticides such as dichloro diphenyl trichloroethane (DDT) and endosulfan, which are detrimental to both the environment and human health.

The concept behind augmenting soil nutrition and thus increasing crop productivity is also as old as agriculture. Ancient people used mulch, manure and guano as a means of fertilizer. Nutritional requirements of plants can be broadly divided into macronutrients, primary among them being, nitrogen (N), phosphorus (P) and potassium (K), collectively called NPK requirement. Micronutrients include a multitude of minerals and elements, which are needed in much smaller dosage and are plant specific. Fertilizers can be classified on the basis of different number of nutrients present in the product. Single or straight fertilizers contain either N, P or K. Binary fertilizers contain two nutrients of any of the above type, whereas NPK fertilizer contains all the three (Erisman et al. 2008). Although fertilizers, unlike pesticides, are non-toxic and nutrient rich, yet they pose a considerable

risk to the environment. Ecologically, they are responsible for disturbing the delicate nutrient cycle via nutrient loading, as well as being a source for pollutant by-products. Most of the fertilizers used are washed away with surface water from rain or irrigation and end up in water bodies. Fertilizers contain a large amount of phosphate, which coincidentally is a limiting factor in such aquatic ecosystems. This causes an exponential increase in the cyanobacterial and algal population, which is detrimental to the aquatic ecosystem, as it prevents sunlight from penetrating deeper into the water body. The eventual death of cyanobacterial and algal population causes significant oxygen depletion in the water body in a process known as eutrophication. Nutrient loading from estuaries into oceans can also cause oxygen depletion in a similar manner and lead to dead zones or areas of significant lower dissolved oxygen with reduced biodiversity. Cyanobacterial blooms can release toxins, which can accumulate and magnify within the ecosystem (Schmidt et al. 2013).

The fertilizer industry is also considered a source of radionuclide and heavy metals such as mercury, cadmium, arsenic, lead, copper and nickel which lead to contamination and accumulation in soil and plant biota such as fruits and vegetables, from where they can affect humans (Atafar et al. 2010; Savci 2012). Out of all the nitrogen fertilizers widely used, only 50% is used by the plants, while 2–20% is evaporated, 15–25% reach with soil organic compounds, whereas remaining 10% contaminates groundwater (Savci 2012). In case of such fertilizers, they are converted to nitrates via microbial nitrification. These excess nitrates can percolate into ground water or leach into any water bodies in the catchment areas. Physiologically an excess of nitrates cause methemoglobin, affecting infants, aged and sick.

The effect of chemical fertilizers on soil is not immediately obvious, due to complex chemical and microbial profile, which offers it a buffering capability. However, prolonged fertilizers misuse or overuse can overwhelm this mechanism (Savci 2012). Particularly, fertilizers high in sodium and potassium have a negative impact on soil profile, pH and prevent uptake of micronutrients by plants (Savci 2012). Excessive fertilizers use can cause a breakdown between the microbial symbiotic relations with plant roots. Volatilization or decomposition of fertilizers has also been linked with emission of nitrogen oxides into the atmosphere, which is not only a potent greenhouse gas, but can also cause acid rain, thereby affecting the soil pH even more. Figure 1 is a very simplistic model of the nitrogen cycle and how nutrient loading may lead to most of it being unassimilated and reaching aquatic bodies unintentionally, where it might lead to algal blooms.

Fig. 1 Nitrogen cycle and the effect of nutrient loading by fertilizer on the same

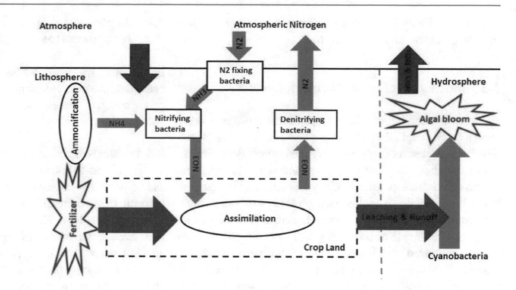

2   Role of Nanotechnology in Agriculture

Nanotechnology helps to promote agricultural practices and also offers sustainable development. Conventional agrochemicals are usually applied through spraying or broadcasting. As a result of this, small amount of agrochemicals reach the target sites of plants, which is very less than the minimum concentration required for plant growth. Agrochemicals are greatly lost due to chemical leaching, hydrolysis, photolysis degradation and also due to microbial degradation. This largely affects the production of crops and thus requires better technologies to enhance the crop yield (Shang et al. 2019). Nanotechnology-based strategies are used for precise farming and also to meet the food demand for the increasing population. Improved and environmental-friendly ENPs are used in nanotechnology to overcome the loss of nutrients and also to increase the crop yield. These ENPs are applied in agriculture in the form of nano-biotechnology, nano-toxicology, nano-fertilizers, nano-pesticides, livestocks, hydroponics, etc. Nanotechnology is proved to be the alternative method to revolutionize the agricultural practices in the current scenario (Elemike et al. 2019). Here, we have broadly discussed the role of these ENPs especially nano-fertilizers and nano-pesticides to increase the plant growth and also their advantages over conventional agrochemicals.

3   Engineered Nanoparticles-Based Nano-fertilizers

3.1   Need for Nano-fertilizers

The application of fertilizers in soil is necessary to increase the soil fertility for improved crop production. Chemical fertilizers cause severe damage of the environment and also destroy the health of the soil. Another problem associated with conventional fertilizers is the accumulation of large part of these fertilizers into soil, which leads to environmental pollution that affect the growth of plants. Due to this reason, a new cost-effective and eco-friendly technique is a need of hour for better crop production. For this purpose, ENPs-based nano-fertilizers are used nowadays, which carry out controlled release of nutrients in soil and also, they prevent loss due to chemical fertilizers (Tripathi et al. 2018). ENPs-based nano-fertilizers are also proved to be helpful in overcoming severe conditions of eutrophication and also to enhance concentration of macro- and micronutrients in order to alleviate the efficiency of nutrients (Shukla et al. 2019).

3.2   Properties of Engineered Nanoparticles-Based Nano-fertilizers

Engineered nanoparticles-based nano-fertilizers play significant role in agriculture to increase the crop yields and nutrient efficiency by decreasing excessive use of chemical fertilizers. The most important properties of ENPs-based nano-fertilizers, which are responsible for their efficacy, include eco-friendly nature, requirement of small quantities of nano-fertilizers and their composition having one or more macro- or micronutrients. Nano-fertilizers ensure controlled release of nutrients into the soil through site-targeted delivery. They also exhibit reduced toxicity and increased utilization of nutrients through delivered fertilizers. Their unique properties also include ultra-high absorption by plants, enhanced photosynthesis, increased crop production and remarkable expansion in the leave's surface area. These properties enhance plant performance, which further results into the rise in crop production. The controlled release of nutrients by ENPs-based nano-fertilizers leads to prevention of eutrophication and water pollution.

3.3 Mode of Action of Nutrient Delivery

Engineered nanoparticles-based nano-fertilizers have been shown to be adaptable to foliar application. It is considered as one of the best methods to rectify the nutrient deficiencies, to increase the quality and yield of crops and also to minimize the environmental pollution. Conventional fertilizers faced several barriers, when applied through foliar method, such as the penetration into the inner tissue becomes difficult, due to the pore size of cell wall that ranges between 5 and 20 nm. The nano-coated fertilizers enhance the penetration via stomata. Nanoparticles having diameter less than the cell wall pore size can easily enter through it and reach up to the plasma membrane (Mahil and Kumar 2019). The nanoparticles applied through foliar method get easily transported to the heterotrophic cells from the site of application, through the plasmadesmata (having diameter of 40 nm) (Etxeberria et al. 2016). The uptake of nanoparticles is done by binding to carrier protein aquaporin, ion channels and endocytosis.

Nanoparticles can also enter the plant cell wall by forming complexes with membrane transporters (Rico et al. 2011). Hong et al. (2014) observed that CeO_2 nanoparticles can enter from atmosphere into the leaf stomata in cucumber leaves, followed by their redistribution to different parts of the plant (Hong et al. 2014). Similarly, calcium oxide nanoparticles were also observed to enter plant cell wall through phloem tissue of groundnut. On the other hand, nanoparticles are found in phloem tissues in wheat plants as investigated by transmission electron microscope (TEM). In *Vicia faba*, polymeric nanoparticles of 43 nm diameter were found to penetrate through stomatal leaf pores, whereas the particles of 1.1 µm were not able to penetrate. These results were obtained with the help of confocal microscopy by Eichert et al. (2008). Another group led by Wang et al. (2013) utilized watermelon plant having large stomata and vessels to study the effect of several nanoparticles such as TiO_2, MgO, Fe_2O_3 and ZnO. These nanoparticles initially had diameter of 27.3–46.7 nm, which increased remarkably in the suspension but reduced during the spraying treatment. They observed that those nanoparticles can easily penetrate the stomata, whose size does not exceed 100 nm, from where they are redistributed to stems via the phloem sieve elements (Wang et al. 2013). After entering the plant system, nanoparticles move from one cell to another through plasmodesmata and are carried by aquaporins, ion channels or endocytosis (Mahil and Kumar 2019).

3.4 Types and Applications of Engineered Nanoparticles-Based Nano-fertilizers

Engineered nanoparticles-based nano-fertilizers can be classified into macronutrient, micronutrient and non-nutrient nano-fertilizers based on the requirements of different nutrients by plants.

3.4.1 Macronutrient Engineered Nanoparticles Nano-fertilizers

This type of ENPs-based nano-fertilizers have the potential to fulfil the requirement of large amount of nutrients by plants, such as N, P, K, Mg, Ca and S. Nano-fertilizers help in decreasing the loss of nitrogen due to leaching, emissions and soil microorganisms. These nanoparticles are also efficient in decreasing the toxic effects caused by the overuse of chemical fertilizers (Vishwakarma et al. 2018). Nano-enabled urea-modified hydroxyapatite and urea-coated zeolite chips were utilized to achieve controlled release of macronutrients (Chhipa 2017). A nanocomposite of urea-modified hydroxyapatite efficiently releases nitrogen under pressure into *Gliricidia sepium*. It was observed that this nanocomposite release approximately 78% more nitrogen, as compared to conventional fertilizer. The slow release of nitrogen results in the increased uptake efficiency, which further lead to remarkably improved plant yield (Kottegoda et al. 2011). Another macronutrient nanocomposite involving urea-hydroxyapatite nanohybrid (6:1) with carbonyl and amine functional groups was used for slow release of nitrogen (Kottegoda et al. 2017). The foliar application of NPK-nano-chitosan composite onto wheat significantly results into shortened plant lifecycle and enhanced grain yield in comparison to conventional fertilizers (Aziz et al. 2016). The effect of P-K-Fe nano-fertilizer was investigated on saffron plants grown on a silty-loam soil. This nano-fertilizer results into increased dry biomass, when exposed through the leaves (Amirnia et al. 2014).

Nano-$CaCO_3$ increases the water-content and dry biomass, when applied to *Vigna mungo*. Similarly, the foliar application of nano-CaO onto peanuts enhances the accumulation of Ca and development of plant roots relative to bulk CaO and $CaNO_3$ (Adisa et al. 2019). The extensive use of conventional fertilizers results into increased accumulation of N, P, K, Mg, Ca and S, which is extremely harmful for agro-ecosystems. These macronutrients cause pollution, when enter the water body. The use of ENPs-based nano-fertilizers diminish the overall environmental pollution, with benefits of increased crop yield via direct delivery, and targeted release of nutrients.

3.4.2 Micronutrient Engineered Nanoparticles-Based Nano-fertilizers

As the name suggests, micronutrients ENPs-based nano-fertilizers supply required nutrients in smaller quantities, generally less than $10 \ mg \ kg^{-1}$ of soil. This type of nano-fertilizers help in enhancing the metabolism of plants and thereby promoting the plant growth and nutritional value. Improved growth of rice was observed under aerobic as well as submerged conditions, due to the presence of nanosized Mn-carbonate hollow core shell system, which favours regulated release of Zn. Foliar application of Mn nano-fertilizers to mung beans (*Vigna radiata*) increases the length of its roots by 52%, of shoots and biomass by 38% as compared to treatment with bulk $MnSO_4$ (Pradhan et al. 2013). Similarly, CuO nano-fertilizers are reported to increase the growth of maize by 51% (Adhikari et al. 2016). The activity of nitrate reductase in soybean was improved by utilizing SiO_2–TiO_2 nanoparticles combination as fertilizer, which further results into better nutrient uptake. FeO nano-fertilizer, when released to black-eyed peas (*Pisum sativum*) and soybeans (*Glycine max*), increases the content of chlorophyll in leaves. Maghemite (Fe_2O_3) nanoparticles were used as fertilizer for peanuts (*Arachis hypogaea*). Fe-based nano-fertilizers increase chlorophyll content and photosynthetic activity. They also significantly increase the concentration of gibberellins and zeatin-riboside, which are growth promoting hormones (Poddar et al. 2018). Both macronutrient as well as micronutrient ENPs-based nano-fertilizers show the potential to increase the biomass or grain yields of plants.

3.4.3 Chitosan Engineered Nanoparticles-Based Nano-fertilizers

Chitosan is a naturally occurring, biodegradable cationic biopolymer, which promotes plant growth, and has antimicrobial and agrochemical potential. Chitosan is generally prepared in acidic aqueous medium in order to improve its distribution on plant surfaces and also it is dialysed to remove the acidity and salinity. It generally increases the toxicity to the target plant, which further inhibits the antimicrobial activity of chitosan. As compared to bulk form of chitosan, its nanoparticles have high solubility in aqueous medium and also have high positive charge on their surface. The affinity of chitosan nanoparticles towards the biological membranes increases, as a result of positive surface charge (Adisa et al. 2019). Chitosan comprises nearly 9–10% N, due to which it behaves as a good source for delivering macronutrients to plants. Several reports have shown the utilization of chitosan nanoparticles in combination with polymethacrylic acid for loading NPK fertilizers. The colloidal suspension of chitosan–polymethacrylic acid along with NPK was found to be highly stable, due to higher anion charges from the calcium phosphate (Hasaneen et al. 2014).

3.4.4 Non-nutrient Engineered Nanoparticles-Based Nano-fertilizers

This is another class of engineered nanoparticles (ENPs), which do not contain plant nutrients and is also proved to be useful for plant growth. These include carbon nanotubes (CNTs), SiO_2, CeO_2 and TiO_2, nano-Zn, Fe, InP/ZnS core shell quantum dots (QDs), ZnCdSe/ZnS core shell QDs, Mn/ZnSe QDs and gold nanorod (Prasad et al. 2017). Nano-fertilizer such as nano-silica could improve the plant growth under the conditions of high temperature humidity by forming a binary film on the cell wall of bacteria or fungi after absorption of nutrients and they also prevent infections. The growth of a seedling and the development of roots could be improved by utilizing fertilizers based on silicon dioxide nanoparticles (Duhan et al. 2017). Mesoporous aluminosilicate-based nanoparticles show excellent ability in order to achieve controlled delivery of macro- and micronutrients in soil. These ENPs promote the growth and thus enhance the crop yield. CNTs increase the shoot length of date palm (*Phoenix dactylifera*) and also improve the growth of tobacco plant by 55–60%, when applied at 5–500 $\mu g \ mL^{-1}$ (Khodakovskaya et al. 2012). On the other hand, CeO_2 nano-fertilizers enhance the growth by 9% and yield by 36% of wheat (*Triticum aestivum* L.) (Rico et al. 2014).

3.5 Advantages of Nano-fertilizers Over Conventional Fertilizers

Nanotechnology using ENPs can help in the manufacture of better "smarter" fertilizers, nano-fertilizers, in various ways. Another avenue of nano-fertilizer research is designing controlled release of fertilizers (CRF), which is shown in Fig. 2. They are basically conventional fertilizers, which have a nanoscale polymer coating. The thickness and characteristics of this nano-polymer coat determining the release characteristics can be engineered depending on agricultural needs. Osmocote® is one such product, which can release fertilizers over 3–4 to 14–16-month period; however, such

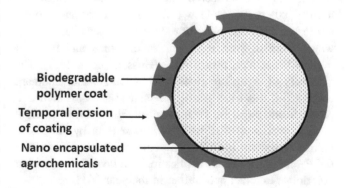

Biodegradable polymer coat

Temporal erosion of coating

Nano encapsulated agrochemicals

Fig. 2 Nano-fertilizer with biodegradable coating

CRF products are three times more expensive than traditional fertilizers (Suppan 2017). This is better than soluble fertilizers because it promotes fertilizer rationalization and while maximizing plant uptake, minimizes its wastage and unintentional release into the environment. CRFs can also be engineered to release their contents in response to particular environmental or plant physiological triggers (Qureshi et al. 2018). Nano-fertilizers have greater nutrient use efficiency, due to a much greater surface area-to-volume or mass ratio; therefore, they are more reactive and thus required in lesser amounts. Nano-fertilizers having particulate size less than the pore size can also be uptaken by the plant roots or leaves via soil application and foliar spray, respectively (Qureshi et al. 2018). Nano-fertilizers can be made to form colloidal suspensions, if the nutrient required has low solubility.

4 Engineered Nanoparticles-Based Nano-pesticides

4.1 Necessity and Limitations of Conventional Pesticides

Pest is usually defined as any animal or plant which is detrimental to human interests, in this case decreasing the productivity of agriculture. They belong to a spectrum of phylum from nematodes, mollusks, arthropods to other plants. These damage crops via direct biomass consumption, root nutrient assimilation or competition for light and water resources. Pests have been around since the advent of agriculture. It has been estimated that a third of all crops are produced globally using pesticides annually. Pests cause the loss of 18% of all crops globally, without the use of pesticides huge losses in production of fruits, vegetables and cereals occur (Oerke 2006; Zhang 2018). The first recorded countermeasure, pesticide, being dusting of elemental sulphur on crops in ancient Sumer around 4500 years ago, followed by the Rig Veda mentioning the extracts of certain poisonous plants, could be used as pesticides around 4000 years ago (Pandya 2018). In the fifteenth century, certain heavy metals such as mercury, arsenic and lead were used to kill pests, while two new natural pesticides pyrethrum and rotenone were introduced in nineteenth century, which were derived from chrysanthemums and roots of tropical vegetables, respectively (Hussaini et al. 2013). The discovery of organochlorides such as DDT was a landmark event and it was used extensively since the 1940s both as a pesticide and a disease vector control. However, in the light of its ecotoxicity, DDT had been phased out by the Stockholm Convention in 2001, on persistent organic pollutants (POP) for agricultural applications, followed by another notorious pesticide, endosulfan in the year 2011.

The problems concerning pesticides and need to develop new pesticides can be better understood, if one studies it from an agroecological point of view. Agricultural fields are artificially selected monocultures of a single species which has been bred or even genetically modified to overexpress a trait which is of economic value to humans. However, these crop fields, although created and maintained by humans, are not isolated from the ecosystem. The pests which continue to feed on the crops are also subject to the same directional evolutionary pressures and are forced to evolve along those lines to ensure survival. The application of pesticides to minimize crop loss only adds to the directional evolutionary forces added on the pests to evolve and adapt.

The second problem in the use and often overuse of pesticides is the unintentional poisoning of the environment and ultimately humans. The average annual global consumption of pesticides stands at 2 million tons, with 3 kg ha^{-1} applied (Devi et al. 2017). It has been estimated that around 3,55,000 people die annually due to overexposure to pesticides (Carvalho 2017). Pesticides are applied on crops in powder, solution or emulsion form where they can be blown off or become volatilized into the atmosphere in a few days, which can lead to a major loss of the applied amount (Aktar et al. 2009). Pesticides can also be leached from the cropland into aquatic bodies. As a general rule, area of unintentional deposition of the pesticide will depend on its volatility. Non-biodegradable synthetic pesticides can persist in the environment for weeks to years and thus classed under persistent organic pollutants (POPs). Figure 3 is a simple representation of the pathway taken by pesticides across ecosystems and trophic levels.

Pesticides achieve pest suppression in several ways, some of the most popular being hormone and neural disruption. However, unintentional targets bearing similar physiology and biochemistry also suffer the ill effects such as bees, shrimps and crabs. The target pest can also be assimilated within the food web, leading to the accumulation of pesticides in subsequently higher trophic levels, and this phenomenon is called as "biological magnification". Overuse of pesticides can also cause detriment to soil microecosystem, which is responsible for the long-term fixing of nutrients in the soil (Aktar et al. 2009). Ultimately, pesticide residues can make their way through the food web to humans, where they have been associated with diseases such as endocrine disruption, obesity and cancer (Caravalho 2017). First-generation organochloride pesticides were especially persistent, because of this reason and development of pest resistance urged their replacement by organophosphates, which are less persistent in the environment (Caravalho 2017). One often overlooked aspect during pesticides discussion is about required allied chemicals. Many pesticides are lipophilic, hence need solvents for dispersing in water.

Fig. 3 Summarized
environmental pathways and fate
of applied pesticides

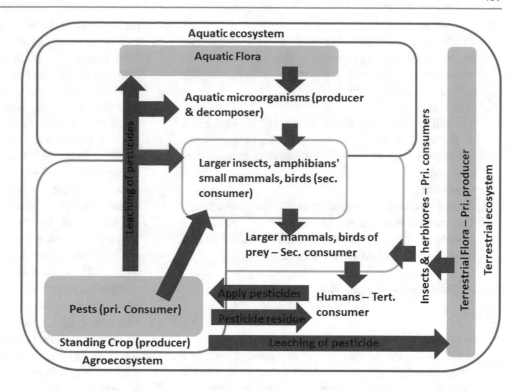

These solvents can be flammable and can pose a serious environmental and health risk.

The hallmark of the good pesticides is that these should be stable, selective, potent and degradable. The limitations of conventional pesticides in the above regard can be supplemented by the research, synthesis and mass production of nano-pesticides.

4.2 Types of Engineered Nanoparticles-Based Nano-pesticides

4.2.1 Nano-carrier-Based Pesticides

A. **Nano-emulsion pesticides**

One major issue with pesticides is of their insolubility in water, because most of them are lipophilic. A solution to this has been the large-scale use of organic solvents to prepare an emulsion (Hayles et al. 2017). However, in addition to the economic and environmental cost of these solvents, the prepared pesticide emulsions are often unstable and prone to separation via flocculation, creaming, etc. This warrants constant agitation and hence attention is being payed to the pesticide mix, which is difficult to make in rural surroundings by farmers, who lack technical knowledge. This results in excessive application of the pesticides, which harm the environment and human health.

Another solution is the use of surfactants to create an interfacial layer and lower the surface energy, resulting in a dispersed solution. Such pesticide dispersed particles are in the 1–20 μm diameter range (Hayles et al. 2017). Nano-emulsions are formed, when the miscles formed by the pesticide, surfactant and water interaction are in the nanometer range. They are formed due to the properties and concentration of the constituents used and are advantageous due to their simple preparations and high stability (Hayles et al. 2017). The advantages of nano-emulsions of pesticides are: increased bioavailability and uptake by pests due to larger surface area, resistant to physiochemical degradation, slow and controlled release, etc.

B. **Polymer-based nano-pesticides**

Polymer-based nano-pesticides offer many of the same advantages provided by the above-mentioned nano-dispersed pesticides including: greater bioavailability, increased stability and controlled release etc. Such nano-carriers can be designed to carry one or more pesticides (Hayles et al. 2017). The pesticide molecules may be physically adsorbed onto the polymeric carrier, covalently linked via chemical crosslinkers, entrapped within a polymeric matrix or a number of other formulations.

C. **Nano-capsule-based pesticides**

Nano-capsules are vesicle shaped nano-carriers with a polymer shell and an inner cavity, which is occupied by the pesticides (Balaure et al. 2017).

D. **Nano-spheres**

Unlike nano-capsules where there exists a central cavity, in polymeric nano-spheres, the pesticide is dispersed throughout the nano-sphere along with the polymer as a solid sphere.

E. **Nano-gel**

The active pesticide is dispersed in a gel, which consists of nanoscale building blocks. This type of formulation gives the active ingredient protection from premature evaporation.

F. **Nano-fibres**

Active pesticide is loaded onto rod like nano-fibres. This method has a benefit that nano-fibres have a higher loading efficiency, meaning that the pesticides can be more densely packed (Balaure et al. 2017).

4.2.2 Non-carrier-Based Nano-pesticides

A. **Solid nanoparticle pesticides**

Solid nanoparticles themselves are applied on crops either in solution or in dry form to act as pesticides. They act on the pests directly and achieve control via disruption of normal physiological functions.

B. **Metals and metal oxide-based nano-pesticides**

Metals and metal oxide nanoparticles, when used as nano-pesticides, can harm the pest in several ways; photocatalytic damage and release of superoxide radicals, membrane lysis leading to leaking of cellular contents and uptake of metal ions, lead to disruption of normal cytological processes and eventual death of the pest. Silver, copper, copper oxide and titanium oxide have been investigated for this purpose.

C. **Non-metal-based nano-pesticides**

Some nanoparticles are naturally toxic to pests such as silica and alumina, which damage the wax protective coating on cuticle of insects (Hayles et al. 2017).

4.3 Mechanism of Action of Pesticides

Pesticides may be classified based on several criteria: class of pests they act on, bio-degradability, solubility, chemical nature, etc. Here, we will consider and classify them according to their mode of action on pests. Broadly, on the basis of mode of action, pesticides may be classified as ionic pumps, neurotransmitter, neural disruptors, hormonal disruptors, juvenile hormone mimics and others.

A. **Ionic pump**

Ionic pumps such as sodium and potassium are necessary to maintain the ionic homeostasis of the neurons, which enables them to polarize and depolarize and thereby transmit impulses. Altering the natural state of these pumps changes their permeability to the ions and thus causes the neurons and by extension, the nerve fibres to continuously fire. This produces convulsions, controlled twitching, loss of coordination and eventual death in the pest. Organochloride chemical class of pesticides is included in this mode of action. The main problem associated with this class is that they are indiscriminate and hence act on mammals as well. Also, these are stable, hence making them persist in the environment (Das 2013).

B. **Neurotransmitter**

This class of pesticides act on the junction between two neurons, where the signal is bridged by a neurotransmitter. They include chemical classes such as organophosphates and carbamates, which act on the enzyme, cholinesterase (ChE), which removes the neurotransmitter Acetylcholine (ACh) from the neural or neuromuscular junction (Das 2013). As a result, accumulation of acetylcholine causes uncontrolled contractions, twitching and eventual death of the pest.

C. **Muscle disruptors**

This includes the chemical classes of pesticides, such as diamedes, which bind to and open calcium channels in the muscle and cause uncontrolled spasms followed by death, much like neurotransmitter disruptors.

D. **Hormonal disruptors**

Among pests, especially of insects, life cycles are heavily regulated by hormones. Therefore, disruption in this hormonal cycle can inhibit pest action. One benefit of exploiting the hormonal system of insects is that unlike the neural system, they are more unique to the pests or at least insects and therefore have less unintentional physiological effects on mammals and by extension humans.

E. **Juvenile Hormone Mimics**

For the immature larvae to metamorphosize into an adult, the concentration of juvenile hormone needs to be decreased. The decrease of this juvenile hormone, prompting transition into adulthood is governed by several physiological, nutritional and environmental conditions. Juvenile hormone mimics, when used as a pesticide, suppress metamorphosis and hence disrupt the life cycle of the pest (Das 2013).

F. **Others**

- **Chitin inhibitors**

  Chitin is a long-chain polysaccharide, which is the main constituent in the exo-skeleton of insects. Chitin synthesis inhibitors act by inhibiting an enzyme called chitin synthases, thereby inducing chitin deficiency in the pest. Insects exposed to this class of pesticides typically die in the moulting position unable to discard their exuviae. Inorganic substances such as silica gels can also act on the waxy cuticle of pests to dehydrate them and induce death.

- **Respiratory inhibitors**

  This includes the class amidino hydrazones, which inhibit the mitochondrion from carrying out cellular respiration. The result of this is that cells are unable to carry out the biochemical functions needed for survival and the pest dies.

4.4 Herbicides

Till now, we have discussed animal pests, however, plants also can be pests, in which case, they are termed as weeds. The term weed is of no botanical significance. Any plant, which is of economic significance in one circumstance may be a weed in another, if it is growing in the cropland is unwanted. Weeds damage crops by directly competing for resources such as sunlight, water and nutrients. The earliest methods of weed control probably involved manually removing them from the fields. However, with the advent of chemical fertilizers and pesticides, which are input costs in agriculture, there was a greater need to remove weeds, which decreased profitability of farmlands. Herbicides have a specific target, mostly enzyme in the plant cells, which they inhibit. Herbicides are able to selectively affect weeds over crop plants based on fundamental biochemical differences, e.g. some herbicides only affect broad leaves, whereas others depend on the fact that some crop plants are able to metabolize and detoxify the herbicide quicker than the weed (Hall et al. 1999).

Much like pesticides, herbicides can also be classified based on several categories; chemical class, mode of application, mode of action, spectrum of use, however, we will focus on the mode of action only, as the basis of classification. Based on the mode of action or mode of toxicity, herbicides can be classified under the following categories:

A. **Lipid synthesis inhibitors**

   Lipids or fats form an integral part of any cell, animal or plant during the formation of the lipid bilayer membrane. Herbicides from the chemical classes like, aryloxy-phenoxy-propionate, cyclohexanedione and phenylpyrazolin, act by blocking the enzyme acetyl coenzyme A carboxylase, which catalyses the first step in fatty acid synthesis and phospholipid production (Sherwani et al. 2015).

B. **Protein synthesis inhibitors**

   These herbicides inhibit the action of the enzyme acetohydroxy acid synthase, which catalyses the first step in the synthesis of branch chained amino acids (Sherwani et al. 2015).

C. **Nucleic acid synthesis inhibitors**

   This class of herbicides is also called as synthetic auxins, as they mimic the activity of indole acetic acid and disrupt nucleic acid synthesis in the cell (Sherwani et al. 2015).

D. **Photosynthetic pigment inhibitors**

   These are also called as carotenoid biosynthesis inhibitors, as these herbicides inhibit the synthesis of photosynthetic pigments and lead to bleaching, wilting and eventual death of pigments. They bind to the Q protein and stop the electron transport chain (ETC) and inhibit carbon dioxide fixation (Das and Mondal 2014).

E. **Reactive Oxygen Species (ROS) formation**

   These herbicides include the Photosystem I (PS I) inhibitor family, which are represented by the bipyridilium family. They accept electrons from PS I and generate herbicide radicals, which in the presence of superoxide dismutase from hydrogen peroxide and hydroxyl radicals generates ROS (Sherwani et al. 2015). ROS thus generated disrupt the cell membrane, finally leading to lysis.

F. **Proto-porphyrinogen oxidase (PPO) inhibitor**

   Proto-porphyrinogen oxidase (PPO) enzyme plays a key role in chlorophyll biosynthesis. Certain herbicides inhibit the action of PPO, leading to its accumulation in the cell. The PPO is then converted to proto-porphyrin, which is toxic and leads to rupture of membranes (Das 2013).

G. **Nano-herbicides**

   Much like pesticides, herbicides also suffer from the same drawbacks of non-selectivity, instability, volatility, etc. Therefore, much like the nano-pesticides, nanotechnology is being used to design and synthesize better nano-herbicides also. Research efforts are being put into polymeric carriers for controlled release of the herbicide as well as nano-coatings to specifically target weed root receptors (Manjunatha et al. 2016). In recent times, some inorganic nanoparticles such as silica nanoparticles (SiNPs) are also being tested as potential herbicides due to selective absorption (Abigail and Chidambaram 2017).

4.5 Limitations of Engineered Nanoparticles-Based Nano-Pesticides

Nano-pesticides offer a promise of augmenting or even replacing the current conventional pesticide regime, due to their perceived advantages of high stability, selectivity and degradability. However, their research and application must be proceeded with good amount of caution, regarding their effects on the environment and human health in the long run.

- **Environmental interaction and fate**
  The environmental and geological cycling of ENPs is still being intensively researched. This includes the fate of nano-carriers after unloading of the pesticides and solid nano-pesticides must be studied to determine the eventual sink and pathways taken by nano-pesticides over time. Realistic environmental research into the fate of these nano-pesticides remain scare and existing environmental models, lacking accurate parameters, cannot predict the environmental fate and effects of nano-materials (Kah and Hofmann 2014; Kookana et al. 2014). The additional environmental parameters, which need to be determined for nano-pesticides and nano-pollutants in general over and beyond conventional pesticides or pollutants, are their concentration, particle density, particle size, particle shape, surface porosity, bound or free, and agglomeration, etc. (Kookana et al. 2014).

- **Nano-toxicity**
  The toxicity of ENPs like nano-fertilizers and nano-pesticides is still not fully known. This is important to note for both, the solid nano-pesticides as well as conventional pesticides, that their properties and thus toxicity might change, when they are in their nano-form (Hayles et al. 2017). Also, there needs to be more research regarding the combined toxicity of the active ingredients, pesticides and the nano-carriers (Sun et al. 2019).

5 Outlook and Future Directions

The often-conflicting objectives of feeding the ever-growing human population and preserving the environment and by extension human health itself are leading to research and application of novel agrochemicals and agricultural methods. Our current use of agrochemicals, primarily conventional fertilizers and pesticides are not only unsustainable but also detrimental to human health. Overuse of agrochemicals like fertilizers has led to several major environmental issues such as disturbing the nutrient cycles, overloading of natural nutrient sinks and unbalancing of the soil pH. Likewise, non-specific toxicity, biological magnification of toxicants, disturbance of native soil microflora and fauna, etc., are major concerns in case of pesticides. Engineered nanoparticles (ENPs) used as nano-agrochemicals in the form of nano-fertilizers and nano-pesticides help overcome many of the shortcomings of conventional agrochemicals. Most important issues include

non-targeted delivery, physicochemical degradation and soil contamination by conventional fertilizers, which may be solved by nano-fertilizers and nano-pesticides by exploiting unique properties of nanoscale materials to increase the specificity, stability, decrease in toxicity and dosage required. However, much more research is needed to study the environmental fate and ecotoxicity of these ENPs-based nano-agrochemicals before their widespread adoption and applications.

Acknowledgements MK would like to thank the research and development grant of University of Delhi, Delhi.

Conflict of Interest None to declare.

References

Abigail E, Chidambaram R (2017) Nanotechnology in herbicide resistance. In: Seehra M (ed) Nanostructured materials: fabrication to applications. IntechOpen, London, pp 207–212

Adhikari T, Sarkar D, Mashayekhi H, Xing B (2016) Growth and enzymatic activity of maize (*Zea mays* L.) plant: solution culture test for copper dioxide nano particles. J Plant Nutr 39:99–115

Adisa IO, Pullagurala VLR, Peralta-Videa JR, Dimkpa CO, Elmer WH, Gardea-Torresdey JL, White JC (2019) Recent advances in nano-enabled fertilizers and pesticides: a critical review of mechanisms of action. Environ Sci Nano 6:2002–2030

Aktar W, Sengupta D, Chowdhury A (2009) Impact of pesticides use in agriculture: their benefits and hazards. Inter Discip Toxicol 2:1–12

Amirnia R, Bayat M, Tajbakhsh M (2014) Effects of nano fertilizer application and maternal corm weight on flowering at some saffron (*Crocus sativus* L.) ecotypes. Turk J Field Crops 19:158–168

Atafar Z, Mesdaghinia A, Nouri J, Homaee M, Yunesian M, Ahmadimoghaddam M, Mahvi AH (2010) Effect of fertilizer application on soil heavy metal concentration. Environ Monit Assess 160:83–89

Aziz HMA, Hasaneen MN, Omer AM (2016) Nano chitosan-NPK fertilizer enhances the growth and productivity of wheat plants grown in sandy soil. Span J Agric Res 14:17

Balaure PC, Gudovan D, Gudovan I (2017) Nanopesticides: a new paradigm in crop protection. In: Grumezescu AM (ed) New pesticides and soil sensors. Academic Press, Cambridge, Massachusetts, United States, pp 129–192

Carvalho FP (2017) Pesticides, environment, and food safety. Food Energy Secur 6:48–60

Chhipa H (2017) Nanofertilizers and nanopesticides for agriculture. Environ Chem Lett 15:15–22

Das SK (2013) Mode of action of pesticides and the novel trends–a critical review. Int Res J Agri Sci Soil Sci 3:393–401

Das SK, Mondal T (2014) Mode of action of herbicides and recent trends in development: a reappraisal. Int J Agric Soil Sci 2:27–32

Devi PI, Thomas J, Raju RK (2017) Pesticide consumption in India: a spatiotemporal analysis. Agric Econ Res 29:163–172

Duhan JS, Kumar R, Kumar N, Kaur P, Nehra K, Duhan S (2017) Nanotechnology: the new perspective in precision agriculture. Biotechnol Rep 15:11–23

Eichert T, Kurtz A, Steiner U, Goldbach HE (2008) Size exclusion limits and lateral heterogeneity of the stomatal foliar uptake pathway for aqueous solutes and water-suspended nanoparticles. Physiol Plant 134:151–160

Elemike EE, Uzoh IM, Onwudiwe DC, Babalola OO (2019) The role of nanotechnology in the fortification of plant nutrients and improvement of crop production. Appl Sci 9:499

Erisman JW, Sutton MA, Galloway J, Klimont Z, Winiwarter W (2008) How a century of ammonia synthesis changed the world. Nat Geosci 1:636

Etxeberria E, Gonzalez P, Bhattacharya P, Sharma P, Ke PC (2016) Determining the size exclusion for nanoparticles in citrus leaves. Hort Sci 51:732–737

Hall L, Beckie H, Wolf T (1999) How herbicides work: biology to application. Alberta agriculture, food and rural development. Publishing Branch, Alberta, United States, pp 10–12

Hasaneen MNA, Abdel-Aziz HMM, El-Bialy DMA, Omer AM (2014) Preparation of chitosan nanoparticles for loading with NPK fertilizer. Afr J Biotechnol 13:3158–3164

Hayles J, Johnson L, Worthley C, Losic D (2017) Nanopesticides: a review of current research and perspectives. In: Grumezescu AM (ed) New pesticides and soil sensors. Academic Press, Massachusetts, United States, pp 193–225

Hong J, Peralta-Videa JR, Rico C, Sahi S, Viveros MN, Bartonjo J, Zhao L, Gardea-Torresde JL (2014) Evidence of translocation and physiological impacts of foliar applied CeO_2 nanoparticles on cucumber (Cucumis sativus) plants. Environ Sci Technol 48:4376–4385

Hussaini SZ, Shaker M, Iqbal MA (2013) Isolation of fungal isolates for degradation of selected pesticides. Bull Env Pharmacol Life Sci 2:50–53

Mahil EIT, Kumar BNA (2019) Foliar application of nanofertilizers in agricultural crops—a review. J Farm Sci 32:239–249

Kah M, Hofmann T (2014) Nanopesticide research: current trends and future priorities. Environ Int 63:224–235

Khodakovskaya MV, De Silva K, Biris AS, Dervishi E, Villagarcia H (2012) Carbon nanotubes induce growth enhancement of tobacco cells. ACS Nano 6:2128–2135

Kookana RS, Boxall AB, Reeves PT, Ashauer R, Beulke S, Chaudhry Q, Cornelis G, Fernandes TF, Gan J, Kah M, Lynch I, Ranville J, Sinclair C, Spurgeon D, Tiede K, Brink PJV (2014) Nanopesticides: guiding principles for regulatory evaluation of environmental risks. J Agric Food Chem 62:4227–4240

Kottegoda N, Munaweera I, Madusanka N, Karunaratne V (2011) A green slow-release fertilizer composition based on urea-modified hydroxyapatite nanoparticles encapsulated wood. Curr Sci 73–78

Kottegoda N, Sandaruwan C, Priyadarshana G, Siriwardhana A, Rathnayake UA, Berugoda Arachchige DM, Kumarasinghe AR, Dahanayake D, Karunaratne V, Amaratunga GA (2017) Urea-hydroxyapatite nanohybrids for slow release of nitrogen. ACS Nano 11:1214–1221

Manjunatha SB, Biradar DP, Aladakatti YR (2016) Nanotechnology and its applications in agriculture: a review. J Farm Sci 29:1–13

Oerke EC (2006) Crop losses to pests. J Agric Sci 144:1–43

Pandya IY (2018) Pesticides and their applications in agriculture. Asian J Appl Sci Technol 2:894–900

Poddar K, Vijayan J, Ray S, Adak T (2018) Nanotechnology for sustainable agriculture. In: Singh RL, Mondal S (eds) Biotechnology for sustainable agriculture-emerging approaches and strategies. Woodhead Publishing, Cambridge, United Kingdom, pp 281–303

Pradhan S, Patra P, Das S, Chandra S, Mitra S, Dey KK, Akbar S, Palit P, Goswami A (2013) Photochemical modulation of biosafe manganese nanoparticles on Vigna radiata: a detailed molecular, biochemical, and biophysical study. Environ Sci Technol 47:13122–13131

Prasad R, Bhattacharyya A, Nguyen QD (2017) Nanotechnology in sustainable agriculture: recent developments, challenges, and perspectives. Front Microbiol 8:1014

Qureshi A, Singh DK, Dwivedi S (2018) Nano-fertilizers: a novel way for enhancing nutrient use efficiency and crop productivity. Int J Curr Microbiol App Sci 7:3325–3335

Rico CM, Majumdar S, Duarte-Gardea M, Peralta-Videa JR, Gardea-Torresdey JL (2011) Interaction of nanoparticles with edible plants and their possible implications in the food chain. J Agr Food Chem 59:3485–3498

Rico CM, Lee SC, Rubenecia R, Mukherjee A, Hong J, Peralta-Videa JR, Gardea-Torresdey JL (2014) Cerium oxide nanoparticles impact yield and modify nutritional parameters in wheat (Triticum aestivum L.). J Agr Food Chem 62:9669–9675

Savci S (2012) An agricultural pollutant: chemical fertilizer. IJESD 3:73

Schmidt J, Shaskus M, Estenik J, Oesch C, Khidekel R, Boyer G (2013) Variations in the microcystin content of different fish species collected from a eutrophic lake. Toxins 5:992–1009

Shang Y, Hasan M, Ahammed GJ, Li M, Yin H, Zhou J (2019) Applications of nanotechnology in plant growth and crop protection: a review. Molecules 24:2558

Sherwani SI, Arif IA, Khan HA (2015) Modes of action of different classes of herbicides. In: Price A, Kelton J, Sarunaite L (eds) Herbicides: physiology of action and safety. InTech Open, Rijeka, Croatia, pp 165–186

Shukla P, Chaurasia P, Younis K, Qadri OS, Faridi SA, Srivastava G (2019) Nanotechnology in sustainable agriculture: studies from seed priming to post-harvest management. Nanotech Environ Eng 4:11

Sun Y, Liang J, Tang L, Li H, Zhu Y, Jiang D, Song B, Chen M, Zeng G (2019) Nano-pesticides: a great challenge for biodiversity? Nano Today 28:100757

Suppan S (2017) Applying nanotechnology to fertilizer: rationales, research, risks and regulatory challenges. The Institute for Agriculture and Trade Policy works locally and globally, Brazil. https://www.iatp.org/sites/default/files/2017-10/2017_10_10_Nanofertilizer_SS_f.pdf

Tripathi M, Kumar S, Kumar A, Tripathi P, Kumar S (2018) Agro-nanotechnology: a future technology for sustainable agriculture. Int J Curr Microbiol App Sci 7:196–200

Vishwakarma K, Upadhyay N, Kumar N, Tripathi DK, Chauhan DK, Sharma S, Sahi S (2018) Potential applications and avenues of nanotechnology in sustainable agriculture. In: Tripathi DK, Ahmad P, Sharma S, Chauhan DK, Dubey NK (eds) Nanomaterials in plants, algae, and microorganisms. Academic Press, Massachusetts, United States, pp 473–500

Wang WN, Tarafdar JC, Biswas P (2013) Nanoparticle synthesis and delivery by an aerosol route for watermelon plant foliar uptake. J Nanopart Res 15:1417

Zhang W (2018) Global pesticide use: profile, trend, cost/benefit and more. Proc IAEES 8:1

Bio-nanosensors: Synthesis and Their Substantial Role in Agriculture

Shailja Dhiman, Swati Gaba, Ajit Varma, and Arti Goel

Abstract

Nanotechnology is a recent emerging area having vast potential in almost every field of science due to their small size and larger surface area as compared to bulk phase materials. Synthesis of nanoparticles can be done from physical and chemical methods, but these days, bio-nanotechnology is in demand that associate principles of biology with physical and chemical methods to synthesize nanomaterials having precise functions. In bio-nanotechnology, the nanoparticles are synthesized from biological means such as plants or microbes also called as plant-microbe-engineered nanoparticles (PM-ENPs). PM-ENPs are more efficient, less toxic and cost effective as compared to physical and chemically synthesized nanoparticles. Plant-microbe-engineered nanoparticles have good anti-microbial activity because of electrostatic interaction with cell membrane of microorganisms and electrostatic interaction build-up inside the cell cytoplasm. The PM-ENPs such as zinc oxide (ZnO) and sliver (Ag) are helpful in increasing the growth of plant by guaranteeing that the nutrients are used in controlled manners by the plants. Plant-microbe-engineered nanoparticles as bio-nanosensors have confirmed their possibility of success in agriculture. Bio-nanosensors can be used for monitoring of crop health, pests attack, environmental stressors and plant diseases. The bio-nanosensors can be used in pathogen detection, sensing food eminence, adulterants, dye, vitamins, fertilizers, taste, smell and pesticides. Therefore, plant-microbe-engineered nanoparticles have significant role in advancement of agriculture. This chapter will pave the path for the possibility of synthesis of nanomaterials by biological means such as by different plant parts and microbes. Also, the role of different metal nanoparticles in making of different types of bio-nanosensors and their substantial role in agriculture advancement have also been emphasized.

Keywords

Agriculture • Bio-nanosensors • Nanofertilizers • Plant-microbe-engineered nanoparticles

1 Introduction

Nanotechnology is coming into various fields such as biotechnology, engineering, food technology, agriculture and medical sciences and brought extensive research. It has an impact on all the forms of life because of its enormous use in automobiles, bio-medical sensors, catalyst, electronics, nano-fabrics, packaging, agriculture, bio-engineering, medicines, drug delivery, etc. (Shankar et al. 2004; Song and Kim 2009; Iravani et al. 2011). Nanotechnology is a novel discovery in the field of nanotechnology and changing too fast to cover thoroughly. Richard Feynman an American physicist in 1959 brought the concept of nanotechnology in a conference of the American Physical Society, where he gave the idea of the very vast potential of nanomaterials (Feynman 1960). When bulk materials are engineered into one or two dimensions in nano-range or smaller particles have properties which vary from those of the bulk phase material. Such engineered particles show totally different characteristics from bulk phase materials. The fact on which nanotechnology lies depicts that the reduction of size of the substances in nanometre range changes the properties of substances dramatically (Chattopadhyay and Patel 2016; Ail et al. 2017).

Bulk phase material is reduced to small size nanoparticles through different approaches such as top to bottom and bottom to up (Fig. 1). In top-to-bottom approach, nanoparticles are synthesized by physical and chemical methods which

S. Dhiman · S. Gaba · A. Varma · A. Goel (✉)
Amity Institute of Microbial Technology, Amity University Noida, Noida, U.P., India
e-mail: agoel2@amity.edu

© Springer Nature Switzerland AG 2021
P. Singh et al. (eds.), *Plant-Microbes-Engineered Nano-particles (PM-ENPs) Nexus in Agro-Ecosystems*,
Advances in Science, Technology & Innovation,
https://doi.org/10.1007/978-3-030-66956-0_11

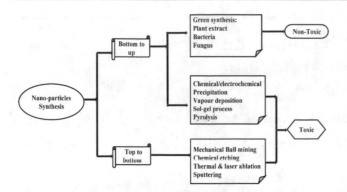

Fig. 1 Different approaches for the synthesis of metallic nanoparticles

mainly include grinding, cutting and etching, while bottom-to-up approach is self-re-arrangement of atom by atom or molecule by molecule for the synthesis of nanoparticles. There are advantages of using these approaches like possibilities to develop nanostructure with more homogenous chemical compositions and less defects. Nanomaterials are based on Gibb's free energy that is why such engineered nanoparticles are in thermodynamic equilibrium state or closure to this state. The top-down approach uses old methods to develop engineered nanoscale materials. Nano-scaled materials have different sizes, in combination with their different behaviour as well as have significant impact on chemical, physical, biological, electrical, mechanical and functional properties (Mukhopadhyay 2014).

This chapter will enhance our knowledge about the synthesis of plant-microbe-engineered nanoparticles (PM-ENPs) and how we can use these nanoparticles for development of particular bio-nanosensors. Besides this, the chapter will give a glance about the newest developments, applications of several nano-biosensors utilized in agriculture and relieving stress of minor population of farmers.

1.1 Synthesis Methods of Plant-Microbe-Engineered Nanoparticles

There are different methods of fabrication or synthesis of nanomaterials such as chemical, physical and biological methods. Methods such as physical and chemical have some disadvantages like high energy use, toxic chemicals and high cost. Therefore, synthesis by biological means has been evolved by use of animal-derived biomaterials, biomolecules of microbial origin and extracts of plant parts. Nanoparticle synthesis through plant extract and microbial means is called as plant-microbe-engineered nanoparticles (PM-ENPs).

Plants have been used as the main natural source for drug preparation and treatment of human illnesses. They are blessed by nature with a magical phenomenon to secrete secondary metabolites which are bioactive. These days,

many natural products are made and used by the humans for treatments of various illnesses. Biosynthesis of nanomaterials is one of the current medicine manufacturing processes from the medicinal plant parts or from non-medicinal too. Nanoparticles possess exceptional properties due to their nanoscale size, morphology and distribution. Biosynthesis of nano-size particles from plant leaves, bark and fruit extract is cheap, environment-friendly and commercialized for large-scale production. There is least requirement of toxic chemicals, temperature and energy (Kharat et al. 2017). Biosynthesis of nanoparticles has become a subject of interest because of choices of reagents which are eco-friendly, whereas chemical synthesis method requires use of harsh chemicals for reduction and stabilization which makes it very expensive and harmful for agriculture (Sabir et al. 2014).

As synthesis process of nanoparticles by microbial cells is reliable, non-toxic and eco-friendly, biological organisms such as bacteria, viruses, algae, yeast and fungi have been used for synthesis of metal nanoparticles. Extracellular synthesis involves enzymes, proteins and organic molecules. Large number of enzymes, e.g. nicotinamide adenine dinucleotide hydrogenase (NADH) dependent reductase, naphthoquinone, anthraquinones and electron shuttle system, is machinery for reduction of toxic metallic ions into non-toxic metal nanoparticles (Patra et al. 2014; Bose and Chatterjee 2016). The mechanisms behind the extracellular and intracellular synthesis of nanomaterials are different among different microorganisms (Mandal et al. 2005; Hulkoti and Taranath 2014). In the intracellular synthesis, the positively charged metal ions are transported through the cell wall and interacted with negatively charged ions of the cell wall. However, in case of fungi, nanoparticle synthesis is extracellularly mediated by nitrate reductase in the presence of enzyme nitrate reductase helping in reduction of metal ions into nano-sized particles (Hulkoti and Taranath 2014).

1.2 Plant-Microbe-Engineered Nanoparticles Based Bio-nanosensors

Biosensors are devices which use biological or living entities for conversion of biological signal into electrical waves for general analysis, and the processor helps in quantification of signal. Biosensors have three functional units named as interactive sensor used for recognition, transducer for signal transfer and the processor which processes the signal transferred from the transducer. There are different types of biosensors such as immunosensors that uses antibody–antigen (Ab–Ag) reactions as a recognition model acting as binary mode. Other types of immunosensors are analytical immunosensor using Ag–Ab as the recognition molecule, in which Ab acts as a recognition entity for an antigen molecule and forms a stable reaction. There are a large number of applications of novel biosensors in research and development, food safety inspection agencies, food industry, food producers and policymakers who take an account for security and food safety (Prasad et al. 2017). However, there are some disadvantages while using traditional methods for detection of food quality such as expensiveness, time consumption which requires multiple steps for sample preparation before food quality detection and requirement of skilled technician and complex instruments which are not accessible to peoples of rural areas (Koedrith et al. 2014). The field of nanotechnology possesses capability to have strong impact in multiple fields such as energy, water, health, agriculture and food, and this is very profound technology among the new technologies.

Nanotechnology is the change of dimensions of bulk phase materials, system and devices at atomic and molecular level, in 1–100 nm range for developing new characteristics (Otles and Yalcin 2012; Prasad et al. 2015). Utilization of nanomaterials in construction of biosensors helps in overcoming the problems associated with old methods. There are advantages associated with nanosensors in place of old methods: they are highly sensitive and highly specific, offer accurate and rapid detection, and they are eco-friendly as well. Bio-nanosensor technology has the potential to detect analytes in low amounts (e.g. chemical or biological materials) that are dangerous to animals, humans and plants at a very low concentration, with very less preparation of sample and handy instrumentation. In field of agriculture, nano-based biosensors can offer opportunities for pesticide detection, drug residues, food-borne pathogens, heavy metal ions and toxic contaminants in foods in a very less time span. Also, nanosensors monitor crop stress, antibiotic resistance, soil conditioning, growth in plants, food quality and nutrient contents (Teodoro et al. 2010; Tarafdar et al. 2013; Prasad 2014; Prasad et al. 2014, 2015, 2017).

2 Types and Roles of Bio-nanosensors

Narayanan and Sakthivel (2010) has documented the large number of nanoparticles such as silver, cadmium, gold, magnetite, silica, titania dioxide, selenium, gold–silver alloy, copper, cobalt and platinum nanoparticles for formation of different types of nanosensors. All the metal nanoparticles and noble metals used are resistant due to corrosion, and hence, they are used for development of different types of utility nanosensors named as acoustic wave biosensors, magnetic biosensor, electrochemical biosensors, nanotube-based sensors, nanowire-based sensors. The functionality of every nanosensors is different.

- **Acoustic nanosensor**: It is used for amplification of the sensing responses and improving the preciseness of the detection limit.
- **Magnetic nanosensor**: It uses ferrite materials with transition metals. Electroactive species are monitored by electrochemical biosensors that are consumed or produced with activity of biological components.
- **Electrochemical nanosensor**: They are divided into potentiometric biosensors and amperometric biosensors. Potentiometric biosensors are not used frequently for checking food quality when compared with amperometric nanosensors having potential to monitor wide range of target analytes. Potentiometric nanosensors has been reported to detect monophenolase activity in apple juice (Dutta et al. 2001), and sucrose concentration detected in drinks (Rotariu et al. 2002), measuring fruit juices for isocitrate concentration (Kim and Kim 2003) and determines urea levels in milk (Verma and Singh 2003).
- **Calorimetric nanosensor**: It gives results of biochemical reactions in the form of heat absorption or production. Calorimetric transduction sensors detect heat consumed or generated in a biological reaction by using heat detection devices. They are used in detection of food quality and metabolites produced.

3 Biosynthesis of Bio-nanosensors Using Metal Nanoparticles

Various categories of metal nanoparticles are involved in the formation of metal oxide nanoparticles including magnetic and nonmagnetic, metal sulphide alloy, gold and silver nanoparticles. All these nanostructures can exist in diverse shapes such as nanoparticles, nanosheet, nanocomposites, nanotubes, nanorods, nanoconjugates, nanowires, etc.

4 Forms of Nanomaterials as Nanosensors

As nanosensors have high sensitivity and quick response, different types of nanomaterials allow quick penetration of fertilizers and nutrient for plant growth promotion, also act as nanosensors for quick monitoring of crop status and hence used in agricultural field such as pesticide detection, pathogen detection, insecticide detection, monitoring the crop biotic and abiotic stress and regulating plant growth. Predominantly, graphene oxide, multiwalled carbon nanotubes, multiwalled chitosan nanocomposite and ZnO chitosan nanocomposite are used enormously in all the mentioned application.

In pesticide detection, carbon nanotubes, gold nanoparticles, and nanocomposites and quantum dots with different polymers have been used (Cesarino et al. 2012; Liu et al. 2012). ZnO chitosan nanocomposite membrane was used for the detection of *Trichoderma harzianum* (Raskar and Laware 2014). Graphene oxides further detect content of nitrate in soil (Pan et al. 2016). Further, carbon nanotubes having single wall (SWNTs) when inserted into chloroplast of plant cells increased the photoabsorption (Wong et al. 2016). Kwak et al. (2017) demonstrated that nanobionic approaches helped in crop improvement and monitoring of environment by inserting nanoparticles into plant cell by improving imaging.

Fang et al. (2017) have reported that glutathione has nanoparticles (Au-NPs) for acetylcholinesterase (AChE) activity detection by means of fluorescence and toxic and heavy metal Cd^{2+} in water samples. Application of nanosensors in agriculture is a promising tool which provides the assurance of development by monitoring soil and crop health. However, the large number of records of research in this area, regarding the performance of reliable nanosensors, is surprisingly insufficient in field, opening a window for research in future.

5 Application of Nanosensors in Agriculture

Nanosensors have more advantages in comparison with conventional sensors because of higher sensitivity, quick response, reliable results, large surface-to-volume ratio and high stability. Detection range is small in gram/mole range or lower than that which is found in several matrixes and facilitates fast electron transfer kit. Nanosensor-based system on global positioning level has been used for monitoring of cultivated fields at real time level in the growing season. All these applications assure the monitoring of crop growth at real time level and high-quality data which could be effective and further provide chances for management practices and ignoring large dose of agricultural inputs.

There are different metal nanomaterials such as quantum dots (QD), carbon nanotubes (CNT), gold nanoparticles (AuNP) and nanocomposites with polymers used in construction of nanosensors for the detection purpose of pesticides, insecticides, acting as a disease detection tool, providing smart agricultural practices, etc. (Zheng et al. 2011; Cesarino et al. 2012; Liu et al. 2012).

A. Nanosensors Provide Smart Agricultural Practices

Society is more dependent on the conventional agriculture practices and is transforming itself into smart agriculture in which the main contribution came from nanosensors, and it facilitates in progress of crop growth, detects pest attack on crop in field condition, detects diseases in various crops and reduces environmental stress (Chen and Yada 2011). Real-time monitoring with nanosensors prevents the use of fertilizer and pesticides and reduces environmental contaminants as well as the cost of the product.

Some of the activities included in smart agriculture are as follows:

(a) Fertilizer or pesticide delivery system facilitated by nanoformulations increases the wettability and dispersion of nutrients.
(b) Fertilizer or pesticide residues are detected by nanosensors.
(c) Disease incidence and crop growth were monitored by remote sensor systems.

B. Nanosensors Detect the Soil Humidity

Ganeshkumar et al. (2016) showed that nanofibers in one dimension made up of potassium niobate ($KNbO_3$) are sensing the humidity because of their large surface-to-volume ratio. Humidity nanosensors produce a result in the form of log value in linear form dependant on conductance versus relative humidity at the interval of two seconds. Results showed an increase in conductance from $10 - 10\mho$ to $10 - 6\mho$ for relative humidity range from 15 to 95% at room temperature.

C. Nanosensors and Crop Improvement

Bionic plants are developed through concise farming with the insertion of nanoparticles into the plant and chloroplast cells for imaging the presence of different objects in environment. Self-powering of plants is enhanced by communication of infrared devices or light sources (Ghorbanpour and Fahimirad 2017; Kwak et al. 2017). In one study, Giraldo et al. (2014) and Wong et al. (2016) reported that in in vivo conditions, single-walled carbon nanotubes (SWNTs) when inserted in plant system increase the photoabsorption. SWNTs suppress the reactive oxygen species (ROS) generation in chloroplast, and near-infrared fluorescence light-harvesting capacity is increased which results in photosynthetic efficiency and yield of plants. Hence, nanobionic approaches help in crop improvement and monitoring of environment.

D. Nanosensors Used for Herbicide Detection

Nanosensors composed of TiO_2 nanotubes were used for atrazine detection in soil reported by Yu et al. (2010). Chitosan composites and carbon nanotubes which are multiwalled were used in detection of methyl parathion in acetylcholinesterase enzyme modified with glassed electrode made up of carbon and detected in small amount in water and soil (Dong et al. 2013). Inhibitory effect of acetylcholinesterase enzyme was used for detection of methyl parathion. An amino-containing phosphorus such as glufosinate and glyphosate herbicide in soil was detected by nanofilm-modified pencil graphite electrode in range of 0.19–0.35 mg mL^{-1}, respectively.

Herbicide chlortoluron was detected by enzymatic nano-biosensor (Haddaoui and Raouafi 2015). The application of herbicide in agriculture provides prevention of growth of weed in cereal fields. The nanostructured method utilizing ZnO nanoparticles and modified carbon electrodes having screen-printed (SPCEs) which allows the detection of an activity of enzyme inhibition in tyrosinase and herbicide level of chlortoluron in part per billion (ppb) is also detected. This nano-biosensor has a 0.47 nano-mole (nM) detection limit. Herbicide chlortoluron induces tyrosinase inhibition in range from 1 to 100 nM.

E. Detection of Pesticide and Insecticide by Aptamer-Based Nanosensor

Aptamers are made up of peptide molecules or single-stranded nucleic acid having size less than 25 kDa with natural or artificial origin, also known as antibodies which are used for recognition element of aptasensors. Aptasensors are further used in pesticide and insecticide detection. They are also used for further detection of heavy metals such as Hg^{2+}, As^{3+} and Cu^{2+}. Even antibiotic kanamycin, tetracycline, oxytetracycline and cocaine were also detected by aptasensors. Acetamiprid in soil ranging from 75 nM to 7.5 µM is detected by nano-biosensor made up of nanoparticles with gold aptamer containing acetamiprid-binder.

Some pesticides such as monocrotophos and organophosphate were detected by an electrochemical biosensor including injection having novel flow (Norouzi 2017). Integrated results of chitosan–gold nanoparticle film are produced with technique called as fast Fourier transform continuous cyclic voltammetry (FFTRCCV). The purpose is to use chitosan–gold nanoparticles for increasing the immobilization level, and the results are obtained in less than 70 s, and further increased sensitivity has 10 nm detection limit; structure obtained was more stable having more than 50-day storage stability.

F. Nanosensors in Insecticide Detection

Another biosensor called as amperometric immunosensor was used for carbofuran detection which is a broad-spectrum insecticide used in agriculture. Gold nanoparticles are immobilized with monoclonal antibody specific to carbofuran on the glutathione. Carbon nanotubes are having multi-walled and sheet of grapheme made up with polyethyleneimine polymer–gold nanocomposites via self-assembly and modified on to the surface of a glass carbon electrode. And further, this antibody conjugates with gold nanoparticles detected by the immunosensing method in detection limit 0.03 ng mL^{-1}. This simultaneous immunological and electrochemical strategy provides highly reproducible, high stability, more specific and good regeneration capability nanosensors (Zhu et al. 2013). Another study on quantum dots (QDs) was done for methyl parathion detection by chronoamperometric sensor in the presence of substrate called as ATCl before and after inhibition with different concentrations of methyl parathion producing results variation in oxidation current which gives the concentration of methyl parathion involved in a reaction. QDs are nanocrystals known for their fluorescence spectra, full wavelength absorbance, high photo-stability and fluorescence emission in controlled manner. QDs have all these properties which provide them a property for imaging and sensing purpose. Construct prepared in the study was made up of ZnSe quantum dots attached with graphene–chitosan nanocomposites electrostatically and casted on a glassy carbon electrode, and acetylcholinesterase with mercaptophenyl boronic acid-functionalized was quantitatively detecting methyl parathion in 0.2 nM in the form of electrochemical signals.

G. Nanosensors As Disease Detection Tool

a. Nano-biosensor made up of chitosan and ZnO nanoparticles nanocomposite with electrode made up of gold is developed to detect fungal pathogen *Trichoderma harzianum* (Siddiquee and Suryani 2014).

b. *Polymyxa betae*, causal agent of necrotic yellow vein virus, which is detected by quantum dots consisting of fluorescence resonance energy transfer (FRET) nano-biosensor is used for identification of disease in sugar beet named as *Rhizomania* (Safarpour et al. 2012).

c. Bakhori et al. (2013) have reported that FRET is used for identification of *Ganoderma boninense* with oligonucleotide in which sensor is made up of deoxyribonucleic acid (DNA) probes and quantum dots.

d. Gold nanoparticle tagged with horse radish peroxidase for bacterial detection such as *Pantoae stewartii* (Zhao et al. 2014a, b).

e. Label-free gold nanorods were used for Odontoglossum ringspot virus detection and Cymbidium mosaic virus detection in 42 and 48 pg mL^{-1}, respectively.

f. A bacterial plant pathogen *Xanthomonas axonopodis* pv. vesicatoria in solanaceous crops was detected with fluorescent silica nanoprobes tagged with secondary antibody of goat anti-rabbit Ig (Yao et al. 2009).

g. Karnal bunt disease in wheat was detected with immunosensor of nano-gold (Singh et al. 2010).

H. Nanosensors Detect Nutrient concentration

Soil suffers from a loss of nutrient concentration, and it is important to analyse soil requirement for conditioning and productivity increase, and components which are in excess suffer from leaching. Some of the nanosensors which are used in nutrient detection are as follows:

Graphene oxide-based nanosensors were useful in nitrate detection, and nanosensors such as nanofibres made up of graphite oxide sheet and compound poly (3, 4-ethylene dioxythiophene) were for detection of nitrate (Pan et al. 2016; Ali et al. 2017).

I. Nanosensors Detect Fertilizer Activity

In the current society, fertilizer estimation is increasing with the help of nanosensors, further helping in cost management of fertilizers for reducing the pressure for farmer and saving fertilizers which are unutilized. Some of the nanoparticle-based biosensors were used to determine the urea, urease inhibition and urease activity which are as follows:

Urea, urease inhibition and urease activity were recognized by nanosensor made up of gold nanoparticle-3, 3′, 5, 5′-tetramethylbenzidine-H–O (Deng et al. 2016). Gold nanoparticle acts as a detection tool and produces yellow colour and has detection limit for recording urease activity (1.8 unit per L) in soil.

J. Nanosensors As An Agent for Promotion of Sustainable Agriculture

Nanofertilizers deliver nutrients to crops in the form of a product encapsulated with nanoparticle. Advantage of using nanofertlizers is reducing nitrogen loss due to emissions and leaching (De Rosa et al. 2010)

There are three ways of encapsulation:

(a) Nanoporous materials or nanotubes can contain nutrients or coating with thin film made up of polymers and delivering an emulsions or nanoparticles.

(b) Carbon nanotubes have penetrated in tomato seeds (Khodakovskaya et al. 2009).

(c) Nanoparticles made up of ZnO enter the ryegrass root tissue (Lin and Xing 2008).

Studies suggested that delivery system of nutrients explores the porous domains in nanoscale range on plant surfaces which release nutrients and prevent their changing state into gaseous or chemical forms whose further absorption cannot occur by plants. In order to attain the absorption, biosensor is equipped with nanofertilizers and allows controlled delivery of nutrients. Soil nutrient and environmental conditions also improves the quality of soil by reducing toxic effects caused by fertilizers.

K. Nanosensors in Regulation of Plant hormones

McLamore et al. (2010) demonstrated the use of MWCNTs helped in the study of plant growth by hormone regulation especially auxin and helped to understand the mechanism of plant roots acclimatization in the environment in marginal soils.

6 Conclusion

The use of nano-biosensors in agriculture enabled for improvement in detection capacity of microorganisms contaminants which are toxic and detect pesticide and insecticide residues. The support of nanomaterials to biosensor technology provides a better device, which can be handled easily and more sensitive and helps in improvement of detection speed. In addition, it has capability for sensing single analyte which gives the information of toxic contaminants present in agriculture. Sensing system increases the selective or specific method of detection for pathogens during antigen–antibody interactions.

Nano-biosensors are still in its development stage in rural small-scale farms, but the support of different type of nanomaterials is effective for biosensors because the cost is less, highly sensitive, user-friendly, high specificity and no technician requirement. Therefore, this nano-biosensor technology will be effective in increasing the crop production for fulfilling the increasing demands for food and provide novel devices for farms in rural or remote areas in order to give benefits for early monitoring of crop.

In a nutshell, the use of bio-nanosensors provides smart agricultural practices and has large number of applications in agriculture such as a detection tool for diseases for quick identification of pathogens and therefore help in managing plant diseases. It also helped in hormones delivery such as auxin and gibberellin with the help of multiwalled carbon nanotubes and promoting plant growth and detects activity of fertilizer, nutrient concentration, insecticides and pesticide residues.

References

Ali MA, Jiang H, Mahal NK, Weber RJ, Kumar R, Castellano MJ, Dong L (2017) Microfluidic impedimetric sensor for soil nitrate detection using graphene oxide and conductive nanofibers enabled sensing interface. Sens Actuators B. Chem 239:1289–1299

Bakhori N, Yusof N, Abdullah A, Hussein M (2013) Development of a fluorescence resonance energy transfer (FRET)-based DNA biosensor for detection of synthetic oligonucleotide of ganoderma boninense. Biosensors 3:419–428

Bose D, Chatterjee S (2016) Biogenic synthesis of silver nanoparticles using guava (*Psidium guajava*) leaf extract and its antibacterial activity against *Pseudomonas aeruginosa*. Appl Nanosci 6:895–901

Cesarino I, Moraes FC, Lanza MR, Machado SA (2012) Electrochemical detection of carbamate pesticides in fruit and vegetables with a biosensor based on acetylcholinesterase immobilised on a composite of polyaniline–carbon nanotubes. Food Chem 135:873–879

Chattopadhyay DP, Patel BH (2016) Synthesis, characterization and application of nano cellulose for enhanced performance of textiles. J Text Sci Eng 6:184–215

Chen H, Yada R (2011) Nanotechnologies in agriculture: new tools for sustainable development. Trends Food Sci Technol 22:585–594

Deng HH, Hong GL, Lin FL, Liu AL, Xia XH, Chen W (2016) Colorimetric detection of urea, urease, and urease inhibitor based on the peroxidase-like activity of gold nanoparticles. Anal Chim Acta 915:74–80

DeRosa MC, Monreal C, Schnitzer M, Walsh R, Sultan Y (2010) Nanotechnology in fertilizers. Nat Nanotechnol 5:91

Dong J, Fan X, Qiao F, Ai S, Xin H (2013) A novel protocol for ultra-trace detection of pesticides combined electrochemical reduction of Ellman's reagent with acetylcholinesterase inhibition. Anal Chim Acta 25:78–83

Dutta S, Padhye S, Narayanaswamy R, Persaud KC (2001) An optical biosensor employing tiron-immobilised polypyrrole films for estimating monophenolase activity in apple juice. Biosens Bioelectron 16:287–294

Fang A, Chen H, Li H, Liu M, Zhang Y, Yao S (2017) Glutathione regulation-based dual-functional upconversion sensing-platform for acetylcholinesterase activity and cadmium ions. Biosens Bioelectron 87:545–551

Feynman R (1960) There's plenty of room at the bottom. Eng Sci 23:22–36

Ganeshkumar R, Sopiha KV, Wu P, Cheah CW, Zhao R (2016) Ferroelectric KNbO₃ nanofibers: synthesis, characterization and their application as a humidity nanosensor. Nanotechnology 27:395607

Ghorbanpour M, Fahimirad S (2017) Plant nanobionics a novel approach to overcome the environmental challenges. In: Medicinal plants and environmental challenges. Springer, Berlin, pp 247–257

Giraldo JP, Landry MP, Faltermeier SM, McNicholas TP, Iverson NM, Boghossian AA, Reuel NF, Hilmer AJ, Sen F, Brew JA, Strano MS (2014) Erratum: corrigendum: plant nanobionics approach to augment photosynthesis and biochemical sensing. Nat Mater 13:530

Haddaoui M, Raouafi N (2015) Chlortoluron-induced enzymatic activity inhibition in tyrosinase/ZnO NPs/SPCE biosensor for the detection of ppb levels of herbicide. Sens Actuators B: Chem 219:171–178

Hulkoti NI, Taranath TC (2014) Biosynthesis of nanoparticles using microbes—a review. Colloids Surf, B 121:474–483

Iravani S (2011) Green synthesis of metal nanoparticles using plants. Green Chem 13:2638

Kharat M, Du Z, Zhang G, McClements DJ (2017) Physical and chemical stability of curcumin in aqueous solutions and emulsions: impact of pH, temperature, and molecular environment. J Agric Food Chem 65:1525–1532

Khodakovskaya M, Dervishi E, Mahmood M, Xu Y, Li Z, Watanabe F, Biris AS (2009) Retraction notice for carbon nanotubes are able to penetrate plant seed coat and dramatically affect seed germination and plant growth. ACS Nano 6(8):7541

Kim M, Kim MJ (2003) Isocitrate analysis using a potentiometric biosensor with immobilized enzyme in a FIA system. Food Res Int 36:223–230

Koedrith P, Thasiphu T, Tuitemwong K, Boonprasert R, Tuitemwong P (2014) Recent advances in potential nanoparticles and nanotechnology for sensing food-borne pathogens and their toxins in foods and crops: current technologies and limitations. Sens Mater 711

Kwak SY, Giraldo JP, Wong MH, Koman VB, Lew TT, Ell J, Weidman MC, Sinclair RM, Landry MP, Tisdale WA, Strano MS (2017) A nanobionic light-emitting plant. Nano Lett 17(12):7951–7961

Lin D, Xing B (2008) Root uptake and phytotoxicity of ZnO nanoparticles. Environm Sci Technol 42:5580–5585

Liu D, Chen W, Wei J, Li X, Wang Z, Jiang X (2012) A highly sensitive, dual-readout assay based on gold nanoparticles for organophosphorus and carbamate pesticides. Anal Chem 84:4185–4191

Mandal D, Bolander ME, Mukhopadhyay D, Sarkar G, Mukherjee P (2005) The use of microorganisms for the formation of metal nanoparticles and their application. Appl Microbiol Biotechnol 69:85–92

McLamore ES, Diggs A, Calvo Marzal P, Shi J, Blakeslee JJ, Peer WA, Murphy AS, Porterfield DM (2010) Non-invasive quantification of endogenous root auxin transport using an integrated flux microsensor technique. Plant J 63:1004–1016

Mukhopadhyay SS (2014) Nanotechnology in agriculture: prospects and constraints. Nanotechnol Sci Appl 63

Narayanan KB, Sakthivel N (2010) Biological synthesis of metal nanoparticles by microbes. Adv Coll Interface Sci 156:1–13

Norouzi P (2017) A novel admittometric sensor for determination of theophylline using FFT coulometric admittance voltammetry and flow injection analysis. Int J Electrochem Sci 10057–10070

Otles S, Yalcin B (2012) Review on the application of nanobiosensors in food analysis. Acta Scientiarum Polonorum Technologia Alimentaria 11:7–18

Pan P, Miao Z, Yanhua L, Linan Z, Haiyan R, Pan K, Linpei P (2016) Preparation and evaluation of a stable solid state ion selective electrode of polypyrrole/electrochemically reduced graphene/glassy carbon substrate for soil nitrate sensing. Int J Electrochem Sci 11:4779–4793

Patra CR, Mukherjee S, Kotcherlakota R (2014) Biosynthesized silver nanoparticles: a step forward for cancer theranostics? Nanomedicine 9(10):1445–1448

Prasad R (2014) Synthesis of silver nanoparticles in photosynthetic plants. J Nanopart 1–8

Prasad R, Pandey R, Barman I (2015) Engineering tailored nanoparticles with microbes: quo vadis? Wiley Interdisc Rev Nanomed Nanobiotechnol 8:316–330

Prasad R, Bhattacharyya A, Nguyen QD (2017) Nanotechnology in sustainable agriculture: recent developments, challenges, and perspectives. Front Microbiol 8:10–14

Raskar SV, Laware SL (2014) Effect of zinc oxide nanoparticles on cytology and seed germination in onion. Int J Curr Microbiol App Sci 3:467–473

Rotariu L, Bala C, Magearu (2002) Yeast cells sucrose biosensor based on a potentiometric oxygen electrode. Analytica Chimica Acta 458:215–222

Sabir S, Arshad M, Chaudhari SK (2014) Zinc oxide nanoparticles for revolutionizing agriculture: synthesis and applications. Sci World J 1–8

Safarpour H, Safarnejad MR, Tabatabaei M, Mohsenifar A, Rad F, Basirat M, Shahryari F, Hasanzadeh F (2012) Development of a quantum dots FRET-based biosensor for efficient detection of polymyxa betae. Can J Plant Path 34:507–515

Shankar SS, Rai A, Ahmad A, Sastry M (2004) Rapid synthesis of Au, Ag, and bimetallic Au core–Ag shell nanoparticles using Neem (Azadirachta indica) leaf broth. J Colloid Interface Sci 275:496–502

Siddiquee S, Suryani S (2014) Nanoparticle-enhanced electrochemical biosensor with DNA immobilization and hybridization of Trichoderma harzianum gene. Sens Biosens Res 2:16–22

Singh R, Singh D, Singh D, Mani JK, Karwasra SS, Beniwal MS (2010) Effect of weather parameters on karnal bunt disease in wheat in Karnal region of Haryana. J Agrometeorol 12:99–101

Song JY, Kim BS (2009) Rapid biological synthesis of silver nanoparticles using plant leaf extracts. Bioprocess Biosyst Eng 32:79

Tarafdar JC, Sharma S, Raliya R (2013) Nanotechnology: interdisciplinary science of applications. Afr J Biotech 12:219–226

Teodoro S, Micaela B, David KW (2010) Novel use of nano-structured alumina as an insecticide. Pest Manag Sci

Verma N, Singh M (2003) A disposable microbial based biosensor for quality control in milk. Biosens Bioelectron 18:1219–1224

Wong MH, Giraldo JP, Kwak SY, Koman VB, Sinclair R, Lew TT, Bisker G, Liu P, Strano MS (2016) Nitroaromatic detection and infrared communication from wild-type plants using plant nanobionics. Nat Mater 16:264–272

Yao KS, Li SJ, Tzeng KC, Cheng TC, Chang CY, Chiu CY, Liao CY, Hsu JJ, Lin ZP (2009) Fluorescence silica nanoprobe as a biomarker for rapid detection of plant pathogens. Adv Mater Res 79–82:513–516

Yu Z, Zhao G, Liu M, Lei Y, Li M (2010) Fabrication of novel atrazine biosensora and its subpart-per trillion levels sensitive performance. Envirom Sci Technol 44:7878–7883

Zhao L, Peralta-Videa JR, Rico CM, Hernandez-Viezcas JA, Sun Y, Niu G, Servin A, Nunez JE, Duarte-Gardea M, Gardea-Torresdey JL (2014a) CeO_2 and ZnO nanoparticles change the nutritional qualities of cucumber (Cucumis sativus). J Agric Food Chem 62:2752–2759

Zhao Y, Liu L, Kong D, Kuang H, Wang L, Xu C (2014b) Dual amplified electrochemical immunosensor for highly sensitive detection of Pantoea stewartii subsp. stewartii. ACS Appl Mater Interfaces 6:21178–21183

Zheng Z, Li X, Dai Z, Liu S, Tang Z (2011) Detection of mixed organophosphorus pesticides in real samples using quantum dots/bi-enzyme assembly multilayers. J Mater Chem 21:16955–16962

Zhu Y, Cao Y, Sun X, Wang X (2013) Amperometric immunosensor for carbofuran detection based on MWCNTs/GS-PEI-Au and AuNPs-antibody conjugate. Sensors 13:5286–5301

Engineered Nanoparticles and Microbial Interaction

Interaction of Nanoparticles with Microbes

Sudhir S. Shende, Vishnu D. Rajput, Andrey V. Gorovtsov, Harish,
Pallavi Saxena, Tatiana M. Minkina, Vasiliy A. Chokheli,
Hanuman Singh Jatav, Svetlana N. Sushkova, Pawan Kaur,
and Ridvan Kizilkaya

Abstract

Nanotechnology is a rising area emerged after the amalgamation of the different advanced scientific fields of physics, chemistry and biology, and it has resulted in engineering of nanoparticles (1–100 nm) and their applications. These nanoparticles have an extensive utility in electronic circuits, biochemical sensors, pharmaceuticals, agriculture, cosmetic industry, therapeutic medical science, garment, food industry, etc. The market of nanoparticles is growing substantially, and many different types of nanoparticles and nanoparticle-based products have launched in the recent past. At the same time, unprecedented increases in the usage of nanoparticles have raised concerns over their ultimate release in the ecosystem, posing serious health hazards and environmental impact. The consequences may be more pronounced because of higher surface area against the mass ratio for the nanoparticles than bulk chemistry bestowing them unique physicochemical, electrical, optical and biological properties. Interaction of nanoparticles to the microbes is, therefore, vital to interpret the influence of nanoparticles on the aquatic bodies and soil health. In this regard, it is crucial to know the stability of nanoparticles, and better to understand the interaction and resulting toxicity mechanisms of nanoparticles to the microbes. In the present chapter, we have discussed these aspects with critical insights. Further, antimicrobial and antifungal properties of the nanoparticles are elaborated with a focus on the toxicity mechanism. The impact of nanoparticles could be influenced by the concentration, size, shape, etc. The toxicity mechanisms include inactivation of enzymes because of the interaction of thiol group, oxidative stress leading to surge in reactive oxygen species, restricted nutrient availability due to the aggregation of nanoparticles on the microbial surfaces, ultrastructural membranes, subcellular organelles and DNA damage. Understanding the complex nature of the interaction between the consortium of diverse microorganisms with nanoparticles is thoroughly debated in this chapter.

S. S. Shende · V. D. Rajput (✉) · A. V. Gorovtsov · T. M. Minkina · V. A. Chokheli · S. N. Sushkova
Academy of Biology and Biotechnology, Southern Federal University, Rostov-on-Don, Russia
e-mail: rajput.vishnu@gmail.com

S. S. Shende
Nanobiotechnology Laboratory, Department of Biotechnology, Sant Gadge Baba Amravati University, Amravati, 444602, Maharashtra, India

Harish · P. Saxena
Department of Botany, Mohan Lal Sukhadia University, Udaipur, Rajasthan 313001, India

H. S. Jatav
Sri Karan Narendra Agriculture University, Jobner, Rajasthan 303329, India

P. Kaur
Centre of Excellence in Agriculture for Nanotechnology, Teri-Deakin Nanobiotechnology Centre, The Energy and Resources Institute (TERI), New Delhi, India

R. Kizilkaya
Ondokuz MayısÜniversitesi, Samsun, Turkey

Keywords

Ecosystem • Interaction • Mechanism • Microbes • Nanoparticles • Toxicity

1 Introduction

Nanotechnology mainly deals with the studies involving the fabrication, manipulation and utilization of nanoparticles (NPs; size between 1 and 100 nm) in different areas such as medical science, pharmaceuticals, electronics, textile, biochemical sensors and other allied areas. Several chemical and physical methods are developed for NPs synthesis with some merits and demerits. Chemical methods involve the use of solvents as reducing agent for NPs synthesis, but

© Springer Nature Switzerland AG 2021
P. Singh et al. (eds.), *Plant-Microbes-Engineered Nano-particles (PM-ENPs) Nexus in Agro-Ecosystems*,
Advances in Science, Technology & Innovation,
https://doi.org/10.1007/978-3-030-66956-0_12

hazardous by-products are major environmental concern while, physical methods have high energy consumption (Huang et al. 2011). However, much attention is paid towards the biological synthesis approach, which has many advantages like eco-friendly nature, reliability, biocompatibility and low production cost (Roy et al. 2013; Emeka et al. 2014). NPs synthesized from plants, microbes and other biological resources are therefore considered as preferred way for synthesis of NPs (Khandel and Shahi 2018).

Nanotechnology is progressing rapidly in various fields due to widespread applications and substantial success. Nanotechnology seems to be a suitable option for the protection of plants from various agents of biotic stress (Rodríguez-Cutiño et al. 2018). Recently, interventions of nanotechnological applications in agriculture sciences have been studied like preparation of nanoscale fertilizers (Xu et al. 2015; Jahagirdar et al. 2020), pesticides (Grillo et al. 2016; Adisa et al. 2019), plant disease diagnosis (Prasad et al. 2017; Chen et al. 2019). It can be stated that agricultural productivity would be increased using better varieties and crop plant protection. The significant antimicrobial potential of NPs against plant pathogens has been widely investigated for advanced agricultural applications (Baker et al. 2017; Verma et al. 2018; Chen et al. 2019; Fu et al. 2020). Further use of nano-agro-particles is considered as a valuable alternative against several fungal pathogens. For instance, Ghasemian et al. (2012) demonstrated the role of CuNPs to ward off filamentous fungi *Penicillium chrysogenum*, *Alternaria alternata*, *Fusarium solani* and *Aspergillus flavus*. Giannousi et al. (2013) studied the efficacy of three types of copper oxide NPs against *Phytophthora infestans*, a pathogen for tomato crop. The authors reported, all the Cu-based NPs, which were tested, showed a significant inhibitory effect against the tested pathogen (Giannousi et al. 2013). At the same time, expected massive usage of NPs in the upcoming future poses serious environmental concerns, and therefore, the interaction of NPs to the microbes is of utmost necessity to formulate a sustainable release policy of the NPs, taking into these concerns.

2 Main Sink of Nanoparticles, Their Production, Applications and Environmental Concerns

The use of NPs raises major concerns for agro-ecosystems, and the soil and water are considered as their main sink. Some NPs be present naturally in the environment; however, the concentrations of these NPs are extremely low with the negligible impacts (Remedios et al. 2012). If the NPs release is inevitable, the objective must be to reduce the NPs release, which might pose a noteworthy threat to agro-ecosystems or human health (Yadav et al. 2014). Main sinks of NPs, i.e.

NPs in soil and water, are described in detail in later sections.

2.1 Nanoparticles in Soils

From the starting of the Earth's history, NPs have naturally existed, and it is a known fact that they are not human innovation (Handy et al. 2008). Soils are considered as a source of natural NPs, as it is a multifaceted matrix with different colloidal mineral particles. At the same time, the pollutants immobilization in the soil matrix has a major concern, which greatly outweighs any anthropogenic production as its exposure to natural NPs (Sharma et al. 2015). With reference to the techniques of NPs formation, there are a number of mechanisms that are able to produce NPs in the environment, like geological and biological. Geological way of synthesis involves autogenesis, or the neo-formation found in the soils, physicochemical weathering, as well as the volcanic explosion activity. Typically, the mentioned geological processes are capable of producing inorganic particles, whereas, in biological mechanisms, organic nano-molecules could be produced, even though some organisms are capable of yielding in the cell the minerals granules (Handy et al. 2008).

In the soils, the movement of NPs is explained by Brownian motion and gravity has no role in this. Consequently, solitary NPs could be entering into micropores and unless they get absorbed on mobile colloids, the mobility is greatly improved, while the aggregates of NPs stay remnant in macropores, while the mobility was introverted, when they are adsorbed on particles which are non-mobile.

The NPs and the soil molecules attachment are depending upon collector and the NPs shape as well as onto the diverse properties, which transform NPs surrounding environment. Thus, the stipulations of soils are capable of improving or inhibiting the NPs mobility in soils. The aquifers and the humic acids present in soils could considerably manipulate NPs mobility of different metal oxides (Ben-Moshe 2010), which could persuade NPs composition monitoring and the soil nutrients fate, contaminants and pollutants as well (Ben-Moshe 2010; Mura et al. 2013).

A powerful rising significance in the exploitation of NPs for the various applications for soil is documented in various researches by several researchers (Pan and Xing 2012; Priester et al. 2012; Jośko and Oleszczuk 2013; Suppan 2013; Fernández et al. 2014; Garner and Keller 2014; Jośko et al. 2014; Conway et al. 2015; Schaumann et al. 2015a, b; Watson et al. 2015; Rabbani et al. 2016). Ge et al. (2011) observed the effects of NPs on bacterial communities' present in soil, in which reduced biomass, diversity of microbes and soil enzyme activity is impacted by the action of ZnO NPs. The aggregation and immobilization of NPs in the soil showed phytotoxicity which ultimately leads to decreased root length and biomass (Kim et al. 2011). Authors reported the toxic effect of ZnO NPs on maize and ryegrass, in which the inhibition in the germination was observed. In another experiment, Ma et al. (2010) reported that, when the aluminium oxide and rare element oxide NPs were applied to the plants, such as carrots, cabbage, cucumber, soybeans and maize, the toxic effect was demonstrated, as they act as an inhibitor for elongation of roots.

The field of soil science is related to all materials science, which are commonly found in soils. These matrices can provide the nutrition for organisms along with those microflora and fauna that assist these processes. This is a composite mixture of chemicals as well as organisms, from which some are pre-arranged at the nano-level while the others are unable to do so (Belal and El-Ramady 2016). The scope of the nanotechnology has been extended from the early phase of preliminary innovations of capability to progress and situate atoms (Belal and El-Ramady 2016). Soil is a composite mix of particles homing in size from millimetres (mm) to nanometres scale (nm). By means of some highly sophisticated techniques such as transmission electron microscopy (TEM) and atomic force microscopy (AFM), it may perhaps be promising to recognize these soils makeup. These preceding methods are capable to demonstrate the association of colloid materials in soils like humic acids and phyllosilicates, and the detection of novel material like iron oxides NPs. Thus, nanotechnology is able to offer additional possibility in classifying single cells, proteins, DNA, genes, as well as other biological structures in soils (Dasgupta et al. 2016a).

With reference to soil, nanotechnology is of vital significance, since a number of constituents of the soils have nanoscale features or are nanoparticulate (Mura et al. 2013). At the nanoscale level, interactions are either conquered by stronger polar and electrostatic interactions, weak Van der Waals forces, or covalent bonding. The particulars of interaction forces of nanoparticle-nanoparticle as well as interactions between nanoparticle-fluid are of major significance for illustrating the chemical and physical processes along with time-lapse progression of free NPs (Mura et al. 2013). Also, in soil, different nanomaterials (NMs) can be found such as nanominerals ranging from nanoparticle to nanosize NPs of mineral but larger sized particles are also present (Maurice and Hochella 2008). Sharma et al. (2015) reviewed the natural inorganic NPs formation, their fate as well as its toxicity issue (Sharma et al. 2015). Additionally, variable NPs are also found in soil matrix, bacterial appendages, clay minerals, amorphous substances as well as other nanominerals (Mura et al. 2013).

Manufactured or fabricated or engineered NPs (ENPs) may be present in soils, but these NPs may perhaps leach out in the surroundings deliberately in diverse forms, which include the metal oxides like CeO_2, TiO_2, ZnO NPs; metals with zero valency such as Au, Ag and Fe NPs; as well as metal salts like ceramics and nano-silicates; carbon derived NMs such as carbon nanotubes; nano-polymers, e.g. polystyrene and latex; and semiconductor materials like CdSe, CdTe; or accidentally by-products combustion or corrosion (Belal and El-Ramady 2016). Because of the distribution of NPs in soils, an alteration in their aggregated size, the stability of a suspension, transport as well as bioavailability could be perceived. Hence, the research on the ENPs is indispensable to comprehend their destiny along with associated danger (Philippe and Schaumann 2014; Sharma et al. 2015). The sol of these NPs is able to be exaggerated by conditions of soil such as ionic strength, the amount of dissolved organic matter as well as the biological and chemical reactions (Li et al. 2016). The NPs coated by dissolved organic matter, have their surface properties altered. These properties include pore size, organic contaminants sorption parameter, surface area, aggregation property and the toxicity mechanisms (Li et al. 2016).

According to Wang and Keller (2009), attributable to complexity, no particular property is able to apply as a common interpreter of the deposition as well as transport of ENPs. Therefore, it is significantly essential to illustrate quantitatively the transfer of ENPs in columns of soil (Pan and Xing 2012). Hence, in conclusion, applications for the environment and ENPs risk assessment in the soil significantly not independent on the appreciative of the interaction between the NPs with the various components of soil ENPs possibly will be functional for remediation of soil (Belal and El-Ramady 2016). By reason of the soil system complexity as well as the so primary stage of research of NPs in soils, the appreciative of behaviour of NPs in this system is exceptionally restricted.

2.2 Nanoparticles in Water

Nanoparticles are of different types like natural or engineered or incidental. Natural NPs include lunar dust, volcanic dust, soil particles and these natural NPs are present on the earth since its birth (Belal and El-Ramady 2016). Incidental NPs are formed by human economic activity like coal usage, fumes of iron welding, machinery in industries and vehicle emission (Smita et al. 2012). ENPs are designed and fabricated for their unique physicochemical property for different applications. Different shapes and types of NPs are made like metal-based NMs, carbon-based NMs, nanocomposites and dendrimers (Handy et al. 2008; Yadav et al. 2014). However, ultimately, all of these NPs are discharged into aquatic bodies (Sharma et al. 2015). The term colloid is sort of a generic term usually applied for particles having size between 1 nm and 1 μm. In aquatic bodies, these NPs form colloidal complex after interaction with organic materials like proteins, humic and fulvic acids, and inorganic species notably hydrous manganese, iron oxides. Further, to interpret the future usage of nano-fertilizers, a huge amount of N and P in nanoform is going to be released in the water bodies, which may affect the ecosystem and human health. Consequences of these interactions are completely unknown (Ma et al. 2014; Johnson et al. 2014; Yang et al. 2015). It is estimated that occurrence of NPs in aquatic system is quite low in comparison with the natural NPs (Delay and Frimmel 2012). Therefore, the movement and translocation of NPs within waters are a budding issue. Nano-pollution in aquatic body is a major cause of concern, and there are few reports available which have specifically dealt to remediate the nano-contamination in the water bodies. In one study published recently, it is reported that iron oxide nanoparticles can be accumulated inside the green algae *Coelastrella terrestris*. In this way, remediation of water containing excess NPs is possible. The accumulation factor reported in this study was found to be about 2.9, which means, about 2.9 times iron oxide NP is accumulated inside the algal cell than ambient environment (Saxena et al. 2020). It has been reported that NPs affect the life of aquatic ecosystem by inhibiting growth and nitrogen fixing capacity, increasing the level of ROS and MDA, decreasing the pigment content in photosynthesis organisms, negatively influencing antioxidant enzymes. Further, physical damage to subcellular organs like membrane damage, cell wall damage and intra-thylakoidal damage are also reported (Saxena and Harish 2018).

3 Toxicity Mechanisms of Nanoparticles

3.1 Proposed Mode of Antibacterial Action of Metal Nanoparticles

It is widely known that metal NPs such as AgNPs and CuNPs have significant antibacterial activity (Table 1), but the mechanism of their action is yet not known. There is some literature available on metal NPs mode of action, but until now, the mode of action is very unclear. Das et al. (2010) reported that CuNPs are capable of entering the cell because of their smaller size and subsequently takes place their protein or enzyme inactivation, producing hydrogen peroxide that results in the death of bacterial cells. In another report, it has been stated that the protein inactivation occurs because of the CuNPs and –SH group of proteins interaction with each other (Schrand et al. 2010). Likewise, metal NPs can disturb the DNA helical structure and degrade it. The cell membrane integrity is decided by the electrochemical potential, since according to Deryabin et al. (2013), CuNPs are responsible to reduce the cell membranes

electrochemical potential, that eventually affected the cell membrane integrity. It was also understood that metal NPs liberate their respective ions, and these heavy metal ions are found to have unfavourably affected the cells of bacteria (Cioffi et al. 2005). Metal NPs and metal ions accumulation on surface of cell cause the formation of pits in the membrane, which mainly leading to the outflow of components from bacterial cells ultimately causing the cells death. The next significant reason for the bacterial cell death has been proposed is the oxidative stress development due to the action of NPs (Deryabin et al. 2013). Considering all these possibilities, Shende et al. (2015) have proposed a hypothetical mechanism of action of CuNPs in bacteria; in a similar way, metal NPs could impact the bacterial cells during the bactericidal action (Fig. 1).

3.2 Proposed Mechanism of Antifungal Action of Metal Nanoparticles

The mechanistic action of metal NPs as a fungicidal agent is still unclear; however, there are many ways by which metal NPs could serve as an antifungal agent depending on their mode of action. The probable antifungal action of metal NPs could be correlated with the commercial fungicides available in the market (Fig. 2).

Although the commercially available antifungal agents are target specific and are mainly limited to the plasma membrane and cell wall, which are the targets (Ngo et al. 2016; Scorzoni et al. 2017), the metal NPs which were capped with proteins in case of biogenic synthesize could get attached to the fungal cell wall and initiate a sustainable release of the metal ions inside the cell, which can act on the fungi by different ways.

A lipid responsible for membrane fluidity is ergosterol and essential for cell viability (Tatsumi et al. 2013; Song et al. 2016). A few antifungal agents generally target ergosterol, either by restraining its biosynthesis or binding to it, resulting in the formation of the pores in the membrane, this may be similar to the metal NPs or metal ions. The composition of fungal cell wall primarily constituted chitin, mannans, glycoproteins and glucans, essential for adhesion and pathogenesis of fungi and also provides a protective barrier, limiting the admittance of molecules to the plasma membrane (van der Weerden et al. 2013).

The two most important modes of action of antifungal agents targeting the cell wall are associated with the inhibition of chitin and β-glucan synthesis. Thus, metal NPs may perhaps target chitin synthase, which is responsible for the chitin chain elongation. Another mechanism is inhibition of nucleic acids, protein and microtubule synthesis. There are some antifungals, which may cause more than one effect on the fungal cells under adverse conditions, like in the

Table 1 Recent studies of using application of different nanoparticles for antimicrobial properties

Nanoparticle(s) studied	Microbes investigated	Approach	References
ZnO NPs	*Staphylococcus aureus, E. coli*	ZnONPs coated textile fabrics are tested for antibacterial property	Singh et al. (2020)
Silver nanoparticles (AgNPs)	*E. coli, Enterococcus faecalis* and *Salmonella typhi*	Aqueous leaf extract of *Cestrum nocturnum* is used to synthesize the NPs. Bactericidal activity was checked using growth inhibition assay	Keshari et al. (2020)
Silver nanoparticles (AgNPs)	*S. aureus, S. dysenteriae* and *S. typhi*	*Penicillium oxalicum* mediated synthesis of NPs. Antibacterial activity was evaluated using well diffusion method and spectrophotometric method	Feroze et al. (2020)
Silver and copper oxide NPs-decorated graphene oxide	*S. aureus, E. coli*	Incorporation of silver and copper oxide NPs through graphene oxide nanosheets is found suitable for clinical treatment	Menazea and Ahmed (2020)
Iron oxide nanoparticles (FeONPs)	Six human pathogenic strains including *E. coli* and *S. aureus*	Aqueous extract of leaf of *Psidium guajava* (PG) is used for synthesis of NPs	Madubuonu et al. (2020)
Silver nanoparticles (AgNPs)	*S. aureus* and *Pseudomonas aeruginosa*	Marine macroalgae *Padina* sp. is used for synthesis of NPs and	Bhuyar et al. (2020)
Chitosan encapsulated silver nanoparticles	*Bacillus cereus, S. aureus, Listeria monocytogenes, E. coli* and *Salmonella enterica*	Leaf extract of *Gynura procumbens* and chitosan is used for NPs synthesis	Sathiyaseelan et al. (2020)
MgO nanoparticles	*Bacillus cereus*	Fabrication of cubic structure of MgO nanoparticles showing antibacterial activity	El-Shaer et al. (2020)
Silver nanoparticles/activated carbon co-doped titania nanoparticles	*E. coli* and *S. aureus*	Zones of inhibition comparable to streptomycin were observed with zone of inhibition of 7 mm	Parvathi et al. (2020)
Fe_3O_4 nanoparticles	*S. aureus, Corynebacterium, P. aeruginosa* and *Klebsiella pneumoniae*	Synthesis of NPs using medicinal plants *Malva sylvestris*	Mousavi et al. (2020)
Silver nanoparticles (AgNPs)	Pathogens in Fish such as *Vibrio harveyi, Vibrio parahaemolyticus, Vibrio alginolyticus* and *Vibrio anguillarum*	NPs synthesis by red algae *Portieria hornemannii* and antibacterial activity against pathogens in fish	Fatima et al. (2020)
Iron oxide, Tobramycin, iron nitride conjugated nanoparticles	*P. aeruginosa*	Synthesis of iron oxide NPs capped with alginate. NPs found to have the potential to cross the bacterial biofilm barrier	Armijo et al. (2020)
Silver nanoparticles embedded guar gum/gelatin nanocomposite	*S. aureus, E. coli* and *P. aeruginosa*	Synthesis of NPs is done via in situ method by maltose sugar reduction	Khan et al. (2020)
V_2O_5 nanoparticles	*S. aureus, E. coli, P. aeruginosa* and *P. vulgaris*	Ultrasound assisted synthesis of NPs. NPs was found to useful in dye degradation and biomedical applications	Karthik et al. (2020)
$ZnTiO_3$ and Ag-doped $ZnTiO_3$ perovskite nanoparticles	*S. aureus* and *Vibrio* sp.	NPs were synthesized via the sol–gel method and found to have antibacterial activity	Abirami et al. (2020)

presence of UV light and oxidants, the mitochondria produce free radicals in large quantity, causing the damage to DNA, proteins and lipids, which leads to cell death due to reactive oxygen species (ROS) generation (Ferreita et al. 2013; Mesa-Arango et al. 2014), a similar mode of action by metal ions against filamentous fungi is reported by Vincent et al. (2018). There may be inhibition of heat shock protein 90 (Hsp 90) which has been associated with the fungal pathogenicity, phase transition, regulation of other heat shock proteins and antifungal resistance (Jacob et al. 2015; Scorzoni et al. 2017), due to the leaching of metal ions from metal NPs. On the other hand, Yang et al. (2011) reviewed the different action mechanisms of fungicides as well as their probable impacts on non-target microbes. During their study,

Fig. 1 Graphical representation of hypothetical mechanism of action of metal nanoparticles (NPs). (1) Metal NPs accumulation on the surface of cell, formation of pits in membrane; (2) Interaction of metal NPs with cell membrane, affects membrane integrity; (3) DNA damage; (4) Interaction of metal ions with sulfhydryl (-SH) groups of proteins, leads to protein inactivation; (5) Metal NPs and metal ions entry, oxidative stress development, leads to cell death

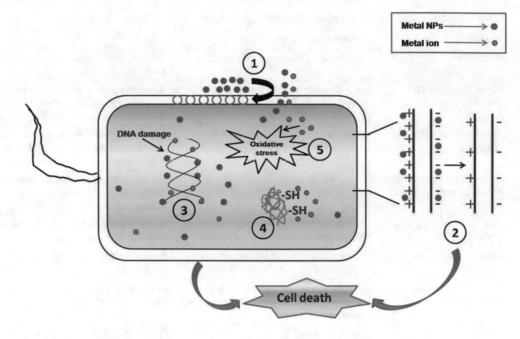

Fig. 2 Schematic representation of hypothetical mechanism of action of metal nanoparticles on fungi

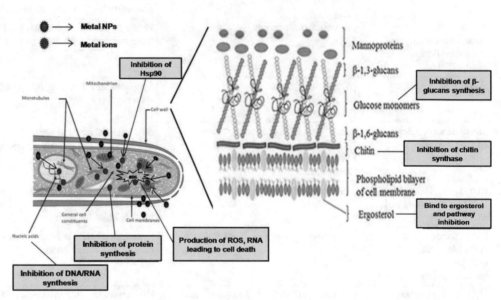

they concluded that the fungicides affected target fungi in the variable ways such as affecting the synthesis of lipids, sterols and other components of membrane, amino acids and protein synthesis, respiration, signal transduction, cell division, and mitosis, multisite fungicidal activity, etc. (Yang et al. 2011). The metal NPs mechanism of action for fungicidal activity is still unclear, but from the above hypothetical mechanism it has been revealed that the metal NPs leaches the metal ions, which ultimately affected the growth and metabolism leading to the growth inhibition of fungus.

4 Effects of Nanoparticles on Soil Microbial Community

More than the last decades, the NPs discipline have progressed as an area of interdisciplinary studies that have fascinated the scientific community as it is very much interesting and challenging (Belal and El-Ramady 2016). Undoubtedly the field of NMs science is pertinent to the structure and composition examination of soil.

Nevertheless, nano-biology related to biology of soil as well as tools intended for distinguishing the substances at their nano-quantities, which are appropriate for soil processes and also significant as are different facets of NPs applications in the environmental sciences (Belal and El-Ramady 2016). It has been documented that the ENPs could possibly be fabricated with single elements such as carbon or silver or with combinations of elements or molecules. These NPs could be categorized depending on their size, their chemical composition or morphological properties. It may also describe these NPs keen on subsequent clusters involving—metal ENPs (elemental Ag, Au, Fe, Se, etc.), metal and non-metal oxides (Al_2O_3, CeO_2, CuO, FeO_2, SiO_2, TiO_2, ZnO), complex compounds such as Co–Zn–Fe oxide, fullerenes and polymer-coated quantum dots for instance cadmium selenide (CdSe) as well as organic polymers similar to polystyrene (Dinesh et al. 2012).

In terms of global biogeochemical cycles, microbes are to be considered as drivers as they are deeply involved in C, N, S and P cycling. Because they are exceptionally sensitive to changes in environmental conditions, the structure, as well as abundance of the microbial community, is likely to change towards the foreign NMs (Ge et al. 2011; Kumar et al. 2011). Since microbes facilitate the regulation and maintenance of overall health of the ecosystem and its function, microbial community alteration will enormously affect the whole ecosystem. Consequently, an improved understanding of how microbes act in response to NPs and/or NMs is able to facilitate our handling of environmental as well as health concerns brought with reference to the manufacturing as well as the application of these NMs (He et al. 2014). Alternatively, it is well documented that, a number of NPs have previously reported for their antimicrobial potential that is why they shown the direct effect on microbes. To date, no standard and established techniques for measuring the NPs toxicity on various soil microbes and microbial diversity.

4.1 Interaction of Soil Contaminants with Soil Microbes

The contaminants effect upon the microbe's community present inside the soil might be evaluated through various methods like viability count, carbon utilization patterns, molecular-based methods, along with fatty acid methyl ester analysis. It has been reported in the literature that, an interaction among the NPs as well as the bacterial cells leads to cytotoxic effect, which has assumed to include the mechanism having two steps (Kumari et al. 2014). The first step involves the oxidative damage by the NPs to the cell membrane, which results in loss of membrane integrity devoid of noteworthy decrease in viability of the cells. The

step second is involving the outflow of the internal cellular components, which leads to the consequence of reduced viability and internalization of the NPs, thus causing cell organelles damage, e.g. the nucleus (Kumari et al. 2014).

The majority of microbes have produced efficient mechanisms at their molecular level as well as explicit pathways for the biochemical reactions for detoxification, efflux, along with to accrue the metal ions greatly previous to it was discovered by plants. In addition, microbes are again competent for the volatilization of a number of metal ions to dispose of their acute toxicity. Even though microorganisms have developed resistance as well as a prevention mechanism, further belowground level studies are essential in views to advantageous microorganisms present in soil like phosphate solubilizers, N_2-fixing, arbuscular mycorrhizal fungi (AMF) to set up the mechanisms of uptake as well as consequences for the soil and microbes (Thul and Sarangi 2015).

Many researchers have published the reviews on the interactions between NPs and microbes, which correlate the physicochemical properties of ENPs (metal and metal oxides) to their biological response (Dinesh et al. 2012; Ge et al. 2012; Pawlett et al. 2013; Holden et al. 2013, 2014; Tilston et al. 2013; Dimkpa 2014; Jośko et al. 2014; Burke et al. 2015; García-Gómez et al. 2015; Judy et al. 2015; Simonin and Richaume 2015; Sillen et al. 2015; Xu et al. 2015; Van Aken 2015; Aliofkhazraei 2016; Sirbu et al. 2016). Moreover, from the above discussion, in conclusion it could be mentioned that the specific toxicity towards the specific species can be attributed to shape and size of NPs. However, the coatings of the materials on the surface, which could be altered importantly by conditions of environment, that can ameliorate or accelerate toxicity to the microbes (Suresh et al. 2013; Thul et al. 2013).

Recent literature was reviewed, and it can be concluded that there are quite a lot of impacts of NPs on soil microorganisms; those involved in the soil enzyme activities, nitrogen cycle, iron metabolism processes, antibiotic and phytohormone production (Dimkpa 2014). These effects are considered to be either positive or negative and the results being dependent on the particular type of NPs, the charge on the surface, size, species of microbes or plant to be examined, dose tested, as well as test medium whether agar, soil, liquid or other used solid media. These communally published results have figured out that NT poses a substantial threat to soil microorganisms and proven that the agricultural processes are driven by microbes. However, it could be demonstrated that there is a prospective for soil and plant microbes to alleviate the NPs bioreactivity (Dimkpa 2014). While roots of all of the terrestrial plants are colonized by microorganisms, a number of studies of NPs interactions with microbes and plants are performed independently. A very few studies in real plant/microbes' systems established the NPs effects onto the implementation of arbuscular mycorrhizal fungi (AMF), nitrogen fixation, and on the fabrication of microbial siderophores in the plant rhizosphere. Hence, it might be recommended that, for a better understanding of the agro-ecological NPs implications, would necessitate additional exhaustive interactive studies in collective plant /microbes/nanoparticles system (Dimkpa 2014).

Regarding the microbes in soil, the comprehensible and metal NPs specific effect was observed on microflora in the soil. For instance, the TiO$_2$ NPs showed an impact on symbiosis of *Rhizobium*-legume in garden peas and Rhizobium (*R. leguminosarum* bv. *vivae* 3841). It was also found that TiO$_2$ NPs put forth morphological modifications in the cells of bacteria. Moreover, Fan et al. (2014) also reported that whenever there the interaction between these two organisms takes place, they disturbed the formation of root nodules and the succeeding postponement in the nitrogen fixation commencement. The immediate application of NPs keen on treated biosolids or soils having transportable NPs might interact with the microbes in the soil. These soil microbes are also competent towards the adsorption and accumulation in one or the other form of NMs that in turn begins the NMs mobilization and is capable to alter communities encompassing the populations of plants, animals and finally humans through the food chains (Holden et al. 2013; Ranjan et al. 2014; Thul and Sarangi 2015).

Conversely, plants, in general, get mineral nutrients from the soils with the help of the soil bacteria and fungi. A study discovered that the AgNPs, which are a popular microbicidal agent, negatively impact the plants growth and eradicates the microbes in the soil that maintain them. Not just microorganisms, but the several soil enzymes activity, e.g. soil peroxidase, catalase, as well as protease, was established to considerably diminished by TiO$_2$ NPs (Du et al. 2011). Furthermore, inorganic TiO$_2$, SiO$_2$ as well as ZnO had found to put forth a lethal effect on bacteria. In the presence of light, the toxicity of these elements further significantly increased (Adams et al. 2006). There are the variety of reports that have been spotted light upon the interactions between NPs-microbe's for associating the ENPs (metal and metal oxides) physicochemical properties and their responses in the biological systems. Additionally, in conclusion, the species-specific toxicity of NPs could be attributed to its shape and size. Research on the ecologically significant species of bacteria, e.g. *Bacillus subtilis*, *Escherichia coli*, *Pseudomonas putida* and other, has noticeably indicated microorganisms be able to take up NPs (Thul and Sarangi 2015; Załęska-Radziwiłł and Doskocz 2015).

In the terrestrial and aquatic ecosystems, bacteria are essential elements as they act as decomposers of organic matter as well as key bases for numerous webs of foods (Thul et al. 2013). Because the dependency of plants on the

fungi and bacteria present in soil and air to get their nutrients, the antimicrobial and cellular toxicity effects of NPs for instance, Ag, TiO$_2$ and Au NPs and nano-emulsions as well might show the effect on the environment (Thul et al. 2013; Dasgupta 2016b, c; Jain et al. 2016; Maddineni et al. 2015; Ranjan et al. 2016). Hajipour et al. (2012) have examined the NPs for their antibacterial properties in a very illustrative manner. It has also been demonstrated that soil microbes, that are plentiful and flexible catalysts, are capable to adsorb and disband the aggregates of ENPs (Horst et al. 2010). It has been reported that the addition of nanoscale zerovalent iron leads to perturbation in soil bacterial community composition, as well as condensed the chloroaromatic mineralizing activity of microbes (Tilston et al. 2013).

4.2 Interaction of Engineered Nanoparticles with Soil Microbiota

The ENPs were also established to considerably modify the bacterial communities in a dose-dependent approach, and NPs are known to influence the dynamics of the microbial community (Ge et al. 2011). In order to this, Priester et al. (2012) reported the uptake of ENPs of CeO$_2$ into the soybean roots and root nodules, which reduced the nitrogen fixation potentials along with the damaged growth of crop plants (Priester et al. 2012). Further studies about the beneficial soil microbes, such as nitrogen fixers, AM fungi, phosphate solubilizers, have demonstrated the uptake mechanisms of the NPs as well as the significance to accumulate in the soil and microorganisms (Ge et al. 2011; Thul et al. 2013). The ENPs mobility in soils is totally dependent on their size, though that is the agglomerates size, not the primary size that is concerned with the transportability of them. There are several aspects, those organize the transfer of these ENPs in the soils; however, charge, size and the rate of agglomeration in the transport medium are prognostic of the mobility of these ENPs in the soils. The metal NPs survival as well as speciation in the soil solution and the understanding on interaction among soil solution or other ions and their active sites is significant for getting a better knowledge about the interactions between metal NPs and soil microbes. Nevertheless, the solution chemistry of metal NPs is somewhat restricted, and thermodynamic data like reaction constants and solubility of NPs are not available. In addition to this, the additional data is requisite to distinguish the effect of ENPs on the soil microbial community in a variety of soils having different physicochemical features and soils from the diverse ecosystem (Dinesh et al. 2012).

In conclusion, a number of novel ENPs from both environmental and industrial applications and resulting from various activities of human as by-products, act as xenobiotics and find their own way to enter into the soil. Thus, the fortification of microbial biomass and diversity present in the soil is most important challenging issue for sustainable resource utilization, since advanced higher levels of microbial diversity as well as biomass indicate higher turnover of nutrient. Very little studies have been performed and reported the toxicity of ENPs to soil microorganisms due to the complex nature of soil through which the organisms are rendered to these ENPs inside diverse phases of soil. For understanding the complete effects of ENPs on different soil organisms under different environmental conditions, more studies are required that detect the different parameters of soil, which influence the bioavailability in addition to the toxicity of ENPs.

5 Future Perspectives and Challenges

Nanoparticles from the environment and the ENPs interact with the microbes in the soil and agro-ecosystems. The NPs form of chemicals, metal (ions), smoke, etc., in air, water and soil cause the environmental pollution when occurring over a quantity of forbearance limit for living animals that is a problem from an age-old. Inappropriate as well as excessive utilization of pesticides and fertilizers has augmented nutrients as well as toxins in surface waters and groundwater, incurring health and water purification expenses, and lessening fish farming as well as recreational opportunities (Mukhopadhyay 2014). Moreover, the soil quality is degraded due to different practices in agriculture, which leads to the eutrophication in the aquatic habitats and may perhaps require the disbursement of augmented fertilization, irrigation and energy to maintain productivity on tarnished soils (Mukhopadhyay 2014; Belal and El-Ramady 2016). These preceding practices could also destroy beneficial soil microbes, insects and other wildlife.

It is well understood that the nanotechnology's application in the field of agriculture might be triumphant, whenever the naturally occurring processes are stimulated within huge articulation of science or sophistication intended for booming accomplishment. For example, the objective may be to build the soil extremely competent to advance the nutrient usage in efficient manner for productivity boosting and superior security of environment. Consequently, the nutrient management in the nanotechnology frame should be based on some imperative parameters, which includes (1) in the soil system, ions of nutrient should be available as an obtainable forms for the plant, and (2) within plant and soil systems, transport of nutrient relies on exchange of ions, desorption and adsorption and the precipitation or solubility reactions, as well as (3) NMs should ease the process that would guarantee the nutrients accessibility for the plants in the rate and manner as per their requirement (Mukhopadhyay 2014; Belal and El-Ramady 2016).

Nanotechnology provides a number of modern approaches or strategies that could employed for water management, fertilizers, pesticides, sensors and restrictions in the application of chemically prepared pesticides, and the NMs potential in the agriculture management in sustainable way (Prasad et al. 2014). There are a number of publications, which have determined the agriculture sustainability beneath the nanotechnology's roof and effect of NPs on the terrestrial environments (Mura et al. 2013; Mukhopadhyay 2014; Prasad et al. 2014; Sekhon 2014; Takeuchi et al. 2014; Ditta et al. 2015; Patil et al. 2016; Salamanca-Buentello and Daar 2016; Rajput et al. 2018a, b, 2020a, b). From these reports, it has been clearly noticed that nanotechnology will participate a progressively more significant role in the agriculture field. Moreover, the last decade researches demonstrated that the potential nanotechnology's applications in transforming the field of agriculture with the revolution in the fields such as regulators for plant growth, biosensors, smart delivery systems for drugs, plants and animals genetic improvement, food additives, pesticides and fertilizers transformed into nano-pesticides and nano-fertilizers (Hong et al. 2013). Hong et al (2013) suggested that for thwarting the probable unfavourable effects from the nanotechnology application in the agriculture sector, research on the issues like in the ecosystem, the transport and the fate of the NMs, uptake as well as its accumulation in animals and plants, with the NMs toxicity evaluation need to be performed. Risk assessment research should also be executed prior to nano-products application for agriculture, and the effects must be examined.

The prospect of the nanotechnology application in agriculture is extraordinary. The implementation of some novel technologies is an imperative concern in the sustainable development frame, and it is well-documented (Mukhopadhyay 2014). It has been proposed that the nanotechnology application in agriculture may take a timeline period of few decades to shift from the laboratory scale to field, particularly because of the drawbacks experienced to evade with biotechnology. Nanotechnology's application is important, as it provided the global population, who carry on the deficiency in access to safe water, education, health care, trustworthy sources of energy, as well as other basic development needs of human (Belal and El-Ramady 2016).

In conclusion, in the light of sustainability the potential nanotechnology applications needs to be re-assessed, considering the ethical (Salamanca-Buentello and Daar 2016), societal (Roure 2016), economic (Shapira and Youtie 2015) as well as environmental factors (Bottero 2016) and interdependencies. It means, the products based on nanotechnology needs sustainability, must not only while the phase of its manufacturing but also must be considered over the complete life cycle of the product. Thus, as presented in review by Rickerby (2013), an entire life cycle of product must be considered for an assessment of technology as well as it is also essential at preliminary stages of development for accomplishing the accurate balance between the cost-effectiveness as well as the environmental impacts. Besides this, the analysis of life cycle of product is expected to provide important insights regarding this issue. The standard methods, which are existing for the risk assessments may perhaps be insufficient for recognizing meticulous vulnerability related to NMs and tools for nano-specific risk assessments, have to be produced. Suitable recycling and strategies used for recovery in the sustainability frame also have to limit the NMs dispersal in the surrounding environments. Procedures should also be accepted for minimizing the environmental and health risks because of the NMs release at every stage of life cycle of products from its production phase, while its application, and towards the final dumping or recycling. If the trustworthy data as well as ethics do not appear, nevertheless, the most awful circumstances have to be supposed for nanotechnology risk management in the sustainable agriculture and development frame.

6 Concluding Remark

Currently, nanotechnology is expanding in each and every field; there is not a single area that is untouched with the nanotechnology and its pioneering modernizations in scientific manner involving the field of agriculture. The interaction of NPs with the soil microbial community played a vital role in agriculture; hence, the nanotechnology application in the agriculture field has been touched numerous fields involving plant protection, plant nano-nutrition, food industry, nano-fertilizers and nano-pesticides, plant productivity, etc. With reference to the ENPs effect on agro-ecosystem, the providence and performance of the ENPs and the probable toxic inferences to the agricultural crops and the plants, as well as naturally present microbes in the rhizosphere of soil and nano-waste generation in the agricultural ecosystem, etc., are of the major burning concerns. In addition, ENPs negative effects produced via free radicals lead to the DNA damage and lipid peroxidation affect the microflora of soil including bacteria and fungi, soil enzymatic activities, plant productivity and the entire environment. Therefore, there is an urgent need to forecast the ENPs effects on the environment to make their application a predictable opportunity in the direction of sustainable agriculture.

Acknowledgements The authors are thankful for the financial support to the Russian Foundation for Basic Research, project no. 19-05-50097 and the Ministry of Science and Higher Education of the Russian Federation, project no. 0852-2020-0029.

References

Abirami R, Kalaiselvi CR, Kungumadevi L, Senthil TS, Kang M (2020) Synthesis and characterization of ZnTiO$_3$ and Ag doped ZnTiO$_3$ perovskite nanoparticles and their enhanced photocatalytic and antibacterial activity. J Solid State Chem 281:121019

Adams LK, Lyon DY, Alvarez PJ (2006) Comparative eco-toxicity of nanoscale TiO$_2$, SiO$_2$, and ZnO water suspensions. Water Res 40:3527–3532

Adisa IO, PullaguralaVLR P-V, Dimkpa CO, Elmer WH, Gardea-Torresdey JL, White JC (2019) Recent advances in nano-enabled fertilizers and pesticides: a critical review of mechanisms of action. Environ Sci Nano 6:2002

Aliofkhazraei M (2016) Handbook of nanoparticles. Springer International Publishing, Cham. https://doi.org/10.1007/978-3-319-15338-4

Armijo LM, Wawrzyniec SJ, Kopciuch M, Brandt YI, Rivera AC, Withers NJ, Cook NC, Huber DL, Monson TC, Smyth HD, Osiński M (2020) Antibacterial activity of iron oxide, iron nitride, and tobramycin conjugated nanoparticles against *Pseudomonas aeruginosa* biofilms. J Nanobiotechnol 18(1):1–27

Baker S, Volova T, Prudnikova SV, Satish S, Nagendraprasad MN (2017) Nanoagroparticles: emerging trends and future prospect in modern agriculture system. Environ Toxicol Pharmacol 53:10–17

Belal E-S, El-Ramady H (2016) Nanoparticles in water, soils and agriculture. In: Ranjan S, Dasgupta N, Lichtfouse E (eds) Nanoscience in food and agriculture 2, sustainable, agriculture reviews, vol 21, Springer, Berlin, pp 311–358. https://doi.org/10.1007/978-3-319-39306-3_10

Ben-Moshe T (2010) Transport of metal oxide nanoparticles in saturated porous media. Chemosphere 81:387–393

Bhuyar P, Rahim MH, Sundararaju S, Ramaraj R, Maniam GP, Govindan N (2020) Synthesis of silver nanoparticles using marine macroalgae *Padina* sp. and its antibacterial activity towards pathogenic bacteria. Beni-Seuf Univ J Appl Sci 9(1):1–5

Bottero J-Y (2016) Environmental risks of nanotechnology: a new challenge? In: Lourtioz J-M et al. (eds) Nanosciences and nanotechnology, Springer, Berlin, pp 287–311. https://doi.org/10.1007/978-3-319-19360-1_13

Burke DJ, Pietrasiak N, Situ SF, Abenojar EC, Porche M, Kraj P, Lakliang Y, Samia ACS (2015) Iron oxide and titanium dioxide nanoparticle effects on plant performance and root associated microbes. Int J Mol Sci 16(10):23630–23650. https://doi.org/10.3390/ijms161023630

Chen J, Mao S, Xu F, Ding W (2019) Various antibacterial mechanisms of biosynthesized copper oxide nanoparticles against soil borne *Ralstonia solanacearum*. RSC Adv 9:3788–3799. https://doi.org/10.1039/c8ra09186b

Cioffi N, Torsi L, Ditaranto N, Tantillo G, Ghibelli L, Sabbatini L, Bleve-Zacheo T, D'Alessio M, Zambonin PG, Traversa E (2005) Copper nanoparticle/polymer composites with antifungal and bacteriostatic properties. Chem Mater 17:5255–5262

Conway JR, Beaulieu AL, Beaulieu NL, Mazer SJ, Keller AA (2015) Environmental stresses increase photosynthetic disruption by metal oxide nanomaterials in a soil-grown plant. ACS Nano 9(12):11737–11749. https://doi.org/10.1021/acsnano.5b03091

Das R, Gang S, Nath SS, Bhattacharjee R (2010) Linoleic acid capped copper nanoparticles for antibacterial activity. J Bionanosci 4:82–86

Dasgupta N, Shivendu R, Bhavapriya R, Venkatraman M, Chidambaram R, Avadhani GS, Ashutosh K (2016a) Thermal co-reduction approach to vary size of silver nanoparticle: its microbial and cellular toxicology. Environ Sci Pollut Res 23:4149–4163. https://doi.org/10.1007/s11356-015-4570-z

Dasgupta N, Shivendu R, Patra D, Srivastava P, Kumar A, Ramalingam C (2016b) Bovine serum albumin interacts with silver nanoparticles with a "side-on" or "end on" conformation. Chem Biol Int 253:100–111. https://doi.org/10.1016/j.cbi.2016.05.018

Dasgupta N, Shivendu R, Shraddha M, Ashutosh K, Chidambaram R (2016c) Fabrication of food grade vitamin E nanoemulsion by low energy approach: characterization and its application. Int J Food Prop 19(3):700–708. https://doi.org/10.1080/10942912.2015.1042587

Delay M, Frimmel FH (2012) Nanoparticles in aquatic systems. Anal Bioanal Chem 402:583–592. https://doi.org/10.1007/s00216-011-5443-z

Deryabin DG, Aleshina ES, Vasilchenko AS, Deryabin TD, Efremova LV, Karimov IF, Korolevskay LB (2013) Investigation of copper nanoparticles antibacterial mechanisms tested by luminescent Escherichia coli strains. Nanotechnol Russ 8(5):402–408

Dimkpa CO (2014) Can nanotechnology deliver the promised benefits without negatively impacting soil microbial life? J Basic Microbiol 2014(54):1–16. https://doi.org/10.1002/jobm.201400298

Dinesh R, Anandaraj M, Srinivasan V, Hamza S (2012) Engineered nanoparticles in the soil and their potential implications to microbial activity. Geoderma 173:19–27. https://doi.org/10.1016/j.geoderma.2011.12.018

Ditta A, Arshad M, Ibrahim M (2015) Nanoparticles in sustainable agricultural crop production: applications and perspectives. In: Siddiqui MH, Al-Whaibi MH, Mohammad F (eds) Nanotechnology and plant sciences: nanoparticles and their impact on plants. Springer, Berlin, pp 55–75. https://doi.org/10.1007/978-3-319-14502-4

Du W, Sun Y, Ji R, Zhu J, Wu J, Guo H (Du) TiO2 and ZnO nanoparticles negatively affect wheat growth and soil enzyme activities in agricultural soil. J Environ Monit 13:822–828

El-Shaer A, Abdelfatah M, Mahmoud KR, Momay S, Eraky MR (2020) Correlation between photoluminescence and positron annihilation lifetime spectroscopy to characterize defects in calcined MgO nanoparticles as a first step to explain antibacterial activity. J Alloys Compd 817:152799

Emeka EE, Ojiefoh OC, Aleruchi C (2014) Evaluation of antibacterial activities of silver nanoparticles green-synthesized using pineapple leaf (Ananas comosus). Micron 57:1–5

Fan R, Huang YC, Grusak MA, Huang CP, Sherrier DJ (2014) Effects of nano-TiO2 on the agronomically-relevant Rhizobium -legume symbiosis. Sci Total Environ 466–467:503–512. https://doi.org/10.1016/j.scitotenv.2013.07.032

Fatima R, Priya M, Indurthi L, Radhakrishnan V, Sudhakaran R (2020) Biosynthesis of silver nanoparticles using red algae Portieria hornemannii and its antibacterial activity against fish pathogens. Microb Pathog 138:103780

Fernández MD, Alonso-Blázquez MN, García-Gómez C, Babin M (2014) Evaluation of zinc oxide nanoparticle toxicity in sludge products applied to agricultural soil using multispecies soil systems. Sci Total Environ 497:688–696. https://doi.org/10.1016/j.scitotenv.2014.07.085

Feroze N, Arshad B, Younas M, Afridi MI, Saqib S, Ayaz A (2020) Fungal mediated synthesis of silver nanoparticles and evaluation of antibacterial activity. Microsc Res Techniq 83(1):72–80

Ferreira GF, Baltazar LEM, Santos JR, Monteiro AS, Fraga LA, Resende-Stoianoff MA, Santos DA (2013) The role of oxidative and nitrosative bursts caused by azoles and amphotericin B against the fungal pathogen Cryptococcus gattii. J Antimicrob Chemother 68:1801–1811

Fu L, Wang Z, Dhankher OP, Xing B (2020) Nanotechnology as a new sustainable approach for controlling crop diseases and increasing agricultural production. J Exp Bot 71(2):507–519. https://doi.org/10.1093/jxb/erz314

García-Gómez C, Babin M, Obrador A, Álvarez JM, Fernández MD (2015) Integrating ecotoxicity and chemical approaches to compare the effects of ZnO nanoparticles, ZnO bulk, and ZnCl2 on plants and microorganisms in a natural soil. Environ Sci Pollut Res 22 (21):16803–16813. https://doi.org/10.1007/s11356-015-4867-y

Garner KL, Keller AA (2014) Emerging patterns for engineered nanomaterials in the environment: a review of fate and toxicity studies. J Nanopart Res 16:2503. https://doi.org/10.1007/s11051-014-2503-2

Ge Y, Schimel JP, Holden PA (2011) Evidence for negative effects of TiO2 and ZnO nanoparticles on soil bacterial communities. Environ Sci Technol 45:1659–1664

Ge Y, Schimel JP, Holden PA (2012) Identification of soil bacteria susceptible to TiO2 and ZnO nanoparticles. Appl Environ Microbiol 78(18):6749–6758. https://doi.org/10.1128/AEM.00941-12

Ghasemian E, Naghoni A, Tabaraie B, Tabaraie T (2012) In vitro susceptibility of filamentous fungi to copper nanoparticles assessed by rapid XTT colorimetry and agar dilution method. J Mycol Med 22:322–328. https://doi.org/10.1016/j.mycmed.2012.09.006

Giannousi K, Avramidis I, Dendrinou-Samara C (2013) Synthesis, characterization and evaluation of copperbased nanoparticles as agrochemicals against Phytophthora infestans. RSC Adv 3:21743–21752. https://doi.org/10.1039/c3ra42118j

Grillo R, Abhilash PC, Fraceto LF (2016) Nanotechnology applied to bio-encapsulation of pesticides. J Nanosci Nanotechnol 16:1231–1234. https://doi.org/10.1166/jnn.2016.12332

Hajipour MJ, Fromm KM, Ashkarran AA, de Aberasturi DJ, de Larramendi IR, Rojo T, Serpooshan V, Parak WJ, Mahmoudi M (2012) Antibacterial properties of nanoparticles. Trends Biotechnol 30(10):499–511. https://doi.org/10.1016/j.tibtech.2012.06.004

Handy RD, von der Kammer F, Lead JR, Richard Owen MH, Crane M (2008) The ecotoxicology and chemistry of manufactured nanoparticles. Ecotoxicology 17:287–314. https://doi.org/10.1007/s10646-008-0199-8

He X, Aker WG, Leszczynski J, Hwang H-M (2014) Using a holistic approach to assess the impact of engineered nanomaterials inducing toxicity in aquatic systems. J Food Drug Anal 22:128–146. https://doi.org/10.1016/j.jfda.2014.01.011

Holden PA, Klaessig F, Turco RF, Priester J, Rico CM, Arias HA, Mortimer M, PacpacoK,Gardea-Torresdey JL (2014) Evaluation of exposure concentrations used in assessing manufactured nanomaterial environmental hazards: are they relevant? Environ Sci Technol 48(18):10541–10551. https://doi.org/10.1021/es502440s

Holden PA, Nisbet RM, Lenihan HS, Miller RJ, Cherr GN, Schimel JP, Gardea-Torresdey JL (2013) Ecological nanotoxicology: integrating nanomaterial hazard considerations across the subcellular, population, community, and ecosystems levels. Acc Chem Res 46:813–822

Hong J, Peralta-Videa JR, Gardea-Torresdey JL (2013) Nanomaterials in agricultural production: benefits and possible threats? In: Shamim N, Sharma VK (eds) Sustainable nanotechnology and the environment: advances and achievements, vol 1124, ACS symposium series. American Chemical Society, Washington, DC, pp 73–90. https://doi.org/10.1021/bk-2013-1124.ch001

Horst AM, Neal AC, Mielke RE, Sislian PR, Suh WH, Mädler L, Stucky GD, Holden PA (2010) Dispersion of TiO2 nanoparticle agglomerates by Pseudomonas aeruginosa. Appl Environ Microbiol 76:7292–7298

Huang J, Zhan G, Zheng B, Sun D, Lu F, Lin Y, Chen H, Zheng Z, Zheng Y, Li Q (2011) Biogenic silver nanoparticles by Cacumen platycladi extract: synthesis, formation mechanism and antibacterial activity. Ind Eng Chem Res 50:9095–9106

Jacob TR, Peres NT, Martins MP, Lang EA, Sanches PR, Rossi A, Martinez-Ross NM (2015) Heat shock protein 90 (Hsp90) as a molecular target for the development of novel drugs against the dermatophyte *Trichophyton rubrum*. Front Microbiol 6:1241

Jahagirdar AS, Shende S, Gade A, Rai M (2020) Bio-inspired synthesis of copper nanoparticles and its efficacy on seed viability and seedling growth in Mungbean (*Vigna radiata* L.) Curr Nanosci 16 (2):1–7

Jain A, Shivendu R, Nandita D, Chidambaram R (2016) Nanomaterials in food and agriculture: an overview on their safety concerns and regulatory issues. Crit Rev Food Sci 58(2):297–317. https://doi.org/10.1080/10408398.2016.1160363

Johnson CA, Freyer G, Fabisch M, Caraballo MA, Ksel K, Hochella MF (2014) Observations and assessment of iron oxide and green rust nanoparticles in metalpolluted mine drainage within a steep redox gradient. Environ Chem 11(4):377–391

Joško I, Oleszczuk P (2013) Influence of soil type and environmental conditions on the ZnO, TiO_2 and Ni nanoparticles phytotoxicity. Chemosphere 92:91–99

Joško I, Oleszczuk P, Futa B (2014) The effect of inorganic nanoparticles (ZnO, Cr_2O_3, CuO and Ni) and their bulk counterparts on enzyme activities in different soils. Geoderma 232–234:528–537. https://doi.org/10.1016/j.geoderma.2014.06.012

Judy JD, McNear DH Jr, Chen C, Lewis RW, Tsyusko OV, Bertsch PM, Rao W, Stegemeier J, Lowry GV, McGrath SP, Durenkamp M, Unrine JM (2015) Nanomaterials in biosolids inhibit nodulation, shift microbial community composition, and result in increased metal uptake relative to bulk/dissolved metals. Environ Sci Technol 49(14):8751–8758. https://doi.org/10.1021/acs.est.5b01208

Karthik K, Nikolova MP, Phuruangrat A, Pushpa S, Revathi V, Subbulakshmi M (2020) Ultrasound-assisted synthesis of V_2O_5 nanoparticles for photocatalytic and antibacterial studies. Mater Res Innov 24(4):229–234

Keshari AK, Srivastava R, Singh P, Yadav VB, Nath G (2020) Antioxidant and antibacterial activity of silver nanoparticles synthesized by Cestrum nocturnum. J Ayurveda Integr Med 11 (1):37–44

Khan N, Kumar D, Kumar P (2020) Silver nanoparticles embedded guar gum/gelatin nanocomposite: green synthesis, characterization and antibacterial activity. Colloid Interfac Sci 35:100242

Khandel P, Shahi SK (2018) Mycogenic nanoparticles and their bio-prospective applications: current status and future challenges. J Nanostruct Chem 8:369–391

Kim S, Kim J, Lee I (2011) Effects of Zn and ZnO nanoparticles and Zn^{2+} on soil enzyme activity and bioaccumulation of Zn in *Cucumis sativus*. Chem Ecol 27(1):49–55. https://doi.org/10.1080/02757540.2010.529074

Kumar N, Shah V, Walker VK (2011) Perturbation of an arctic soil microbial community by metal nanoparticles. J Hazard Mater 190 (1–3):816–822

Kumari J, Kumar D, Mathur A, Naseer A, Kumar RR, Chandrasekaran PT, Chaudhuri G, Pulimi M, Raichur AM, Babu S, Chandrasekaran N, Nagarajan R, Mukherjee A (2014) Cytotoxicity of TiO_2 nanoparticles towards fresh water sediment microorganisms at low exposure concentrations. Environ Res 135:333–345. https://doi.org/10.1016/j.envres.2014.09.025

Li S, Ma H, Wallis LK, Etterson MA, Riley B, Hoff DJ, Diamond SA (2016) Impact of natural organic matter on particle behavior and phototoxicity of titanium dioxide nanoparticles. Sci Total Environ 542:324–333. https://doi.org/10.1016/j.scitotenv.2015.09.141

Ma R, Levard C, Judy JD, Unrine JM, Durenkamp M, Martin B, Jefferson B, Lowry GV (2014) Fate of zinc oxide and silver nanoparticles in a pilot wastewater treatment plant and in processed biosolids. Environ Sci Technol 48(1):104–112

Ma Y, Kuang L, HeX BW, Ding Y, Zhang Z, Zhao Y, Chai Z (2010) Effects of rare earth oxide nanoparticles on root elongation of plants. Chemosphere 78:273–279

Maddineni SB, Badal KM, Shivendu R, Nandita D (2015) Diastase assisted green synthesis of size-controllable gold nanoparticles. RSC Adv 5:26727–26733. https://doi.org/10.1039/C5RA03117F

Madubuonu N, Aisida SO, Ahmad I, Botha S, Zhao TK, Maaza M, Ezema FI (2020) Bio-inspired iron oxide nanoparticles using *Psidium guajava* aqueous extract for antibacterial activity. Appl Phys A 126(1):1–8

Maurice PA, Hochella MF (2008) Nanoscale particles and processes: a new dimension in soil science. Adv Agron 100:123–138

Menazea AA, Ahmed MK (2020) Silver and copper oxide nanoparticles-decorated graphene oxide via pulsed laser ablation technique: preparation, characterization, and photoactivated antibacterial activity. Nano-Struct Nano-Objects 22:100464

Mesa-Arango AC, Trevijano-Contador N, Román E, Sánchez-Fresneda R, Casas C, Herrero E, Argüelles JC, Pla J, Cuenca-Estrella M, Zaragoza O (2014) The production of reactive oxygen species is a universal action mechanism of Amphotericin B against pathogenic yeasts and contributes to the fungicidal effect of this drug. Antimicrob Agents Chemother 58:6627–6638

Mousavi SM, Hashemi SA, Zarei M, Bahrani S, Savardashtaki A, Esmaeili H, Lai CW, Mazraedoost S, Abassi M, Ramavandi B (2020) Data on cytotoxic and antibacterial activity of synthesized Fe_3O_4 nanoparticles using *Malva sylvestris*. Data Brief 28:104929

Mukhopadhyay SS (2014) Nanotechnology in agriculture: prospects and constraints. Nanotechnol Sci Appl 7:63–71

Mura S, Seddaiu G, Bacchini F, Roggero PP, Greppi GF (2013) Advances of nanotechnology in agro-environmental studies. Ital J Agron 8:127–140

Ngo HX, Garneau-Tsodikova S, Green KD (2016) A complex game of hide and seek: the search for new antifungals. Med Chem Comm 7:1285–1306

Pan B, Xing B (2012) Applications and implications of manufactured nanoparticles in soils: a review. Eur J Soil Sci 63(4):437–456. https://doi.org/10.1111/j.1365-2389.2012.01475.x

Parvathi VP, Umadevi M, Sasikala R, Parimaladevi R, Ragavendran V, Mayandi J, Sathe GV (2020) Novel silver nanoparticles/activated carbon co-doped titania nanoparticles for enhanced antibacterial activity. Mater Lett 258:126775

Patil SS, Shedbalkar UU, Truskewycz A, Chopade BA, Ball AS (2016) Nanoparticles for environmental clean-up: a review of potential risks and emerging solutions. Environ Technol Innov 5:10–21

Pawlett M, Ritz K, Dorey RA, Rocks S, Ramsden J, Harris JA (2013) The impact of zero-valent iron nanoparticles upon soil microbial communities is context dependent. Environ Sci Pollut Res 20 (2):1041–1049. https://doi.org/10.1007/s11356-012-1196-2

Philippe A, Schaumann GE (2014) Interactions of dissolved organic matter with natural and engineered inorganic colloids: a review. Environ Sci Technol 48(16):8946–8962. https://doi.org/10.1021/es502342r

Prasad R, Bhattacharyya A, Nguyen QD (2017) Nanotechnology in sustainable agriculture: recent developments, challenges, and perspectives. Front Microbiol 8:1014. https://doi.org/10.3389/fmicb.2017.01014

Prasad R, Kumar V, Prasad KS (2014) Nanotechnology in sustainable agriculture: present concerns and future aspects. Afr J Biotechnol 13 (6):705–713. https://doi.org/10.5897/AJBX2013.13554

Priester JH, Ge Y, Mielke RE, Horst AM, Moritz SC, Espinosa K, Gelb J, Walker SL, Nisbet RM, An YJ, Schimel JP (2012) Soybean susceptibility to manufactured nanomaterials with evidence for food quality and soil fertility interruption. Proc Natl Acad Sci USA 109: E2451–E2456

Rabbani MM, Ahmed I, Park S-J (2016) Application of nanotechnology to remediate contaminated soils. In: Hasegawa H, RahmanMofizur IM, AzizurRahman M (eds) Environmental remediation technologies for metal-contaminated soils. Springer, Berlin, pp 219–229. https://doi.org/10.1007/978-4-431-55759-3_10

Rajput V, Minkina T, Ahmed B, Sushkova S, Singh R, Soldatov M, Laratte B, Fedorenko A, Mandzhieva S, Blicharska E, Musarrat J (2020a) Interaction of copper-based nanoparticles to soil, terrestrial, and aquatic systems: critical review of the state of the science and future perspectives, In: Reviews of environmental contamination and toxicology. Springer, Berlin, pp 51–96. https://doi.org/10.1007/398_2019_34

Rajput V, Minkina T, Sushkova S, Behal A, Maksimov A, Blicharska E, Ghazaryan K, Movsesyan H, Barsova N (2020b) ZnO and CuO nanoparticles: a threat to soil organisms, plants, and human health. Environ Geochem Health 42:147–158. https://doi.org/10.1007/s10653-019-00317-3

Rajput VD, Minkina TM, Behal A, Sushkova SN, Mandzhieva S, Singh R, Gorovtsov A, Tsitsuashvili VS, Purvis WO, Ghazaryan KA, Movsesyan HS (2018a) Effects of zinc-oxide nanoparticles on soil, plants, animals and soil organisms: a review. Environ Nanotechnol Monit Manag 9:76–84. https://doi.org/10.1016/j.enmm.2017.12.006

Rajput VD, Minkina T, Sushkova S, Tsitsuashvili V, Mandzhieva S, Gorovtsov A, Nevidomskyaya D, Gromakova N (2018b) Effect of nanoparticles on crops and soil microbial communities. J Soils Sediments 18:2179–2187. https://doi.org/10.1007/s11368-017-1793-2

Ranjan S, Nandita D, Arkadyuti RC, Melvin SS, Chidambaram R, Rishi S, Ashutosh K (2014) Nanoscience and nanotechnologies in food industries: opportunities and research trends. J Nanopart Res 16(6):2464. https://doi.org/10.1007/s11051-014-2464-5

Ranjan S, Nandita D, Bhavapriya R, Ganesh SA, Chidambaram R, Ashutosh K (2016) Microwave irradiation-assisted hybrid chemical approach for titanium dioxide nanoparticle synthesis: microbial and cytotoxicological evaluation. Environ Sci Pollut Res 23(12):12287–12302. https://doi.org/10.1007/s11356-016-6440-8

Remedios C, Rosario F, Bastos V (2012) Environmental nanoparticles interactions with plants: morphological, physiological, and genotoxic aspects. J Bot 1–8. https://doi.org/10.1155/2012/751686

Rickerby DG (2013) Nanotechnology for more sustainable manufacturing: opportunities and risks. In: Shamim N, Sharma VK (eds) Sustainable nanotechnology and the environment: advances and achievements. ACS symposium, vol 1124. American Chemical Society, Washington, DC, pp 91–105

Rodríguez-Cutiño G, Gaytán-Andrade JJ, García-Cruz A, Ramos-González R, Chávez-González ML, Segura-Ceniceros EP, Martínez-Hernández JL, Govea-Salas M, Ilyina A (2018) Nanobiotechnology approaches for crop protection. In: Kumar V et al (eds) Phytobiont and ecosystem restitution. Springer, Berlin, pp 1–21. https://doi.org/10.1007/978-981-13-1187-1_1

Roure F (2016) Societal approach to nanoscience and nanotechnology: when technology reflects and shapes society. In: Lourtioz J-M et al (eds) Nanosciences and nanotechnology. Springer, Berlin, pp 357–404. https://doi.org/10.1007/978-3-319-19360-1_17

Roy S, Mukherjee T, Chakraborty S, Kumar das T (2013) Biosynthesis, characterisation and antifungal activity of silver nanoparticles by the fungus Aspergillus foetidus MTCC8876. Digest J Nanomater Biostruct 8:197–205

Salamanca-Buentello F, Daar AS (2016) Dust of wonder, dust of doom: a landscape of nanotechnology, nanoethics, and sustainable development. In: Bagheri A et al (eds) Global bioethics: the impact of the UNESCO international bioethics committee, vol 5, Advancing global bioethics. Springer, Berlin, pp 101–123. https://doi.org/10.1007/978-3-319-22650-7_10

Sathiyaseelan A, Saravanakumar K, Mariadoss AV, Wang MH (2020) Biocompatible fungal chitosan encapsulated phytogenic silver nanoparticles enhanced antidiabetic, antioxidant and antibacterial activity. Int J Biol Macromol. https://doi.org/10.1016/j.ijbiomac.2020.02.291

Saxena P, Harish (2018) Nanoecotoxicological reports of engineered metal oxide nanoparticles on algae. Curr Poll Rep 4(2):128–142

Saxena P, Sangela V, Harish (2020) Toxicity evaluation of iron oxide nanoparticles and accumulation by microalgae Coelastrella terrestris. Environ Sci Poll Res 1–11. https://doi.org/10.1007/s11356-020-08441-9

Schaumann GE, Baumann T, Lang F, Metreveli G, Vogel H-J (2015b) Engineered nanoparticles in soils and waters. Sci Total Environ 535:1. https://doi.org/10.1016/j.scitotenv.2015.06.006

Schaumann GE, Philippe A, Bundschuh M, Metreveli G, Klitzke S, Rakcheev D, Grün A, Kumahor SK, Kühn M, Baumann T, Lang F, Manz W, Schulz R, Vogel H-J (2015a) Understanding the fate and biological effects of Ag- and TiO$_2$-nanoparticles in the environment: the quest for advanced analytics and interdisciplinary concepts. Sci Total Environ 535:3–19. https://doi.org/10.1016/j.scitotenv.2014.10.035

Schrand AM, Rahman MF, Hussain SM, Schlager JJ, Smith DA, Syed AF (2010) Metal-based nanoparticles and their toxicity assessment. WIREs Nanomed Nanobiotechnol 2:554–568

Scorzoni and de Paula e Silva, ACA, Marcos CM, Assato PA, de Melo WCMA, de Oliveira HC, Costa-Orlandi CB, Mendes-Giannini MJS, Fusco-Almeida AM, , 2017.Scorzoni L, de Paula e Silva, ACA, Marcos CM, Assato PA, de Melo WCMA, de Oliveira HC, Costa-Orlandi CB, Mendes-Giannini MJS, Fusco-Almeida AM (2017) Antifungal therapy: new advances in the understanding and treatment of mycosis. Front Microbiol 8:36

Sekhon BS (2014) Nanotechnology in agri-food production: an overview. Nanotechnol Sci Appl 7:31–53. https://doi.org/10.2147/NSA.S39406

Shapira P, Youtie J (2015) The economic contributions of nanotechnology to green and sustainable growth. In: Basiuk VA, Basiuk EV (eds) Green processes for nanotechnology. Springer, Berlin, pp 409–434. https://doi.org/10.1007/978-3-319-15461-9_15

Sharma VK, Filip J, Zboril R, Varma RS (2015) Natural inorganic nanoparticles—formation, fate, and toxicity in the environment. Chem Soc Rev 44(23):8410–8423. https://doi.org/10.1039/c5cs00236b

Shende S, Ingle AP, Gade A, Rai M (2015) Green synthesis of copper nanoparticles by Citrus medica Linn. (Idilimbu) juice and its antimicrobial activity. World J Microbiol Biotechnol 31(6):865–873

Sillen WMA, Thijs S, Abbamondi GR, Janssen J, Weyens N, White JC, Vangronsveld J (2015) Effects of silver nanoparticles on soil microorganisms and maize biomass are linked in the rhizosphere. Soil Biol Biochem 91:14–22. https://doi.org/10.1016/j.soilbio.2015.08.019

Simonin M, Richaume A (2015) Impact of engineered nanoparticles on the activity, abundance, and diversity of soil microbial communities: a review. Environ Sci Pollut Res 22:13710–13723. https://doi.org/10.1007/s11356-015-4171-x

Singh G, Joyce EM, Beddow J, Mason TJ (2020) Evaluation of antibacterial activity of ZnO nanoparticles coated sonochemically onto textile fabrics. J Microbiol Biotechnol Food Sci 9(4):106–120

Sirbu T, Maslobrod SN, Yu AM, Borodina VG, Borsch NA, Ageeva LS (2016) Influence of dispersed solutions of copper, silver, bismuth and zinc oxide nanoparticles on growth and catalase activity of Penicillium funiculosum. In: Sontea V, Tiginyanu I (eds) 3rd international conference on nanotechnologies and biomedical engineering, IFMBE proceedings 55, Springer, Berlin, pp 271–274. https://doi.org/10.1007/978-981-287-736-9_66

Smita S, Gupta SK, Bartonova A, Dusinska M, Gutleb AC, Rahman Q (2012) Nanoparticles in the environment: assessment using the causal diagram approach. Environ Health 11(Suppl 1):S13. https://doi.org/10.1186/1476-069X-11-S1-S13

Song J, Zhai P, Zhang Y, Zhang C, Sang H, Han G, Keller NP, Lu L (2016) The *Aspergillus fumigatus* damage resistance protein family coordinately regulates ergosterol biosynthesis and azole susceptibility. MBio 7:e01919-15

Suppan S (2013) Nanomaterials in soil: our future food chain? The institute for agriculture and trade policy. Published by IATP (www.iatp.org)

Suresh AK, Pelletier DA, Doktycz MJ (2013) Relating nanomaterial properties and microbialtoxicity. Nanoscale 5:463–474

Takeuchi MT, Kojima M, Luetzow M (2014) State of the art on the initiatives and activities relevant to risk assessment and risk management of nanotechnologies in the food and agriculture sectors. Food Res Int 64:976–981. https://doi.org/10.1016/j.foodres.2014.03.022

Tatsumi Y, Nagashima M, Shibanushi T, Iwata A, Kangawa Y, Inui F, Jo Siu WJ, Pillai R, Nishiyama Y (2013) Mechanism of action of efinaconazole, a novel triazole antifungal agent. Antimicrob Agents Chemother 57:2405–2409

Thul ST, Sarangi BK (2015) Implications of nanotechnology on plant productivity and its rhizospheric environment. In: Siddiqui MH, Al-Whaibi MH, Mohammad F (eds) Nanotechnology and plant sciences: nanoparticles and their impact on plants. Springer, Berlin, pp 37–54. https://doi.org/10.1007/978-3-319-14502-3

Thul ST, Sarangi BK, Pandey RA (2013) Nanotechnology in agroecosystem: implications on plant productivity and its soil environment. Expert Opin Environ Biol 2:27. https://doi.org/10.4172/2325-9655.1000101

Tilston EL, Collins CD, Mitchell GR, Princivalle J, Shaw LJ (2013) Nanoscale zero valent iron alters soil bacterial community structure and inhibits chloroaromatic biodegradation potential in Aroclor 1242-contaminated soil. Environ Pollut 173:38–46

Van Aken B (2015) Gene expression changes in plants and microorganisms exposed to nanomaterials. Curr Opin Biotechnol 33:206–219. https://doi.org/10.1016/j.copbio.2015.03.005

Van der Weerden NL, Bleackley MR, Anderson MA (2013) Properties and mechanisms of action of naturally occurring antifungal peptides. Cell Mol Life Sci 70:3545–3570

Verma SK, Das AK, Patel MK, Shah A, Kumar V, Gantait S (2018) Engineered nanomaterials for plant growth and development: a perspective analysis. Sci Total Environ 630:1413–1435

Vincent M, Duval RE, Hartemann P, Engels-Deutsch M (2018) Contact killing and antimicrobial properties of copper. J Appl Microbiol 124(5):1032–1046

Wang P, Keller AA (2009) Natural and engineered nano and colloidal transport: role of zeta potential in prediction of particle deposition. Langmuir 25:6856–6862

Watson JL, Fang T, Dimkpa CO, Britt DW, McLean JE, Jacobson A, Anderson AJ (2015) The phytotoxicity of ZnO nanoparticles on wheat varies with soil properties. Biometals 28(1):101–112. https://doi.org/10.1007/s10534-014-9806-8

Xu C, Peng C, Sun L, Zhang S, Huang H, Chen Y, Shi J (2015) Distinctive effects of TiO_2 and CuO nanoparticles on soil microbes and their community structures in flooded paddy soil. Soil Biol Biochem 86:24–33. https://doi.org/10.1016/j.soilbio.2015.03.011

Yadav T, Mungray AA, Mungray AK (2014) Fabricated nanoparticles: current status and potential phytotoxic threats. In: Whitacre DM (ed) Reviews of environmental contamination and toxicology, vol 230. Springer, Berlin, pp 83–110. https://doi.org/10.1007/978-3-319-04411-8_4

Yang C, Hamel C, Vujanovic V, Gan Y (2011) Fungicide: modes of action and possible impact on nontarget microorganisms. Inter Schol Res Net ISRN Ecol 130289:1–8

Yang Y, Colman BP, Bernhardt ES, Hochella MF (2015) Importance of a nanoscience approach in the understanding of major aqueous contamination scenarios: case study from a recent coal ash spill. Env Sci Technol 49(6):3375–3382. https://doi.org/10.1021/es505662q

Załęska-Radziwiłł M, Doskocz N (2015) DNA changes in *Pseudomonas putida* induced by aluminum oxide nanoparticles using RAPD analysis. Desalin Water Treat 57(3):1573–1581. https://doi.org/10.1080/19443994.2014.996015

Nano-toxicity and Aquatic Food Chain

Deeksha Krishna and H. K. Sachan

Abstract

Toxicity of nanoparticles in the aquatic environment is of serious concern as increasing concentration of nanoparticles potentially affects the aquatic plants and animals living in the aquatic ecosystem. Engineered nanoparticles (ENPs) are derived from anthropogenic sources, which are highly stable and uniform in distribution. In the aquatic environment, there is an alarming situation and indefinite safety use for the ENPs. The ENPs interact with aquatic organisms at trophic levels (lower and upper levels) throughout the aquatic food chain. Advancement is rendered in the evaluation of bioaccumulation in recent years, and the transfer in trophic level of ENPs. While findings of numerous studies carried out in different locations of the world have proved the noxious consequences of nanomaterials upon the organism's in the aquatic environment as well as in what manner they impact food chain resulting in bioaccumulation, affecting marine animals' wellbeing, development, reproduction, and physiology. We are exploring the nanotoxicity in the aquatic food chain and aquatic species, trophic transition, and biomagnification in this chapter. The critical points of the study are that ENPs are able to go up to three trophic stages in the aquatic food chain. Biomagnification of various nanoparticles (quantum dots, nAu, $nCeO_2$ and $nTiO_2$) fit for two trophic levels have a biomagnification ratio greater than one. Not many studies on the third trophic stage nevertheless demonstrated biomagnification. The deposition of ENPs in aquatic plants and animals has also been shown to affect physiological processes of different organisms.

Keywords

Aquatic systems • Bioaccumulation • Environment • Nanoparticles • Phytoplankton • Reactive oxygen species • Trophic levels

1 Introduction

Nanotechnology is reportedly expected to hit a market size of $3 trillion by 2020. In the consumer market, more than 1800 nano-enabled items are now available. There has been a tremendous progress in nanoscience and nanotechnology field over the past decade, with nanomaterials being utilized in a wide-ranging field, including business, chemistry, healthcare, medicine, fabric textiles, forestry, wastewater management electronics as well as communications devices (Walters et al. 2016; Bundschuh et al. 2018). Therefore, unintentional liberation of engineered nanoparticles (ENPs) is initiated around the environment, predominantly in waters. On lower and upper trophic stages, ENPs may communicate with food chain species. Advancement has taken place on bioaccumulation evaluation and trophic transition of ENPs in recent years. The released ENPs from nano-enabled products during their life cycle raised environmental health and safety issues.

Nonetheless, ample evidence can be found in recent research articles upon the ecological influences of ENPs (Adiloğlu et al. 2012; Holden et al. 2016; Zhang et al. 2018; Abbas et al. 2020; Attarilar al. 2020) supporting the forthcoming impacts of ENPs to damage aquatic organisms if existing in abundantly higher concentrations. A previous report by Shi et al. (2013) revealed a massive consumption of ENPs engrained out the toxicological properties in the aquatic ecosystem (Salieri et al. 2015), causing prominent harm to aquatic biota. Studies have shown that the toxicity precisely associated with ZnO-NPs is due to buildup of Zn^{+2} ions in the water environment (Brun et al. 2014; Zhang et al.

D. Krishna (✉) · H. K. Sachan
CAFF, Fiji National University, 1544 Suva, Fiji
e-mail: dikshakrishna@gmail.com

© Springer Nature Switzerland AG 2021
P. Singh et al. (eds.), *Plant-Microbes-Engineered Nano-particles (PM-ENPs) Nexus in Agro-Ecosystems*,
Advances in Science, Technology & Innovation,
https://doi.org/10.1007/978-3-030-66956-0_13

2018). In this chapter, we emphasize on the ecotoxicological consequences of nanoparticles on the aquatic food chain, especially aquatic ecosystems (in plants, marine invertebrates, and fish).

2 Nanotoxicology in the Aquatic Food Chain

Natural nanoparticles have existed in the atmosphere naturally since centuries while man-made NPs due to their specific surface interactions and properties are related with the design of ENPs providing them different physico-chemical and toxicological characters in contrast to naturally occurring NPs (Handy et al. 2008). Nanotoxicology is a modern and developing research field in toxicology on nanomaterials (Walters et al. 2016; Bundschuh et al. 2018). Evaluating toxicological assets of nanoparticles (NPs) to know if it may pose a threat to the atmosphere or society and its extent is covered in nanotoxicology studies. Nanoparticles toxicity is considered to have a significant impact on plants, animals, and marine organisms (Fig. 1). Indeed, many major chemical manufacturers who produce NPs, discharge effluent into the ocean or rivers. Massive damage to humans and the environment is now happening and is expected to grow significantly. Nanotechnology advancement has not succeeded devoid of questions about its prospective detrimental ecological influences. Much is still unclear, however. Altered interactions with ENPs entail sedimentation, degradation, agglomeration, or else chemical transition, in the same way as absorption plus conversion in the food ecosystem (Vázquez Núñez and De la Rosa-Álvarez 2018).

2.1 Sources of Nanomaterials

Broad classes of substances that consist of particulate elements are called nanoparticles (NPs, which are having one dimension less than 100 nm at the minimum (Laurent et al. 2010; Khan et al. 2019). Nanomaterials are classified into naturally produced nanomaterials that are found in the organism's body. Further, NPs can be studied under the subgroup of naturally occurring NPs categories based on

their origin as they are created by the way as a consequence of engineering activities like vehicle engine exhaust, soldering emissions, ignition activities and forest fires etc. Engineered nanoparticles (ENPs) are man-made having properties for desired applications (Jeevanandam et al. 2018). ENPs include numerous metals based NPs such as fumed silica, titanium dioxide, carbon black, iron oxide, carbon nanotubes (CNTs), etc. (Hristozov and Malsch 2009). Metal nanoparticles (MNPs) possess specific characteristics and have size smaller than twenty to thirty nm. This usually creates additional energy on the exterior of the particles, which makes them extraordinarily reactive and thermodynamic. The scale is also a key factor in reactivity, distribution and toxicity of nanoparticles.

2.2 Physicochemical Properties of Engineered Nanoparticles Influencing Their Toxicity

Engineered nanoparticles (ENPs) may have the probability of toxicity risk. Still, it is also dependent on (a) amount and extent of exposure, (b) integral and inherent nanoparticles toxicity, (c) persistence in body of the nanoparticles, and (d) susceptibility of the organism (Dhasmana et al. 2017). Nanoparticle toxicity is primarily based on properties like (Fig. 2): (i) outer layer of the surface can change the physico-chemical properties of nanoparticles and consequently influence noxiousness, (ii) total area of the surface is amplified with the rise in the chemical activity of nanoparticles and is also a significant factor accountable for toxicity, (iii) composition of nanoparticle (chemically) and toxicity is dependent upon the phase of nanoparticles, i.e. the chemistry and crystalline, (iv) size: smaller size from the same material will be more toxic than bulk, and larger particles, (v) interface along with toxins accessible in the water, and (vi) ENPs functional behaviour (Walters et al. 2016; Dhasmana et al. 2017; Mahaye et al. 2017). The active behaviour of ENPs are the dissolution of ENPs and generation of reactive oxygen species (ROS) into metal ions in the water (Mahaye et al. 2017). Ionic structure of metals endures being less lethal than nanomaterials (Bielmyer et al. 2006; Batley et al. 2013). Higher levels of ENPs of nearly one mg L^{-1} was stated as the precise cause for death than small levels of about 5–50 µg

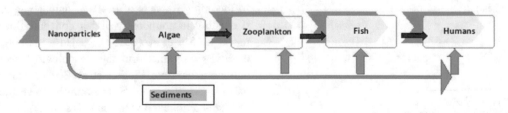

Fig. 1 A schematic representation of the aquatic food chain

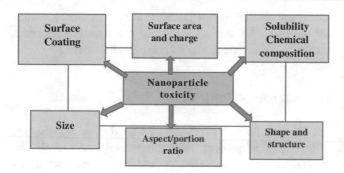

Fig. 2 Properties of nanomaterials influencing their toxicity (based on Turan 2019)

L^{-1} which caused physical alterations, chromosomal modifications and oxidative strain (Bystrzejewska-Piotrowska et al. 2009; Barbara Rasco 2013).

2.3 Transformation Processes

The reliant on the inbuilt assets and on water properties dispersed by nanoparticles are subjected to several conversion pursuits. The key routes are physical, biological or else, chemical alterations that later outline the performance of nanoparticles in the aqueous system (Stone et al. 2010; Lowry et al. 2012). Hetero and homo aggregation, deposition, agglomeration, as well as sedimentation, are some of the physical processes, whereas suspension and redox effects (oxidation, sulfidation) and photochemical reaction are chemical processes. The chief instances of the biological processes are microbial mediated biodegradation and bio-modification activities (Lead et al. 2018). The kind of nanoparticles and factors such as the chemistry of water, pH, the strength of ion and natural organic matter (NOM) make a difference in transformations. Outcome and behaviour in water will be affected by the collaboration of distributed nanoparticles with NOM according to the properties of surface establishing a diverse natural coating (Biswas and Sarkar 2019).

The reduction and oxidation processes are outlined by electron transfer among the chemical moieties in the environment. Reduction and oxidation processes are commenced by silver and iron (Shah et al. 2015). There is ample oxygen in the oxidizing natural environment, e.g. aerated soils as well as natural waters, whereas the reductive ecosystem is drained of oxygen (Lowry et al. 2012). Sunlight-catalyzed redox reactions like photooxidation and photoreduction alter oxidation status of nanoparticles, persistence, ROS, and coating. For example, it was observed that TiO_2 and carbon nanotubes (CNTs) are instinctively photoactive and capable of generating ROS (Chen and Jafvert 2011). Dissolution and sulfidation processes have significant impacts on the surface

properties, persistence and toxicity of the nanomaterials (Levard et al. 2011). Adsorption of inorganic and organic ligands and macromolecules on NPs alter the behaviour and exterior interface of NPs substantially.

A physical change like aggregation is an unalterable process that reduces the surface area, the surge in NPs size altering, in turn, their reactivity, transport, sedimentation and toxicity. Consequently, reduction in surface area of the NPs leads to the decrease in toxicity which in turn alters ROS generation or dissolution (Nichols et al. 2002; Oberdörster et al. 2006; Sellers et al. 2008; Aitken et al. 2010; Lowry et al. 2012; Rist and Hartmann 2018). The photocatalytic reactions in the presence of sunlight resulted in lowering of the pH of the medium which further resulted in high ionic strength and presence of divalent ions (Hartmann et al. 2014; Yin et al. 2015). Porous aggregates can be available as sediment rather than compact ones that remain suspended in water due to erosion and disaggregation processes that create smaller pieces which consumes natural organic matter (NOM) around them (Chekli et al. 2015). The redox reactions change coating, nanoparticles' reactivity, toxicity, surface charging and aggregation state properties which change these transformations (Lowry et al. 2012). Biotransformation on modified NPs of the bioavailable poly (ethylene glycol) (PEG) coatings initiates their aggregation (Kirschling et al. 2011). Due to exclusive change in seawater and freshwater at high dilutions, toxicity of ENPs in all aquatic habitats is not consistent (Renzi and Guerranti 2015; Ju-Nam and Lead 2008).

2.4 Pathway of Exposure in the Aquatic Environment

Fundamental mechanisms of toxicity for numerous nanoparticles are studied in vitro at the cellular level to oxidative stress. Oxidative stress creates reactive oxygen in species (Oberdorster et al. 2005; Nel et al. 2006). Physical injury to cell membranes may cause toxicity (Stoimenov et al. 2002). Route of uptake is by adhesion of nanoparticles to the cell coat and disconnection of soluble toxic species (Klaine et al. 2008). The type of organisms, uni- or multicellular level and its trophic level determines the absorption of nanoparticles and its toxicity in aquatic biota. For example, the mechanism of crossing the cell membrane (viz. direct or via endocytosis) in unicellular organisms remains a significant issue. However, endocytosis has been observed as the preferred pathway for internalization of nanoparticles in eukaryotic organisms (Moore 2006; Nowack and Bucheli 2007). In the case of higher organisms, the nanoparticles might be absorbed by the gill or the external surface epithelia. In contrast, interaction with the aquatic plants may include root surface adsorption, cell wall integration, or intercellular space diffusion (Nowack and Bucheli 2007). Another contaminant uptake pathway is through the food chain, mostly via direct ingestion. The water fleas (*Daphnia magna*) ingested and metabolized the lipid-coated nanotubes present in the aquatic system as its normal feeding behaviour (Roberts et al. 2007). Similarly, Bouldin et al. (2008) reported the absorption of quantum dots in the water fleas (*Ceriodaphnia dubia*) through dietary mechanism from an algal food.

The toxicity of ENPs in aquatic animals is particulate dependent and depends on how they penetrate the cells of the organism (Singh et al. 2011). The technique of the process of entry into the cell starts with their adhesion to the pores of the cell membrane followed by their final entry into the cell by endocytosis or by ion transfer systems (Fig. 3). Interference with the electron transport mechanism or the development of reactive oxygen species (ROS) caused during the entry of ENPs has substantial adverse effects; beginning with cell membrane damage (Ross et al. 2007). The nanoparticles ability to enhance cell damage (by reactive oxygen generation) governs the toxic effects of ENPs in the aquatic system. For example, Smith et al. (2007) demonstrated that the single-walled carbon nanotubes (SWCNTs) increase in oxidative stress and iono-regulatory disturbance in the gut lumen of fish when exposed to sub-lethal concentration for 10-days.

3 Nanotoxicity to Individual Species in Aquatic Food Chain

After the release of nanomaterials in the environment, the aquatic system is the main sink of ENPs. ENPs can influence not only the growth of aquatic species but also the whole ecological equilibrium in the aquatic system. Some studies on nanomaterials and its effect on the aquatic ecosystem have been discussed in the following sub-sections.

3.1 Microbial Toxicity

The consequences of ENPs are of considerable significance in the ecological process. In reaction to high nAg levels, the composition of a bacterial population shifted, while its metabolic processes remained usual (Das et al. 2012). There is significant proof that nanoparticles are moved trophically within the food chain. These hazards were observed in nTiO_2 toxicity, where biofilm-accumulated TiO_2 was relocated to biofilm-exhausting snails which caused trophic harm (Yeo and Nam 2013; Banerjee and Choudhury 2019). Pakrashi et al. (2014) detected related deteriorating consequences on nAu-exposed algae, carboxyl quantity. Biomagnification of these inter-trophic transitions has also not been observed (Laws et al. 2016). Banerjee and Choudhury (2019) emphasize another hypothesis stating that the potential for transferring ENPs across ecosystem boundaries also lies. ENPs can be transported via floods or evolving insects from the aquatic to the terrestrial ecosystem. This perspective requires confirmation by additional studies.

Engineered nanoparticles (ENPs) also seem to be non-toxic to specific populations of microorganisms, because they are trapped within biofilm's extra polymeric material. Lone organisms, such as leaf dwelling bacteria and fungi, are generally immune to nCuO and nAg. These findings indicate the effects of ENPs across microorganisms on the community and evolution (Bundschuh et al. 2016; Banerjee and Choudhury 2019). The absorption of metal ENPs like ZnO and CuO in water depends on the original

Fig. 3 Pathway of exposure in the aquatic environment (based on Walters et al. 2016)

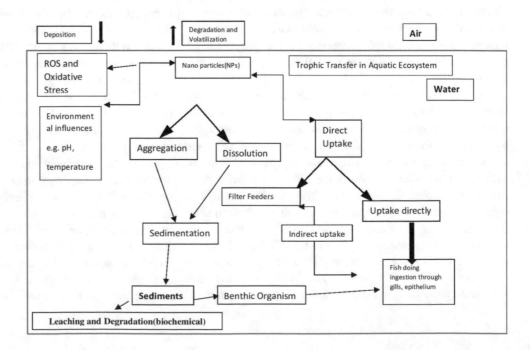

scale of the nanoparticles (Hanna et al. 2013). In the case of nAg, the uptake rate was observed to increase with a change in the size of the ENP (Pan et al. 2012; Banerjee and Choudhury 2019). Zhao and Wang (2012) have noted a contrary reverse trend, however. Related data were well accessible (Handy et al. 2008; Klaine et al. 2008). Various nanomaterials, particularly silver, indicate bactericidal properties (Sondi and Salopek-Sondi 2004; Morones et al. 2005; Banerjee and Choudhury 2019). The titanium dioxide also shows strong antimicrobial activity (Wolfrum et al. 2002).

3.2 Toxicity to Aquatic Plants

Less research has been done on the impact of the ENPs on aquatic plants. Synchrotron-based micro X-ray fluorescence mapping and extended X-ray absorption structure spectroscopy revealed deposits of the fraction of Ag_2S and silver thiol species in the roots of duckweed after exposure to 24 h of ENP (Stegemeier et al. 2017; Banerjee and Choudhury 2019). The development of Ag derivatives in the plant roots was possibly due to the plant molecular defence system to retort the intake of Ag-ENPs (Stegemeier et al. 2017). Kim et al. (2011) reported hindrances to the development of *Lemna paucicostata* plants exposed to Ag-ENP (even at a low concentration of 1 ppm) and TiO_2-ENP (at a higher concentration of 250 ppm).

3.3 Toxicity to Phytoplankton

Phytoplanktons are an essential means of the marine food web system and are the most significant consumers in aquatic habitats. Where ENPs have significant toxic effects on phytoplankton, the whole ecosystem is affected due to phytoplankton toxicity as they hold crucial importance in the aquatic food chain. The toxicity to phytoplankton and reduction in their growth will automatically allow the entire food system to fail or collapse. Therefore, ecotoxicological studies on phytoplankton are of particular importance for aquatic systems. As expected, the algae were the most sensitive group of aquatic organisms to ENP. It was found that ZnO exhibited maximum toxicity in freshwater plankton *Pseudokirchneriella subcapitata* among other metal and metal oxide ENPs, with substantial growth reduction (EC_{50}) at 42 mg l^{-1} (Aruoja et al. 2009; Banerjee and Choudhury 2019). In the marine algae, *Thalassiosa pseudonanathe* EC_{50} for ZnO was found to be 4.6 mg L^{-1} (Wong et al. 2010). Particles of Nano-C_{60} impaired the growth of *P. subcapitata* at a concentration of 90 mg L^{-1} nearly 30%. The C_{60} ENPs' contact with the algal cells has facilitated the entrance into the cells of other contaminants. This stimulated more

significant damage to algal cells and improved cellular apoptosis (Sigg et al. 2014; Banerjee and Choudhury 2019). Toxic effects of NiO ENPs on the alga *Chlorella vulgaris* have been tested. The tests showed 32.28 mg L^{-1} EC_{50} values with 72 h sensitivity to NiO. NiO toxicity of thylakoid systems in *Chlorella vulgaris* has caused plasmolysis, cell membrane damage, and disorder. The most alarming discovery was that the NiO effects could be transmitted to herbivores at a higher trophic level, devouring the NiO effect (Gong et al. 2011; Banerjee and Choudhury 2019).

The exposure of nanomaterials to phytoplankton and the deposition in phytoplankton will directly or indirectly impact the whole marine environment because they are the primary consumers of the nutrient in aquatic environments. Phytoplanktons are the primary producers, so the nanocrystals lying on the exteriors of this biota enter up in the food chain. The iron nanoparticles hamper the growth of marine phytoplankton. There is also inhibition of development of marine phytoplankton species *Isochrysis galban* due to presence of iron nanoparticles (Keller 2012). In photosynthesis of phytoplankton, chlorophyll is of a, b, and c types (Chen et al. 2012). As the Fe_3O_4 nanoparticle intensity enhanced, the chlorophyll a matter tends to decline in *C. vulgaris* (Chen et al. 2012). There has been a significant toxic effect of Fe_3O_4 nanoparticles on CO_2 absorption and the net photosynthetic rate. In lipid peroxidation and cellular oxidative, the malondialdehyde (MDA) is an important marker in *C. vulgaris*, which steadily rises as the Fe_3O_4 nanoparticle concentration rises (Chen et al. 2012). This has demonstrated that the MDA content in *C. vulgaris* has been increased due to stress-induced by Fe_3O_4 nanoparticles (Chen et al. 2012).

3.4 Fish Nanotoxicity

Fish is a common aquatic vertebrate, serving an essential ecological role in aquatic systems. It is also an important food source for humans—a study on the toxicity and behaviour of ENPs in fish directly related to human safety. To forecast the toxicity of a specific material, different stages in the fish life cycle are studied. Harmful ENPs can be highly toxic to many invertebrate organisms, including fish species which are also part of the aquatic food chain. Marine invertebrates such as *Hediste diversicolor* and *Scrobicularia* sp. had been chosen for the study of their behavioural and biochemical reactions to Cu NPs. Impaired coping habits were found at *Scrobicularia* sp. for Cu-ENPs or Cu soluble as well; however, *H. diversicolor* reflected harmful effects only on soluble Cu. All species showed no variations in their cholinesterasic behaviour, demonstrating that either the Cu-ENPs or the soluble Cu did not induce neurotoxicity (Buffet et al. 2011; Banerjee and Choudhury 2019). When nanoparticles enter their digestive glands and gills, ENPs injure suspension-fed invertebrates and detritivores. ENPs typically reach cells along endocytotic pathways, causing damage to large tissues, particularly in tissues that contain highly phagocytic cells (Moore 2006; Banerjee and Choudhury 2019). The bivalve mollusk is another important invertebrate that can be used for research into the effects of ENPs in both fresh and coastal waters.

The oxidative stress in fish causes toxic possessions in the liver and gills (Aschberger et al. 2011). Cu-ENP in zebrafish triggered damage to gills and may cause severe, dangerous effects (Griffitt et al. 2007). Gills and liver were having Ag-ENPs as well as Cu-ENPs of main targets for accumulation as investigated by histological tests, interpreting these nanoparticles tremendously poisonous to zebrafish as the concentration of LC_{50} was 1.5 mg L^{-1} for 48 h (Bilberg et al. 2010; Sigg et al. 2014). ZnO-ENPs and ZnO microparticles showed a dose-dependent effect in the degree of injury, though Al_2O_3 and TiO_2-ENPs did not cause any substantial harm (Sigg et al. 2014). Nano-C_{60} and nano-C_{70} particles in zebrafish embryos also showed the same impacts (Usenko et al. 2008; Vieira et al. 2009; Sigg et al. 2014). Nanoparticle ecotoxicity on fish is significant since fish are the primary species in the aquatic environment as well as potent bioindicators of environmental waste and toxicology studies. *Daphnia magna* can filter and feed on synthesized particles ranging from 0.4 to 40 µm (e.g. algal cells, bacteria, and other organic or inorganic particles) (Xu et al. 2019).

Thus, it is inevitable that NPs may enter into the body of *D. magna* as food. Indeed, uptake of NPs has been found in many reports. Another comparative study reported was done on nanotoxicity of metals on zebrafish. 48 h exposure of Cu on zebrafish eggs revealed deformity and late hatching, although no teratogenic effects for a similar time under Au–NPs was observed. The indicator for toxicant contact is done on model water fleas of genus *Daphnia* members. In supplement, to the entire accessibility of the comprehensive genome sequence, *Daphnia* has a significant fraction of genes familiar with humans (Sá-Pereira et al. 2018). With the surge in TiO_2-NP concentration, there has been a growth in mortality rate when TiO_2-NP was exposed to *D. magna*.

3.5 Toxicity to Human Health

Severe threat to human health may arise due to the direct or indirect contact of ENPs. Due to the contact with water comprising the residue of ENPs leads to direct contact, which usually occurs by the use of industrial effluents released into aquatic systems. Breathing of water aerosols, skin, inhalation or ingestion or intake of polluted and contaminated drinking water is some of the immediate interaction practices (Daughton 2004). Predicted environmental concentrations (PECs) of nanoparticles regularly used in aquatic systems have been outlined in the following Table 1.

Table 1 Predicted environmental concentrations (PECs) of nanoparticles regularly used in aquatic systems

NMs	Compartments	Concentrations	Regions/Countries	References
TiO_2	Water (ng L^{-1})	400–1400	Europe	Sun et al. (2014)
		540–3000	Switzerland	Sun et al. (2014)
		≈200	Los Angeles, US	Liu and Cohen (2014)
		380–11,500	Surface water, Europe	Sun et al. (2014)
		Photostable TiO_2: 0.6–100 Photocatalytic TiO_2: 0.05–7	Freshwater, Denmark	Gottschalk et al. (2015)
		Photostable TiO_2: 0.04–1 Photocatalytic TiO_2: 0.004–0.099	Seawater, Denmark	Gottschalk et al. (2015)
		0–30	Rhône River, France	Sani-Kast et al. (2015)
		80–9000	Europe	Meesters et al. (2016)
		240–2700	Ireland	O'Brien and Cummins (2010), Musee (2011)
		2.7–270	Johannesburg City, South Africa	
		10^8 particles/m³	Rhine River, France	Praetorius et al. (2012)
	Sediment (mg kg^{-1})	≈7	Los Angeles, US	Liu and Cohen (2014)
		0–2.7	Rhône River, France	Sani-Kast et al. (2015)
		62.9–186	Europe	Sun et al. (2016)
		Photostable TiO_2: 0.2–2.8 Photocatalytic TiO_2: 0.017–2.6	Freshwater, Denmark	Gottschalk et al. (2015)
		Photostable TiO_2: 0.049–1.3 Photocatalytic TiO_2: 0.0043–0.12	Seawater, Denmark	
		0.09–30	Europe	Meesters et al. (2016)
		10^{13} particles/m³	Rhine River, France	Praetorius et al. (2012)

(continued)

Table 1 (continued)

NMs	Compartments	Concentrations	Regions/Countries	References
Ag	Water (ng L^{-1})	0.87–7.84	Surface water, Europe	Sun et al. (2016)
		0–6 × 10^{-4}	Seawater, Denmark	Gottschalk et al. (2015)
		0–0.044	Freshwater, Denmark	
		≈ 1 × 10^{-4}	Los Angeles	Liu and Cohen (2014)
		0	Seawater	Giese et al. (2018)
		0.03–2.79	Freshwater	
		0.002–0.3	Europe	Dumont et al. (2015)
		3.3–58.9	Ireland	O'Brien and Cummins (2010)
		0.51–0.94	Europe	Sun et al. (2014)
		0.37–0.73	Switzerland	Sun et al. (2014)
		2.80–619	Johannesburg, City South Africa	Musee (2011)
	Sediment (mg kg^{-1})	0–0.016	Freshwater, Denmark	Gottschalk et al. (2015)
		0–7 × 10^{-4}	Seawater, Denmark	
		≈4 × 10^{-5}	Freshwater, Los Angeles	Liu and Cohen (2014)
		0.053–0.125	Europe	Sun et al. (2016)
		2 × 10^{-5} to 0.47065	Freshwater	Giese et al. 2018)
CNT	Water (ng L^{-1})	0.1–1.82	Surface water, Europe	Sun et al. (2016)
		≈0.31	Los Angeles, US	Liu and Cohen (2014)
		0.17–0.35	Surface water, Europe	Sun et al. (2014)
		0.27–0.56	Surface water, Switzerland	Sun et al. (2014)
		2 × 10^{-4}-0.015	Freshwater, Denmark	Gottschalk et al. (2015)
		2 × 10^{-5} to 2 × 10^{-4}	Seawater, Denmark	
	Sediment (mg kg^{-1})	1.25 × 10^{-2} to 2.66 × 10^{-2}	Europe	Sun et al. (2016)
		≈0.015	Los Angeles, US	Liu and Cohen (2014)
		1 × 10^{-4} to 5.6 × 10^{-3}	Freshwater, Denmark	Gottschalk et al. (2015)
		0–2 × 10^{-4}	Seawater, Denmark	
Cu or CuCO$_3$	Water (ng L^{-1})	≈0.04	Los Angeles, US	Liu and Cohen (2014)
		0.02–0.07	Seawater, Denmark	Gottschalk et al. (2015)
		0.1–6	Freshwater, Denmark	
	Sediment (mg kg^{-1})	≈3.5 × 10^{-3}	Los Angeles, US	Liu and Cohen (2014)
		0.043–2.1	Freshwater, Denmark	Gottschalk et al. (2015)
		0.025–0.083	Seawater, Denmark	

4 Conclusion

The existence of nanoparticles influences aquatic life. ENPs toxicity can be initiated or mitigated by the occurrence of chemical stressors and DOM (dissolved organic matter). This chapter assesses ENPs properties on the aquatic environment with ecotoxic effects due to its event. To improve nanoparticle risk assessment, there is a need for more research. Toxicity evaluation needs to be begun on formulations of nanoparticles evaluated involving at least five species from different trophic levels for extracting the predicted no-effect concentrations (PNECs) for identifying the species sensitivities to other species. Moreover, evaluation of the toxicity to different natural and artificial aquatic and soil systems should also be performed to demonstrate the toxicity of nanomaterials at a holistic scale.

References

Abbas Q, Yousaf B, Amina AMU, Munir MAM, El-Naggar A et al (2020) Transformation pathways and fate of engineered nanoparticles (ENPs) in distinct interactive environmental compartments: a review. Environ Int 138:105646. https://doi.org/10.1016/j.envint.2020.105646

Adiloğlu SI, Yu C, Chen R, Li JJ, Li JJ et al (2012) We are IntechOpen, the world's leading publisher of Open Access books Built by scientists, for scientists TOP 1%. Intech, i(tourism):13. https://doi.org/10.1016/j.colsurfa.2011.12.014

Aitken RJ, Peters SAK, Jones AD, Stone V (2010) Regulation of carbon nanotubes and other high aspect ratio nanoparticles: approaching this challenge from the perspective of asbestos. In: International handbook on regulating nanotechnologies. https://doi.org/10.4337/9781849808125.00020

Aruoja V, Dubourguier HC, Kasemets K, Kahru A (2009) Toxicity of nanoparticles of CuO, ZnO and TiO2 to microalgae *Pseudokirchneriella subcapitata*. Sci Total Environ 407:1461–1468

Aschberger K, Micheletti C, Sokull-Klüttgen B, Christensen FM (2011) Analysis of currently available data for characterizing the risk of engineered nanomaterials to the environment and human health-lessons learned from four case studies. Environ Int 37:1143–1156. https://doi.org/10.1016/j.envint.2011.02.005

Attarilar S, Yang J, Ebrahimi M, Wang Q, Liu J (2020) The toxicity phenomenon and the related occurrence in metal and metal oxide nanoparticles: a brief review from the biomedical perspective, vol 8. https://doi.org/10.3389/fbioe.2020.00822

Banerjee A, Choudhury AR (2019) Nanomaterials in plants, algae and microorganisms concepts and controversies, vol 22019. Academic Press, pp 129–141. https://doi.org/10.1016/B978-0-12-811488-9.00007-X

Barbara Rasco MO (2013) Impact of engineered nanoparticles on aquatic organisms. J Fish Livestock Prod. https://doi.org/10.4172/2332-2608.1000e106

Batley GE, Kirby JK, McLaughlin MJ (2013) Fate and risks of nanomaterials in aquatic and terrestrial environments. Acc Chem Res. https://doi.org/10.1021/ar2003368

Bielmyer GK, Grosell M, Brix KV (2006) Toxicity of silver, zinc, copper, and nickel to the copepod Acartia tonsa exposed via a phytoplankton diet. Environ Sci Technol. https://doi.org/10.1021/es051589a

Bilberg K, Malte H, Wang T, Baatrup E (2010) Silver nanoparticles and silver nitrate cause respiratory stressin Eurasian perch (*Perca fluviatilis*). Aquat Toxicol 96:159–165

Biswas JK, Sarkar D (2019) Nanopollution in the aquatic environment and ecotoxicity: no nano issue! Current Pollut Rep. https://doi.org/10.1007/s40726-019-0104-5

Bouldin JL, Ingle TM, Sengupta A, Alexander R, Hannigan RE, Buchanan RA (2008) Aqueous toxicity and food chain transfer of quantum dots (TM) in freshwater algae and Ceriodaphnia dubia Environ. Toxicol Chem 27(9):1958–1963

Brun NR, Lenz M, Wehrli B, Fent K (2014) Comparative effects of zinc oxide nanoparticles and dissolved zinc on zebrafish embryos and eleuthero-embryos: importance of zinc ions. Sci Total Environ. https://doi.org/10.1016/j.scitotenv.2014.01.053

Buffet PE, Tankoua OF, Pan JF, Berhanu D, Herrenknecht C, Poirier L, Amiard-Triquet C, Amiard JC, Bérard JB, Risso C, Guibbolini M, Roméo M, Reip P, Valsami-Jones E, Mouneyrac C (2011) Behavioural and biochemical responses of two marine invertebrates Scrobicularia plana and Hediste diversicolor to copper oxide nanoparticles. Chemosphere 84:166–174. https://doi.org/10.1016/j.chemosphere.2011.02.003

Bundschuh M, Seitz F, Rosenfeldt RR, Schulz R (2016) Effects of nanoparticles in fresh waters: risks, mechanisms and interactions. Freshw Biol 61:2185–2196

Bundschuh M, Filser J, Lüderwald S, McKee MS, Metreveli G, Schaumann GE et al (2018) Nanoparticles in the environment: where do we come from, where do we go to? Environ Sci Eur. https://doi.org/10.1186/s12302-018-0132-6

Bystrzejewska-Piotrowska G, Golimowski J, Urban PL (2009) Nanoparticles: their potential toxicity, waste and environmental management. Waste Manag. https://doi.org/10.1016/j.wasman.2009.04.001

Chekli L, Zhao Y, Tijing L, Phuntsho S, Donner E, Lombi E, Gao B, Shon H (2015) Aggregation behaviour of engineered nanoparticles in natural waters: characterizing aggregate structure using on-line laser light scattering. J Hazard Mater 284:190–200

Chen CY, Jafvert CT (2011) The role of surface functionalization in the solar light-inducedproduction of reactive oxygen species by single-walled carbon nanotubes in water. Carbon 49(15):5099–5106. https://doi.org/10.1016/j.carbon.2011.07.029

Chen G, Liu X, Su C (2012) Distinct effects of humic acid on transport and retention of TiO$_2$ rutile nanoparticles in saturated sand columns. Environ Sci Technol 46:7142–7150. https://doi.org/10.1021/es204010g

Daughton CG (2004) Non-regulated water contaminants: emerging research. Environ Impact Assess Rev 24:711–732

Das P, Xenopoulos MA, Williams CJ, Hoque ME, Metcalfe CD (2012) Effects of silver nanoparticles on bacterial activity in natural waters. Environ Toxic Chem 31:122–130

Dhasmana A, Firdaus S, Singh KP, Raza S, Jamal QMS, Kesari KK et al (2017) Nanoparticles: applications, toxicology and safety aspects. Environ Sci Eng (Subseries: Environ Sci). https://doi.org/10.1007/978-3-319-46248-6_3

Dumont E, Johnson AC, Keller VDJ, Williams RJ (2015) Nano silver and nano zinc-oxide in surface waters—Exposure estimation for Europe at high spatial and temporal resolution. Environ Pollut 196:341–349

Giese B, Klaessig F, Park B, Kaegi R, Steinfeldt M, Wigger H, et al. (2018) Risks, release and concentrations of engineered nanomaterial in the environment. Sci Rep 8(1):S. 1565. https://doi.org/10.1038/s41598-018-19275-4

Gong N, Shao KS, Feng W, Lin ZZ, Liang CH, Sun YQ (2011) Biotoxicity of nickel oxide nanoparticlesand bio-remediation by microalgae *Chlorella vulgaris*. Chemosphere 83:510–516

Gottschalk F, Lassen C, Kjoelholt J, Christensen F, Nowack B (2015) Modeling flows and concentrations of nine engineered nanomaterials in the Danish environment. Int J Environ Res Pub Health. https://doi.org/10.3390/ijerph120505581

Griffitt RJ, Weil R, Hyndman KA, Denslow ND, Powers K et al (2007) Exposure to copper nanoparticlescauses gill injury and acute lethality in zebrafish (*Danio rerio*). Environ Sci Technol 41:8178–8186

Handy RD, Von Der Kammer F, Lead JR, Hassellöv M, Owen R, Crane M (2008) The ecotoxicology and chemistry of manufactured nanoparticles. Ecotoxicology. https://doi.org/10.1007/s10646-008-0199-8

Hanna SK, Miller RJ, Zhou DX, Keller AA, Lenihan HS (2013) Accumulation and toxicity of metaloxide nanoparticles in a soft-sediment estuarine amphipod. Aquat Toxicol 142:441–446

Hartmann NIB, Skjolding LM, Hansen SF, Baun A, Kjølholt J, Gottschalk F (2014) Environmental fate and behaviour of nanomaterials: new knowledge on important transfomation processes

Holden PA, Gardea-Torresdey JL, Klaessig F, Turco RF, Mortimer M, Hund-Rinke K et al (2016) Considerations of environmentally relevant test conditions for improved evaluation of ecological hazards of engineered nanomaterials. Environ Sci Technol. https://doi.org/10.1021/acs.est.6b00608

Hristozov D, Malsch I (2009) Hazards and Risks of engineered nanoparticles for the environment and human health. Sustainability. https://doi.org/10.3390/su1041161

Jeevanandam J, Barhoum A, Chan YS, Dufresne A, Danquah MK (2018) Review on nanoparticles and nanostructured materials: history, sources, toxicity and regulations. Beilstein J Nanotechnol. https://doi.org/10.3762/bjnano.9.98

Jing H, Zhou Y, Wang C, Li S, Wang X (2017) Toxic effects and molecular mechanism of different types of silver nanoparticles to the aquatic crustacean *Daphnia magna*. Environ Sci Technol 51 (21):12868–12878. https://doi.org/10.1021/acs.est.7b03918

Ju-Nam Y, Lead JR (2008) Manufactured nanoparticles: an overview of their chemistry, interactions and potential environmental implications. Sci Total Environ 400:396–414

Keller AA, Garner K, Miller RJ, Lenihan HS (2012) Toxicity of nano-zero valent iron to freshwater and marine organisms. PLoS One 7(8):e43983

Khan R, Inam MA, Khan S, Park DR, Yeom IT (2019) Interaction between persistent organic pollutants and ZnO NPs in synthetic and natural waters. Nanomaterials 9. https://doi.org/10.3390/nano9030472

Kim E, Kim SH, Kim HC, Lee SG, Lee SJ, Jeong SW (2011) Growth inhibition of aquatic plant caused by silver and titanium oxide nanoparticles. Toxicol Environ Health Sci 3:1–6

Kirschling TL, Golas PL, Unrine JM, Matyjaszewski K, Gregory KB, Lowry GV, Tilton RD (2011) Microbial bioavailability of covalently bound polymer coatings on model engineered nanomaterials. Environ Sci Technol 45(12):5253–5259. https://doi.org/10.1021/es200770z

Klaine SJ, Alvarez PJJ, Batley GE, Fernes TF, Hy RD, Lyon DY, Mahendra S, Mclaughlin MJ, Lead JR (2008) Nanomaterials in the environment: behavior, fate, bioavailability, and effects. Environ Toxicol Chem 27:1825–1851

Laurent S, Forge D, Port M, Roch A, Robic C, Vander Elst L, Muller RN (2010) Magnetic iron oxide nanoparticles: synthesis, stabilization, vectorization, physicochemical characterizations, and biological applications. Chem Rev 2574–2574

Laws J, Heppell CM, Sheahan D, Liu CF, Grey J (2016) No such thing as a free meal: organotin transferacross the freshwater-terrestrial interface. Freshwater Biol 61:2051–2062

Lead JR, Batley GE, Alvarez PJJ, Croteau MN, Handy RD, McLaughlin MJ et al (2018) Nanomaterials in the environment: behavior, fate, bioavailability, and effects—an updated review. Environ Toxic Chem. https://doi.org/10.1002/etc.4147

Levard C, Reinsch BC, Michel FM, Oumahi C, Lowry GV, Brown GE (2011) Sulfidation processes of PVP-coated silver nanoparticles in aqueous solution: impact on dissolution rate. Environ Sci Technol. https://doi.org/10.1021/es2007758

Liu HH, Cohen Y (2014) Multimedia environmental distribution of engineered nanomaterials. Environ Sci Technol. https://doi.org/10.1021/es405132z

Lowry GV, Gregory KB, Apte SC, Lead JR (2012) Transformations of nanomaterials in the environment. Environ Sci Technol. https://doi.org/10.1021/es300839e

Mahaye N, Thwala M, Cowan DA, Musee N (2017) Genotoxicity of metal based engineered nanoparticles in aquatic organisms: a review. Mutation Res Rev Mutation Res. https://doi.org/10.1016/j.mrrev.2017.05.004

Meesters JAJ, Quik JTK, Koelmans AA, Hendriks AJ, Van De Meent D (2016) Multimedia environmental fate and speciation of engineered nanoparticles: a probabilistic modeling approach. Environ Sci Nano. https://doi.org/10.1039/c6en00081a

Morones JR, Elechiguerra JL, Camacho A, Holt K, Kouri JB, Ramirez JT, Yacaman MJ (2005) The bactericidal effect of silver nanoparticles. Nanotechnology 16:2346–2353

Moore MN (2006) Do nanoparticles present ecotoxicological risks for the health of the aquatic environment? Environ Int 32:967–976

Musee N (2011) Simulated environmental risk estimation of engineered nanomaterials: a case of cosmetics in Johannesburg City. Human Exp Toxicol. https://doi.org/10.1177/0960327110391387

Nel A, Xia T, Madler L, Li N (2006) Toxic potential of materials at the nanolevel. Science 311:622–627

Nichols G, Byard S, Bloxham MJ, Botterill J, Dawson NJ, Dennis A, Diart V, North NC, Sherwood JD (2002) A review of the terms agglomerate and aggregate with a recommendation for nomenclature used in powder and particle characterization. J Pharm Sci 91:2103–2109

Nowack B, Bucheli TD (2007) Occurrence, behavior and effects of nanoparticles in the environment. Environ Pollut 150(1):5–22. https://doi.org/10.1016/j.envpol.2007.06.006

O'Brien N, Cummins E (2010) Nano-scale pollutants: fate in irish surface and drinking water regulatory systems. Human Ecol Risk Assess. https://doi.org/10.1080/10807039.2010.501270

Oberdörster G, Oberdörster E, Oberdörster J (2005) Nanotoxicology: an emerging discipline evolving from studies of ultrafine particles. Environ Health Perspect 113:823–839

Oberdörster E, Zhu S, Blickley TM, McClellan-Green PML (2006) Haasch Ecotoxicology of carbon-based engineered nanoparticles: effects of fullerene (C60) on aquatic organisms. Carbon 44:1112–1120

Pakrashi S, Dalai S, Chandrasekaran N and Mukherjee A (2014) Trophic transfer potential of aluminium oxide nanoparticles using representative primary producer (Chlorella ellipsoides) and a primary consumer (Ceriodaphnia dubia). Aquat Toxicol 152:74–81

Pan JF, Buffet PE, Poirier L, Amiard-Triquet C, Gilliland D, Joubert Y et al (2012) Size dependent bioaccumulation and ecotoxicity of gold nanoparticles in an endobenthic invertebrate: the tellinid clam *Scrobicularia plana*. Environ Pollut 168:37–43

Praetorius A, Scheringer M, Hungerbühler K (2012) Development of environmental fate models for engineered nanoparticles—a case study of TiO2 nanoparticles in the rhine river. Environ Sci Technol. https://doi.org/10.1021/es204530n

Renzi M, Guerranti C (2015) Ecotoxicity of nanoparticles in aquatic environments: a review based on multivariate statistics of meta-data. J Environ Anal Chem 2:149

Rist S, Hartmann NB (2018) Aquatic ecotoxicity of microplastics and nanoplastics: lessons learned from engineered nanomaterials. In: Handbook of environmental chemistry. https://doi.org/10.1007/978-3-319-61615-5_2

Roberts AP, Mount AS, Seda B, Souther J, Qiao R, Lin S, Ke PC, Rao AM, Klaine SJ (2007) In vivo biomodification of lipid-coated carbon nanotubes by Daphnia magna. Environ Sci Technol 41 (8):3025–3029

Ross JRM, Flegal AR, Brown CL, Squire S, Scelfo GM, Hibdon S (2007) Spatial and temporal variations in silver contamination and toxicity in San Francisco Bay. Environ Res 105:34–52

Salieri B, Pasteris A, Baumann J, Righi S, Köser J, D'Amato R et al (2015) Does the exposure mode to ENPs influence their toxicity to aquatic species? A case study with TiO2 nanoparticles and *Daphnia magna*. Environ Sci Poll Res. https://doi.org/10.1007/s11356-014-4005-2

Sani-Kast N, Scheringer M, Slomberg D, Labille J, Praetorius A, Ollivier P, Hungerbühler K (2015) Addressing the complexity of water chemistry in environmental fate modeling for engineered nanoparticles. Sci Total Environ. https://doi.org/10.1016/j.scitotenv.2014.12.025

Sá-Pereira P, Diniz MS, Moita L, Pinheiro T, Mendonça E, Paixão SM, Picado A (2018) Protein profiling as early detection biomarkers for TiO$_2$ nanoparticle toxicity in *Daphnia magna*. Ecotoxicology 27:430–439

Sellers K, Mackay C, Bergeson LL, Clough SR, Hoyt M, Chen J, Henry K, Hamblen J (2008) Nanotechnology and the environment. CRC Press

Shah M, Guo QX, Fu Y (2015) The colloidal synthesis of unsupported nickel-tin bimetallic nanoparticles with tunable composition that have high activity for the reduction of nitroarenes. Catal Commun 65:85–90

Shi H, Magaye R, Castranova V, Zhao J (2013) Titanium dioxide nanoparticles: a review of current toxicological data. Particle Fibre Toxicol. https://doi.org/10.1186/1743-8977-10-15

Sigg L, Behr R, Groh K, Isaacson C, Odzak N et al (2014) Chemical aspects of nanoparticle ecotoxicology. Chimia 68:806–811

Singh S, Dosani T, Karakoti AS, Kumar A, Seal S, Self WT (2011) A phosphate-dependent shift in redox state of cerium oxide nanoparticles and its effects on catalytic properties. Biomaterials 32:6745-6753. https://doi.org/10.1016/j.biomaterials.2011.05.073

Smith CJ, Shaw BJ, Handy RD (2007) Toxicity of single walled carbon nanotubes on rainbow trout, (*Oncorhynchus mykiss*): respiratory toxicity, organ pathologies, and other physiological effects. Aquat Toxicol 82:93–109

Sondi I, Salopek-Sondi B (2004) Silver nanoparticles as antimicrobial agent: a case study 620 on *E. coli* as a model for Gram-negative bacteria. J Colloid Interface Sci 275:177–182

Stegemeier JP, Colman BP, Schwab F, Wiesner MR, Lowry GV (2017) Uptake and distribution of silver in the aquatic plant *Landoltia punctata* (Duckweed) exposed to silver and silver sulphide nanoparticles. Environ Sci Technol 51:4936–4943

Stoimenov PK, Klinger RL, Marchin GL, Klabunde KJ (2002) Metal oxide nanoparticles as bactericidal agents. Langmuir 18:6679–6686. https://doi.org/10.1021/la0202374

Stone V, Nowack B, Baun A, van den Brink N, von der Kammer F, Dusinska M et al (2010) Nanomaterials for environmental studies: classification, reference material issues, and strategies for physico-chemical characterization. Sci Total Environ 408 (7):1745–1754. https://doi.org/10.1016/j.scitotenv.2009.10.035

Sun TY, Bornhöft NA, Hungerbühler K, Nowack B (2016) Dynamic probabilistic modeling of environmental emissions of engineered nanomaterials. Environ Sci Technol. https://doi.org/10.1021/acs.est.5b05828

Sun TY, Gottschalk F, Hungerbühler K, Nowack B (2014) Comprehensive probabilistic modelling of environmental emissions of engineered nanomaterials. Environ Poll. https://doi.org/10.1016/j.envpol.2013.10.004

Turan NB, Erkan HS, Engin GO, Bilgili MS (2019) Nanoparticles in the aquatic environment: usage, properties, transformation and toxicity—a review. Process Saf Environ Prot 130:238–249. https://doi.org/10.1016/j.psep.2019.08.014

Usenko CY, Harper SL, Tanguay RL (2008) Fullerene C60 exposure elicits an oxidative stress response in embryonic zebrafish. Toxicol Appl Pharmacol 229:44–55

Vieira LR, Gravato C, Soares AMVM, Morgado F, Guilhermino L (2009) Acute effects of copper and mercury on the estuarine fish *Pomatoschistus microps*: linking biomarkers to behaviour. Chemosphere 76:1416–1427

Vázquez Núñez E, de la Rosa-Álvarez G (2018) Environmental behavior of engineered nanomaterials in terrestrial ecosystems: uptake, transformation and trophic transfer. Curr Opi Environ Sci Health. https://doi.org/10.1016/j.coesh.2018.07.011

Walters C, Pool E, Somerset V (2016) Nanotoxicology: a review. Toxicol New Aspects Sci Conundrum. https://doi.org/10.5772/64754

Wolfrum EJ, Huang J, Blake DM, Maness PC, Huang Z, Fiest J (2002) Photocatalytic oxidation of bacteria, bacterial and fungal spores, and model biofilm components to carbon dioxide on titanium dioxide–coated surfaces. Environ Sci Technol 36:3412–3419

Wong SWY, Leung PTY, Djurisi AB, Leung KMY (2010) Toxicities of nano zinc oxide to five marine organisms: influences of aggregate size and ion solubility. Anal Bioanal Chem 396:609–618

Xu Z, Liu Y, Wang Y (2019) Application of *Daphnia magna* for nanoecotoxicity study methods in molecular biology book. Nanotoxicity. https://doi.org/10.1007/978-1-4939-8916-4_21

Yin Y, Yang X, Zhou X, Wang W, Yu S, Liu J, Jiang G (2015) Water chemistry controlled aggregation and photo-transformation of silver nanoparticles in environmental waters. J Environ Sci 34:116–125

Yeo MK, Nam DH (2013) Influence of different types of nanomaterials on their bioaccumulation in a paddy microcosm: a comparison of TiO$_2$ nanoparticles and nanotubes. Environ Poll 178:166–172

Zhang W, Xiao B, Fang T (2018) Chemical transformation of silver nanoparticles in aquatic environments: mechanism, morphology and toxicity. Chemosphere 191(7):324–334. https://doi.org/10.1016/j.chemosphere.2017.10.016

Zhao CM, Wang WX (2012) Size-dependent uptake of silver nanoparticles in *Daphnia magna*. Environ Sci Technol 46:11345–11351

Plant-Microbe-Soil Health-Engineered Nanoparticles Nexus: Conclusion

Impact of Engineered Nanoparticles on Microbial Communities, Soil Health and Plants

Akhilesh Kumar, Prashant Kumar Sharma, Saurabh Singh, and Jay Prakash Verma

Abstract

Today, nanoparticles (NPs) have received tremendous attention due to their unusual properties and multiple applications. Engineered nanoparticles (ENPs) are applied in medicine, industries, agriculture, space science, etc. Anthropogenic release of ENPs to the environment poses a potential hazard to soil, plants, and human health. Soil is a major repository of ENPs and its exposure modulates microbial diversity, soil properties, and plant growth. The effects of ENPs on soil result in many anomalies on soil properties and plants. Soil enzymes such as dehydrogenase, urease, and phosphatase are highly affected by ENPs. ENPs exert toxic effects on multiple economically important crops and trigger severe oxidative stress in plants leading to cell death. Due to their unique size, ENPs penetrate plant tissues and translocate from one part to another. Also, uptake, translocation, and accumulation of ENPs in crops pose potential risk to animals and human beings. Thus, in the present scenario, it is necessary to explore the effects of different ENPs on soil physicochemical, microbial community, and plant growth parameters. In this chapter, we will briefly highlight the effects of different ENPs on soil, microbs, and plant responses.

Keywords

Adsorption • Bio-availability • Ecotoxicity • Oxidative stress toxicity • Soil enzymes • Transportation

A. Kumar (✉) · S. Singh · J. P. Verma
Institute of Environment and Sustainable Development, Banaras Hindu University, Uttar Pradesh, Varanasi, 221005, India
e-mail: akhiballia@gmail.com

J. P. Verma
e-mail: verma_bhu@yahoo.co.in

P. K. Sharma
Department of Chemistry, Indian Institute of Handloom Technology, Varanasi, 221005, India

1 Introduction

The global demand for food is predicted to increase by around 70–100% by 2050 (Foley et al. 2011; Muller et al. 2012; WWAP 2012). The present intensive agriculture has resulted stress on ecosystems and natural resources resulting in erosion of soil, soil pollution, loss of biodiversity, and disturbance of global nutrient cycles (Foley et al. 2011). Therefore, the pattern of agricultural practices is changing rapidly by incorporating a sustainable approach and modern innovative technology like nanotechnology. However, the application of nanotechnology to agriculture is still at a nascent stage as compared to their application in energy, water treatment, etc. (Qu et al. 2013; Zhang et al. 2003; Shah et al. 2014). *The 'nano' size has resulted in* large surface-to-volume ratios, unique surface functionalization, plasmon resonance, and photoactivity which can be utilized to improve the agro-food systems. Nanotechnology is applied in the field for the supply of nutrients, monitoring, and suppression of disease (Asli and Neumann 2009).

The commercial products of engineered nanoparticles (ENPs) are rapidly moving from laboratory to market. The widespread applicability has raised significant concerns about the harmful impact of ENPs to the environment. Soil is the primary sink for ENPs which get accumulated through various pathways, such as direct when ENPs containing pesticides, fertilizers, sewage sludge, etc., are used for improved productivity, while the indirect exposure is via atmospheric deposition, landfills, or accidental spills during industrial production (Zhang 2003; DeRosa et al. 2010). In a given environment, ENPs may interact with soil, microbes, and plants. The effect of ENPs on the soil depends on the type, size, composition, concentration together with soil type, and its enzymatic activities. Higher concentrations of ENPs induce a negative effect on dehydrogenase activity (Josko et al. 2014). ENPs also cause detrimental effects on the rate of self-cleansing capacity of soil and nutrients balance. These processes are instrumental in plant nutrition regulation and soil fertility improvement (Janvier et al. 2007; Suresh et al. 2013).

The presence of ENPs in soil has raised notable concern regarding its impact on soil biodiversity (Bondarenko et al. 2013). Various properties of soil such as texture, pH, organic matter, and structure alter the capability of ENPs that have toxic effects on microorganisms (bioavailability) (Fierer and Jackson 2006; Simonin and Richaume 2015). Soil microbes are key indicator of change as they play an important role in biogeochemical cycling (Kandeler et al. 1996; Holden et al. 2014; Kumar and Verma 2018). ENPs mobility in soil is lower (Darlington et al. 2009) as the soil porosity and transport are governed by mucilage, voids, and exudates from hyphae, roots, and bacteria (Oades 1993; Zhao et al. 2012). Therefore, subtle changes in the microbial community induced by ENPs exposure could alter the uptake of nutrients, disease suppression, and development of plants as well as the fate of nanoparticles (NPs). ENPs affect plants at different levels such as physiological, biochemical, and molecular. Unlike animals, plants are sessile and roots absorb NPs along with water and nutrient from the contaminated environment. These ENPs are accumulated in plant products and reach human and animals. Additionally, they also induce toxicity to plant such as inhibition of seed germination, nutrient acquisition, plant growth, and transport (Xiong et al. 2017; Zhao et al. 2017). Thus, the toxicity of NPs considers numerous perspectives from deoxyribonucleic acid (DNA) to physiological level and ecosystem functioning. The quality of soil, water, and the environment is an extremely important issue. Maintaining the ecosystem quality, particularly soil is necessary for proper functioning (Kumar et al. 2021). Therefore, intensive development of nanotechnology, enormous use of ENPs, and their release in the environment are a challenge for the future. Although nanotechnology is relatively a young field in science, its contribution is developing dramatically due to its wider application. Besides, it is expected that the production scale of ENPs will be much higher in a few years. In this chapter, we have briefly discussed about the different types of engineered NPs, their sources, and its impact on microbial community, soil health, and plant responses.

2 Sources of Engineered Nanoparticles

Nowadays, ENPs are extensively being used in a wide range of industrial products for multiple applications. They are released in the environment naturally, intentionally, or accidentally through various means (Fig. 1). In the environment, ENPs may pose a potential threat to soil properties, water, and air due to their small size and easy transportation. The ENPs are mostly used in cosmetics, electronic devices, paint, pigment, which are the potential sources of soil and the environment contamination. The ENPs used in paint, pigment, and cosmetics are released in the environment at the time of use and contaminate soil and surface water (Keller et al. 2013; Tripathi et al. 2017a, b). The sources of ENPs can be broadly classified in point and non-point sources. Point sources include production unit, research laboratory, storage unit, and wastewater producing treatment plants, while non-point sources include cosmetics, paints, electronic devices, and medical waste (Fig. 1).

Soil acts as a major repository of these ENPs which leads toward contamination of soil as the concentration of ENPs is

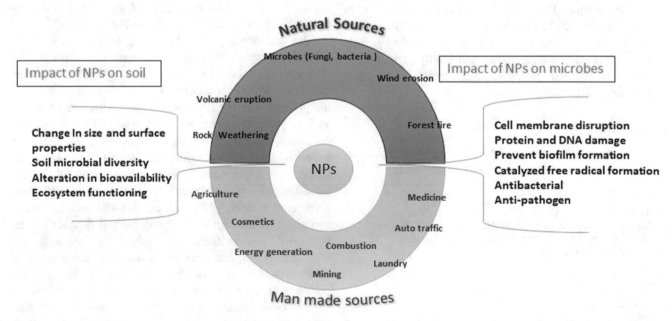

Fig. 1 Sources of engineered nanoparticles and its impact on soil and microbes

predicted to be five times higher as compared to water and air. The major sources of ENPs in soil include sewage sludge used for agricultural purpose (forecasted annual load of 1.01–2380 μg^{-1} engineered nanomaterials (ENMs) kg^{-1} $year^{-1}$ sludge) (Schwab et al. 2016), increased usage of novel pesticides or fertilizers' nanoformulations (Gardea-Torresdey et al. 2014), wastewater effluents (0.001–15 ng L^{-1} $year^{-1}$), and atmospheric sources like dry depositions and rainfall (551 μg m^{-3} $year^{-1}$) (Gottschalk et al. 2009; Hendren et al. 2011). In addition, nanotechnology has been used frequently in the remediation of environmental pollutants (Cecchin et al. 2017; Lv et al. 2019). The use of ENPs to remediate pollution also adds NPs to the soil and the immediate environment. The extensive applications of NPs in different areas have attracted more attention due to their potential environmental risks. Hence the research is increasing for the toxicity of nanomaterials on animals, plants, and microbes (Lee et al. 2012). So, there is an urgent need to assess the potential risks of NPs on soil, bacteria, plants, and the environment.

3 Impact of Engineered Nanoparticles on Environmental Components

The expeditious use of NPs in cosmetics, drug delivery, biosensor, electronics, environmental remediation, and wastewater treatment has an urgent need to evaluate their impact on soil and microbes present in soil and plants. Extensive research has shown both positive and negative impacts of NPs on soil, soil microbes, and plants (Table 1, Fig. 2). However, there are many challenges and unresolved issues related to the impact of NPs (Schwab et al. 2016; Lv et al. 2019). So, it is crucial to understand the biological effects of different types of ENPs and its long-term environmental consequences. Study reveals that ENPs can have a toxic effect on living organism including, viruses, bacteria, fungi, plants (Fig. 3), and animals (Cecchin et al. 2017). The plants and soil microbes are among the most closely associated biotic components in the environment. Alterations in soil microbial biomass are a sensitive indicator of changes in soil properties and health. Therefore, it is essential to explore the fate and transport of engineered NPs in soil, plants, animals, and associated environment.

3.1 Impact of Engineered Nanoparticles on Soil Microbes

The introduction and use of NPs in the environment cause harm to beneficial microbes such as bacteria, fungi, and actinomycetes which are involved in plant growth and development (Kumar et al. 2020). NPs used for different purposes are directly or indirectly released into the environment. These ENPs accumulate in soil and change their properties with time. The major impact of ENPs includes changes in soil enzyme and soil respiration, and ultimately microbial diversity. TiO_2 NPs decrease soil microbial and enzymatic activity in the environment (Peng et al. 2015; Peyrot 2015). Silver (Ag) NPs are produced by many fungi by both intracellular and extracellular pathways (Bhainsa and D'Souza 2006). The Ag NPs are effective to certain pathogenic bacteria such as *Syphilis typhus*, *Vibrio cholera*, *Staphylococcus aureus*, *Pseudomonas aeruginosa* (Siddiqi and Husen 2016). The major mechanism through which ENPs caused bacterial cell death includes oxidative stress leading to damage of lipids, proteins, carbohydrates, and DNA (Table 1). The generation of hydrogen peroxides is one of the major events responsible for antibacterial activity. The green NPs synthesized from *Allium cepa* also have antibacterial activity (Jini and Sharmila 2020).

The Ag NPs widely used in industrial products are released in soil having extensive implications. Soil microbial diversity, biomass, and plant growth is highly affected by these particles. These ENPs easily cross the cell membrane and affect the physiology of bacteria and cause cellular toxicity (Courtios et al. 2019). The ENPs synthesized from *Vitis vinifera* (black grapes) have antibacterial activities against *S. aureus*, *Bacillus cereus*, *P. aeruginosa*, and *Escherichia coli* (Kowsalya et al. 2019). Some NPs show activity against both beneficial as well as harmful microbes. The Cu NPs release copper ions in the soil which may cause considerable loss in counts of both beneficial and pathogenic bacteria (Lofts et al. 2013). The effect of ENPs on microbial community depends on its types, size, and structure. Inorganic NPs have more toxic effect than organic ones (Frenk et al. 2013). A decrease in soil enzyme, nitrogen fixation, and crop productivity occurs due to higher concentration of ENPs. NPs like TiO_2 generate superoxide and hydroxide radicals which show strong antibacterial activity (Table 1) (Peyrot et al. 2014; Xu et al. 2015). NPs cross most biological barriers such as bacterial cell wall and their behaviors on the cell is unpredictable. Hence, as the production of ENPs is increasing, soil microbes facing challenges of toxicity and decreasing microbial diversity.

3.2 Impact of Engineered Nanoparticles on Soil Biophysical Properties

Soil is the main repository of waste products of nanomaterials which are added after use. It is a complex living system that has different assemblages. The properties of soil such as size, surface area, charge, mineral composition, and organic matter are mostly affected by ENPs. NPs may undergo many transformations in soil and alter its properties

Table 1 An overview of the toxicological effects of engineered nanoparticles (ENPs) on different crops

Nanoparticles	Plants	Size	NPs concentration	Mode of application	Growth media and exposure duration	Impacts on plants	References
Carbon-based nanoparticles							
Carbon nano onions	*Cicer arietinum*	20–40 nm	0, 10, 20, and 30 µg mL^{-1} water	Seed	10 days	Increased protein, electrolytes, and micronutrients, size, and weight of mature seeds without ENPs uptake	Tripathi et al. (2017a, b)
Chitin	*Triticum aestivum*	80–200 nm long, 30–50 nm wide	0, 0.002, 0.006, and 0.02 g kg^{-1} sandy soil	Seed and root	Full life cycle	Increased grain protein, Fe, and Zn contents, improved photosynthetic parameters	Xue et al. (2017)
MWCNTs	*Triticum aestivum; Zea mays; Arachis hypogaea; Allium cepa*	10–20 nm	50 µg mL^{-1}	Seed	Over night	Improved and rapid germination, increased biomass accumulation, and water absorption potential of seeds	Srivastava and Rao (2014)
MWCNTs	*Hordeum vulgare; Zea mays; Glycine max*	15–40 nm wide	50 µg mL^{-1} in hydroponics	Root	20 weeks	Increased shoot growth in maize and barley and decreased root biomass in soybean and maize, increased photosynthetic capacity in maize	Lahiani et al. (2017)
Mesoporous carbon NPs	*Oryza sativa*	80 nm	0, 10, 50, 150 mg L^{-1}	Seed	20 days	Decrease in root length and shoot length, phytohormones like BR, IPA, and DHZR in plant shoots increased significantly	Hao et al. (2019)
Metal-based nanoparticles							
Ag	*Oryza sativa*	<20 nm	0, 0.2, 0.5, and 1 µg mL^{-1}, hydroponics	Seed	7 days	Reduced root elongation, shoot and root fresh weights, total chlorophyll and carotenoids contents, ROS production increased	Nair and Chung (2014a)
Ag	*Arachis hypogaea*	20 nm	50, 500, and 2000 mg kg^{-1} sandy soil	Seed and root	98 days	Reduced plant growth parameters and yield, fatty acid composition was adversely affected	Rui et al. (2017)
Ag	*Triticum aestivum*	5.6 nm	20, 200, and 2000 mg kg^{-1} soil	Seed and root	4 months	Reduced plant growth and biomass, Increased grain Ag and reduced grain Fe, Zn, and Cu, reduced yield and grain protein and amino acid contents	Yang et al. (2018)
Ag	*Vigna unguiculata*	20–100 nm	50–100 µg mL^{-1}, Foliar application	Leaves	7 days	Showed no phytotoxicity	Vanti et al. (2019)
Ag	*Hordeum vulgare*	-	366 mg Ag kg^{-1} dry soil	Root	14 days	Smaller shoots and shorter, thick roots	Gonzalez Linares et al. (2020)
Al$_2$O$_3$	*Solanum lycopersicum*	-	400 mg L^{-1}, Foliar application	Leaves	20 days	Increase in photosystem II subsequently increases photosynthesis process and increase plant growth	Shenashen et al. (2017)
Al$_2$O$_3$	*Allium cepa*	>50 nm	0.1, 10, and 100 mg L^{-1}	Root	3 days	Generation of ROS and MDA, chromosomal abrasion	Debnath et al. (2020)
CeO$_2$	*Triticum aestivum*	8 ± 1 nm	0, 1, 125, and 500 mg kg^{-1} of soil	Root	90 days	Decreased root Ce, Al, Fe, and Mn concentrations and adversely affected grain nutrient quality and growth parameters	Rico et al. (2017)
CeO$_2$	*Lactuca sativa*	16.5 nm	0–2000 mg kg^{-1} in sand	Root	3 weeks	Increased oxidative stress, increased nitrate–N level in shoots, inhibited the biomass production	Zhang et al. (2017a, b)

(continued)

Table 1 (continued)

Nanoparticles	Plants	Size	NPs concentration	Mode of application	Growth media and exposure duration	Impacts on plants	References
CeO_2	*Sorghum bicolor*	15 ± 5 nm	0 and 2 mg per plant, Foliar	Leaves	60 days	Lower lipid peroxidation and increased photosynthetic rates and seed yield per plant (31%)	Djanaguiraman et al. (2018)
CeO_2	*Phaseolus vulgaris*	10–30 nm	0, 250, 500, 1000, and 2000 mg L^{-1}, Foliar	Leaves	30 days	NPs application induced membrane damage	Salehi et al. (2018)
$CoFe_2O_4$	*Lycopersicon lycopersicum*	17 nm	17 62.5, 125, 250, and 500 mg l^{-1}; hydroponically	Seed and root	15 days	Increased root and shoot length, no effect on seed germination	López-Moreno et al. (2016)
$CuFe_2O_4$	*Cucumis sativus*	30.7 nm	0.0, 0.04, 0.2, 1, and 5 mg L^{-1}	Root	8 days	Increase in fresh weight, protein content, superoxide dismutase, and peroxidase activities	Abu-Elsaad and Hameed (2019)
CuO	*Oryza sativa*	40 nm	100 mg L^{-1} in hydroponics	Seed	35 days	NPs transported from roots to leaves through apoplastic pathway	Peng et al. (2015)
CuO	*Spinacia oleracea*	10–100 nm	200 mg kg^{-1} soil	Root	60 days	Improved photosynthesis and biomass production	Wang et al. (2016)
CuO	*Triticum aestivum*	<50 nm	3, 10, 30, 300 mg kg^{-1} grown in sand	Root	7 days	Inhibition of root elongation; exposure resulted in root hair proliferation and shortening of the zones of division and elongation	Adams et al. (2017)
CuO	*Oryza sativa*	43 ± 9 nm	50, 100, 500, and 1000 mg kg^{-1} soil	Root	7, 21, 60, and 88 days	Physiological parameters and grain yield adversely affected (500 and 1000 mg/kg)	Peng et al. (2017)
CuO	*Capsicum annuum*	20–100 nm	0, 125, 250, and 500 mg kg^{-1} soil	Root	90 days	Root Cu concentrations were elevated (250 and 500 mg/kg); reduced nutrient uptake to fruits and leaves	Rawat et al. (2018)
Fe_2O_3	*Triticum aestivum*	20–30 nm	0, 100, 500, 1000, 5000, and 10,000 mg L^{-1}	Seed	8 days	Increased germination at lower concentration, reduced seed germination with increasing treatments	Feizi et al. (2013)
Fe_2O_3	*Cucumis melo*	20 nm	0, 100, 200, and 400 mg L^{-1} in Hoagland	Root	4 weeks	Promote plant growth and increase chlorophyll	Wang et al. (2019)
Fe_3O_4	*Triticum aestivum*	6.8 nm	2000 mg l^{-1}	Seed	5 days	Reduce heavy metals uptake and mitigate their toxicity	Konate et al. (2017)
Fe_3O_4	*Brassica juncea*	80–110 nm	500 mg L^{-1}	Root	4 days	Reduce As toxicity, sulfur-related gene transcripts increased	Praveen et al. (2018)
Fe_3O_4	*Cucumis melo*	20 nm	0, 100, 200, and 400 mg L^{-1} in Hoagland	Root	4 weeks	Promote plant growth and increase chlorophyll	Wang et al. (2019)
FeS_2	*Beta vulgaris*	600–700 nm	80–100 μg mL^{-1}	Seed	12–14 h	Increased germination and crop yield	Das et al. (2016)
MgO	*Solanum lycopersicum*	20–200 nm	7–10 μg mL^{-1}	Root	7 days	Controlled bacterial wilt disease	Imada et al. (2016)
MgO	*Citrus maximus*	50–200 nm	0, 250, 500, or 1 000 mg L^{-1}	Seed	40 days	Reduction in chlorophyll content, antioxidant enzymes activity, and root activity	Xiao et al. (2019)
MoO_3	*Oryza sativa*	21.34 nm	100 m L^{-1}	Root	15 days	Insignificant translocation from root to shoot	Sharma et al. (2020a)

(continued)

Table 1 (continued)

Nanoparticles	Plants	Size	NPs concentration	Mode of application	Growth media and exposure duration	Impacts on plants	References
Nd_2O_3	*Cucurbita maxima*	30–45 nm	100 mg L^{-1}	Root	8 days	Inhibition plants growth and the necessary elements uptake was hampered	Chen et al. (2016)
$NiFe_2O_4$	*Hordeum vulgare*	12.25 nm	0, 125, 250, 500, and 1000 mg L^{-1}	Seed	3 weeks	Decrease in plant growth and biomass at concentration higher than 500 mg/L	Tombuloglu et al. (2019)
SiO_2	*Oryza sativa*	-	2.5 mM L^{-1}, Foliar application	Leaves	70 days	Alleviated heavy metal toxicity and improved growth due decreased bio-concentration and translocation in plants	Wang et al. (2016)
SiO_2	*Zea mays*	30 nm	1000 mg L^{-1}	Seed	3 days	Reduced shoot length, shoot fresh weight, and dry root weight, chlorophyll contents, content of carotenoid, MDA production	Ghoto et al. (2020)
TiO_2	*Oryza sativa*	20 nm	0, 25, 50, 150, 250, 500, and 750 mg kg^{-1} in P-deficient soil	Seed and root	Full life-cycle	Increased P uptake and plant growth (50–750 mg/kg) without translocation to grains	Zhang et al. (2015)
TiO_2	*Triticum aestivum*	21 nm	0, 5, 50, 150 mg L^{-1}	Root	20 days	Down regulation of antioxidant enzyme genes encoding catalase, APX, MDA, and dehydroascorbate reductase with more prominence in roots	Silva et al. (2019)
ZnO	*Sorghum bicolor*	18 nm	6 mg kg^{-1} soil	Root	-	Increased grain yield and grain Zn, N, K, and P under all experimental variations	Dimkpa et al. (2017)
ZnO	*Solanum lycopersicum*	<100 nm	3, 20, 100, and 225 mg kg^{-1} acidic (pH 5.4) or calcareous (pH 8.3) soil	Root	90 days	Increased photosynthetic pigments and protein in calcareous soil and higher leaf Zn in acidic soil	García-Gómez et al. (2017)
ZnO	*Triticum aestivum*	18 nm	6 mg kg^{-1} soil	Root	Grown to maturity	Increased in leaf chlorophyll and shoot height, grain yield and Zn content increased	Dimkpa et al. (2018)
ZnO	*Triticum aestivum*	<100 nm	0, 10, 20, 50, 100, 200, 1000 mg L^{-1}	Seed and root	7 days	Lower biomass of seedlings, structural damage to the roots, and significant changes in enzyme activities	Du et al. (2019)

(Beddow et al. 2014; Michels et al. 2015). Thus, understanding the behavior and effects of these ENPs is very topical for a scientific community. Both primary and transformed ENPs are rich in the soil environment (Schwab et al. 2016). The human added ENPs in soil attract special attention because they have the potential to accumulate for a longer time and are generally resistant to degradation. However, ENPs affect many physical, chemical, and biological properties of soil (Table 1). The effects of ENPs depend on soil types, their concentration, and soil enzyme activity. It also depends on the type, shape, size, and concentration of ENPs. A high concentration of ENPs reduces the activity of the dehydrogenase enzyme (Jośko et al.

2014). Microorganisms determine the status of soil biological activity and functions of soil enzyme. The biochemical and biological diversity of soil act as a sensor for soil health. Most of ENPs disturb the balance of these soil parameters and soil properties through different types of interactions. Dehydrogenase, phosphatases, and urease are the most common soil enzymes. These enzymes are involved in soil respiration and cycling of nutrients (Burns et al. 2013). Silver NPs, one of the most produced NPs in terms of quantity, strongly alter the health and physico-chemical properties of soil (Courtois et al. 2019).

Another important aspect of ENPs is its effects on soil nutrient and fertility. The properties of soil such as pH,

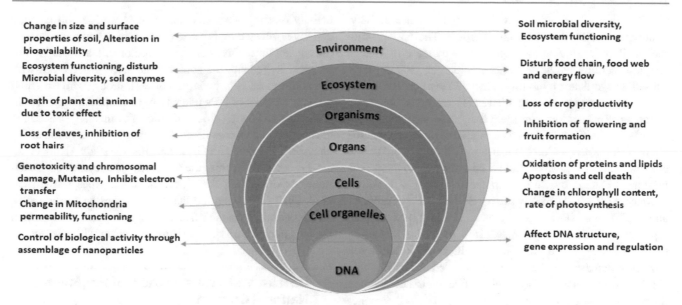

Change In size and surface properties of soil, Alteration in bioavailability

Ecosystem functioning, disturb Microbial diversity, soil enzymes

Death of plant and animal due to toxic effect

Loss of leaves, inhibition of root hairs

Genotoxicity and chromosomal damage, Mutation, Inhibit electron transfer

Change in Mitochondria permeability, functioning

Control of biological activity through assemblage of nanoparticles

Environment

Ecosystem

Organisms

Organs

Cells

Cell organelles

DNA

Soil microbial diversity, Ecosystem functioning

Disturb food chain, food web and energy flow

Loss of crop productivity

Inhibition of flowering and fruit formation

Oxidation of proteins and lipids Apoptosis and cell death

Change in chlorophyll content, rate of photosynthesis

Affect DNA structure, gene expression and regulation

Fig. 2 Impact of engineered nanoparticles on biotic and abiotic components of environment

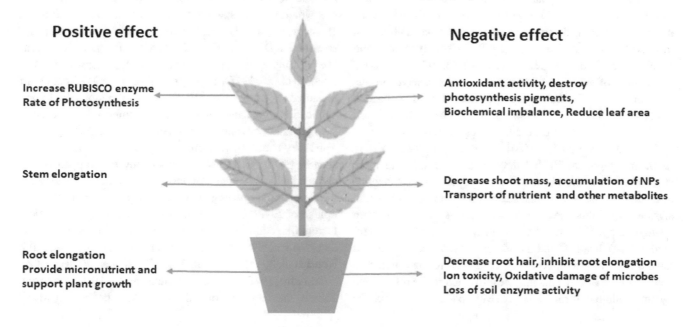

Positive effect

Increase RUBISCO enzyme Rate of Photosynthesis

Stem elongation

Root elongation Provide micronutrient and support plant growth

Negative effect

Antioxidant activity, destroy photosynthesis pigments, Biochemical imbalance, Reduce leaf area

Decrease shoot mass, accumulation of NPs Transport of nutrient and other metabolites

Decrease root hair, inhibit root elongation Ion toxicity, Oxidative damage of microbes Loss of soil enzyme activity

Fig. 3 Impact of engineered nanoparticles on different plant parts

texture, structure, and organic matter are highly affected by ENPs. In addition, size and microbial diversity determine the mobility of ENPs in soil. The small size and high surface area of NPs allow them to interact with soil microorganisms and interfere with their normal metabolism. Therefore, it has strong mobility and antibacterial activity in the soil. Silver NP has growth inhibitory properties with *Nitrosomonas europaea, Nitrosospira multiformis,* and *Nitrosococcus oceani* (Table 1) (Beddow et al. 2014). Wastewater is a major source of ZnO NPs, and its applications in agricultural soil are major causes of soil contamination. However, ENPs

in the environment may undergo photooxidation, dissolution, and sulfidation in a natural way (Ma et al. 2014).

The effect of NPs on soil enzymes is highly variable and significant depending on the type of soil enzyme. While phosphatases and urease are soil enzymes related to nutrient cycling, dehydrogenase activity represents the overall microbial activity. The Ag NPs negatively affect soil dehydrogenase and urease activities (Shin et al. 2012). In a study by Sindhura et al. (2014), the green synthesized Zn NPs was found non-toxic and also enhanced the dehydrogenase and phosphatase activities. In a study by Josko et al. (2014),

Cr NPs showed no effect on the dehydrogenase activity, while the Ni NPs showed a mixed effect. The Ni NPs with the small concentration showed a slight increase (9.2%), in enzyme activity while a concentration of 100 mg kg^{-1} caused a significant inhibitory effect (86.9%) on soil dehydrogenase activity. Kim et al. (2011) observed the toxic effect of Zn NPs on soil dehydrogenase activity which decreased significantly. Josko et al. (2014) observed that the effect of the Zn and Cu NPs caused inhibition of the urease enzyme activity, while Cr and Ni NPs stimulated them. The inhibition caused by Zn NPs in the soil type SL1 ranged between 11.7 and 41.6%, while for SL2, it was between 2.1 and 53.7%. Cu NPs ranged between 0.7 and 44% in both types of soils under experiment. In a study by Luo et al. (2020), it was found that the activities of different soil enzymes increased on the interaction of Ce and Cr NPs with elevated atmospheric CO_2 levels, while their activity was found to decrease on using the NPs alone. The soil enzymes tested under these conditions were soil dehydrogenase, urease, and acid phosphatase. Phosphatase activity is measured in two types, acidic phosphatase, and alkaline phosphatase. The acid phosphatase of the soil was inhibited on the addition of Zn, Cu, and Cr NPs in soil type SL1 and Ni NPs in SL2 soil type (Josko et al. 2014). In SL2 soil type, stimulation of the soil acid phosphatase was observed on the addition of Zn NPs. In a study by Kim et al. (2011), the acid phosphatase activity in the soil decreased upon the addition of different Zn NPs at different concentrations.

In a study by You et al. (2018), the metal oxide NPs were used, and it concludes that the saline-alkali soils are more susceptible than the black soil with respect to their enzyme activities. It also states that the metal oxide NP incubation significantly influences the soil enzyme activities and even changes the soil bacterial community. In a study by Peyrot et al. (2014), it was found that Ag NPs in low concentrations have a higher toxic effect when compared to their increased concentration effect. This may be attributed to the role played by the colloidal form of silver. Multiple factors affect the

activity of enzymes concerned with NPs. The soil enzyme activity is affected by NP size (such as bulk and nano forms), the contact time of NPs and soil, type of soil, and the kind of enzyme. In general, the soil enzyme activity decreases with the addition of NPs in large concentrations, while a small amount addition of NPs tends to have a stimulating effect in some cases. Though this deduction fits in many cases, the exact effect can be concluded only upon its observation as the enzyme NP interaction is a multifactor study (Kim et al. 2011). Apart from this, the intracellular enzymes are much more sensitive than the extracellular enzymes (Asadishad et al. 2017). This may be due to the adhesion of the extracellular enzymes on the surface of clay particles due to which the NPs have a hard time coming in contact with enzymes.

4 Uptake and Translocation of Engineered Nanoparticles in Plants

Interaction of ENPs with plants is dependent on various factors which include species types, transpiration rate, route of exposure, physicochemical properties, exposure duration, and size of NPs (Dietz and Herth 2011; Ma et al. 2015; Duran et al. 2017; Kranjc et al. 2018; Noori et al. 2017; Xiong et al. 2017; Zhao et al. 2017). During agricultural practices, plants are presumably exposed to ENPs due to nanotechnological applications like nanopesticides and nanofertilizers, the use of sewage sludge, and atmospheric depositions (Gardea-Torresdey et al. 2014; Xiong et al., 2017). The fate of atmospheric ENPs on leaves are either they are trapped on the surface by the cuticular wax layer or can enter into the plants through natural openings like stomata (Fig. 4). However, the uptake of ENPs from the soil by roots depends on various factors like cation exchange capacity, pH, rhizospheric exudates, and microorganisms (Du et al. 2017; Huang et al. 2017; Noori et al. 2017; Xue et al. 2017; Rossi et al. 2018).

Initially, ENPs are accumulated on the surface as roots secrete mucilage or organic acids which are negatively

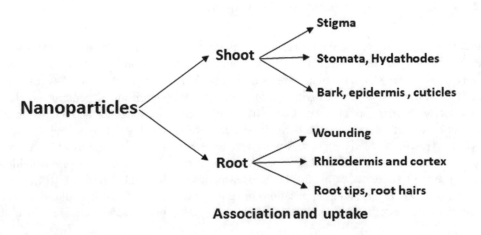

Fig. 4 Uptake and translocation mechanism of engineered nanoparticles in plants

charged (Zhou et al. 2011). These accumulated ENPs have to cross different layers including cuticle, epidermis, cortex, endodermis, and Casperian strips to translocate into shoots by xylem vessels. Likewise, in plant leaves, ENPs get incorporated into cuticles of roots as these are thin. After incorporation into the cuticle, there are two proposed pathways for the uptake of ENPs through epidermis, i.e., apoplastic pathway and symplastic pathway. The apoplastic pathway is the most studied and according to which intercellular transport of ENPs occurs without crossing the cell membrane as these NPs diffuse into the intercellular spaces by penetrating the cell wall (Lv et al. 2019). These ENPs after passing through epidermis, cortex, and endodermis then reaches Casperian strip which is made up of lipophilic compounds and hampers the movement into the vascular system (Schwab et al. 2016). Nonetheless, sometimes these ENPs enters the vascular system as Casperian strip has not yet formed in root tip or through root junctions where Casperian strip is disjointed (McCully 1995; Dietz and Herth 2011; Lv et al. 2015; Schymura et al. 2017). However, the symplastic pathway is a cell-to-cell pathway that occurs in two steps which are the penetration of cell membrane followed by intercellular transfer via plasmodesmata.

According to the literature, the highest feasibility transmembrane pathway for NPs is through endocytosis. After endocytosis, these ENPs reaches vacuoles and get sequestered as it acts as a sink for solutes and also largest organelle in the plant cell (Serag et al. 2011; Bao et al. 2016; Huang et al. 2017). The study on vacuole membrane transport of NPs suggests that dissolved metal NPs can induce anti-oxidative responses in plants. These oxidative response results into the production of thiol-containing glutathione stimulating hormone (GSH) which binds with metal ions or with metal transporters resulting in metal detoxification pathway (Dhankher et al. 2002; Ma et al. 2016). Another mechanism of ENPs toxicity may be due to the Trojan horse mechanism in which these NPs are taken up by plants cell followed by the release of metal ions causing damage to the cellular structure (Singh and Ramarao 2012). Studies of ENPs uptake and translocation by plants are still at a very nascent stage which needs to be studied thoroughly.

4.1 Effects of Engineered Nanoparticles on Plants

The increasing application of ENPs may have an impact on ecosystem functioning and food crops. In this section, major interaction of ENPs are illustrated in terms of some growth and developmental features (seed germination, biomass, yield characteristics, shoot/root growth and leaf production), physiological features (photosynthetic efficiency and effect on various photosystems), biochemical, and molecular features (enzymatic and non-enzymatic components). The response of major crops exposed to different ENPs are presented in Table 1.

4.1.1 Effect of Engineered Nanoparticles on Growth and Developmental Features of Plants

Plant morphological parameters such as seed germination, biomass, leaf area, yield, length, and weight of root and shoot are indicators of plant's health. Exposure of carbon nano onions (CNO) to *Cicer arietinum* seed showed an increase in $\sim 35\%$ in weight, $\sim 24\%$ in length, $\sim 17\%$ in diameter, and $\sim 16\%$ in height over the control when exposed to 30 µg mL^{-1} of CNO (Tripathi et al. 2017a, b). Similarly, studies on effects of multiwalled carbon nanotubes (MWCNTs) (50 µg mL^{-1}) on *A. cepa, Arachis hypogaea, Glycine max, Hordeum vulgare, Triticum aestivum,* and *Zea mays* showed improvement in root and shoot length, biomass, and seed germination by improving water absorption potential (Srivastava and Rao 2014; Lahiani et al. 2017). Yield in terms of crop products is an important parameter for growth assessment which includes the number of spikes, grains per spike, and grain weight. Xue et al. (2017) showed that nanochitin (6 mg kg^{-1})-treated *T. aestivum* led up to 23% increase in yield and other yield parameters. Nanochitin showed a prominent increase in the number of spikelets per spike as apex development and spikelet primordia differentiation was influenced resulting in apex elongation (Xue et al. 2017). In contrast to previous studies, mesoporous carbon NPs (150 mg L^{-1}) showed a reduction in root length (70%), shoot length (57.1%), root fresh weight (34%), and shoot fresh weight (45%) of *Oryza sativa* grown hydroponically (Hao et al. 2019).

Metal-based NPs generally show toxic effects on maize crops due to the release of metal ions (Dimkpa et al. 2012; Mahmoodzadeh et al. 2013; Nair and Chung 2014a). Ag NPs exposed to *A. hypogaea* showed a significant reduction in plant height, biomass, and yield (Rui et al. 2017). Similar results were obtained for *O. sativa* and *T. aestivum* exposed to Ag NPs in hydroponics and soil, respectively (Nair and Chung 2014a; Yang et al. 2018). In contrast, the foliar application showed no toxicity to *Vigna unguiculata* at varied concentrations (Vanti et al. 2019). However, Al$_2$O$_3$ NPs showed improved biomass of *Solanum lycopersicum* during foliar applications (Shenashen et al. 2017; Debnath et al. 2020). CeO$_2$ NPs are generally used in the automobile industry, electronics, and fuel additives (Keller et al. 2013).

Different plants have been exposed to different concentrations of CeO$_2$ NPs to analyze their effect. Zhang et al. (2017a, b) showed that there is no visible impact of CeO$_2$ NPs at a concentration of 500 mg kg^{-1}, while on increasing the concentration (1000–2000 mg kg^{-1}), fresh and dry weight of root and shoot decreased in *Lactuca sativa*.

Conversely, foliar spray of nano Ce on *Sorghum bicolor* (L.) plants under drought showed 31% increase in seed yield per plant as compared to control (Djanaguiraman et al. 2018). Conversely, *Beta vulgaris* when exposed to FeS_2 NPs showed 47% increment in the yield (Das et al. 2016). Similar results of increased root and shoot length and biomass were observed in *Lycopersicon lycopersicum* and *Cucumis melo* when grown with $CoFe_2O_4$ and Fe_2O_3 hydroponically. Some ENPs are used against environmental stresses like heavy metal contamination, drought, or some infectious diseases. *Triticum aestivum* and *Brassica juncea* initially when grown in soil contaminated with Pb, Zn, Cd, Cu, and As showed a reduction in growth and biomass but supplementing the soil with Fe_3O_4 NPs showed improved growth (Konate et al. 2017; Praveen et al. 2018). Similarly, Imada et al. (2016) showed that bacterial wilt in *S. lycopersicum* could be controlled by the application of MgO NPs. Although ENPs may either have a positive or negative impact on plants, the response varies considerably with duration, dose, species, and experimental conditions.

4.1.2 Effects of Engineered Nanoparticles on Physiological Processes of Plants

Physiological responses of plants occured due to exposure of ENPs which induce abiotic stress in plants. Plant biomass indicates phytotoxicity which is mainly affected by a change in photosynthetic efficiency due to change in photosystem I (PS-I) or PS-II. Studies suggest that the higher concentrations of ENPs affect the photosynthesis, which causes suppression or death of crops (Perreault et al. 2014; Da Costa and Sharma 2016). The exposure of 50 $\mu g\ ml^{-1}$ of MWCNT to *Z. mays* showed a 10% increase in the photosynthetic rate, whereas no change in *G. max* in hydroponics was observed (Lahiani et al. 2017). Similar results of increased photosynthetic rate were observed in *T. aestivum* as nanochitins caused an increase in stomatal conductance (Xue et al. 2017). Increased stomatal conductance results in increased diffusion of external CO_2 in the pectin cavity which enhances the CO_2 assimilation rate and hence photosynthetic rate. Stomatal conductance also increases the transpiration rate which causes movement of increase in water uptake, eventually resulting in increased nitrogen and potassium accumulation by *T. aestivum* plant exposed to 6 $mg\ kg^{-1}$ of nanochitin (Xue et al. 2017). Similar results of increased micronutrients (Cu, Fe, Mn, Mo, Ni, and Zn) in the seeds have been shown by CNO treated *C. arietinum* which is due to increased protein content (Tripathi et al. 2017a, b).

Metal-based NPs can affect the photosynthetic apparatus of plants and its productivity, causing acute and chronic effects (Arruda et al. 2015; Da Costa and Sharma 2016). Hydroponically grown *O. sativa* in different concentrations of Ag NPs showed a significant reduction in total chlorophyll and carotenoids due to peroxidation of chloroplast membrane (Nair and Chung 2014a). The decrease in photosynthetic capacity may be due to suppression of fluorescence caused by a decrease in the quantum yield of PS-II and electron transport chain inhibition (Matorin et al. 2013), therefore, resulted in a significant decrease in reducing and total sugar contents (Nair and Chung 2014b). In contrast to previous studies, Al_2O_3 NPs showed an increase in photosynthetic quantum yield of PS-II resulting in increased photosynthesis and plant growth (Shenashen et al. 2017). MoS_2 NPs treatments led to increased chlorophyll-a levels suggesting it to be non-compromising with photosynthetic process in rice (Sharma et al. 2020b). CeO_2 NPs exposure to *L. sativa* has resulted in decreased chlorophyll content of leaves. The reason for decreased chlorophyll is reduced uptake of Fe which acts as an activator of key coenzyme for synthesis of chlorophyll (Terry and Low 1982; Miller et al. 1984). Additionally, it also hampers electron transport during photosynthesis as Fe is an important constituent of enzyme ferredoxin (Arnon 1965). A recent study showed that the transfer of energy from PS-II to the Calvin cycle is disrupted by CeO_2 by electron absorption from PS-II by, or through reactions with reactive oxygen species (ROS) (Conway et al. 2015). However, foliar application of CeO_2 during drought condition improved photosynthetic rate, PS-II quantum yield and stomatal conductance as compared to plant growing in drought with CeO_2 exposure (Djanaguiraman et al. 2018).

Iron (Fe) is an essential element for the synthesis of chlorophyll, and its deficiency would reduce the rate of photosynthesis (Briat et al. 2015). *Cucumis melo* exposed to Fe_2O_3 and Fe_3O_4 initially showed a reduction in photosynthetic ability due to oxidative stress and increased in later phase due to the absorption of Fe (Wang et al. 2019). The presence of heavy metals might disrupt the pigment complex or inhibit enzymes involved in the biosynthetic pathway of chlorophyll. The addition of certain ENPs in soil contaminated with heavy metals showed improvement in photosynthetic activity. For instance, *B. juncea* grown in arsenic-contaminated soil showed 36% improvement in photosynthetic rate when supplemented with Fe_3O_4 NPs (Praveen et al. 2018). TiO_2 NPs were shown to increase the Rubisco activity eventually improving photosynthesis (Sarmast and Salehi 2016; Ghoto et al. 2020). It also improves the electron transport chain by increasing the number and energy of electrons, improving ATP formation and photolysis of water and also by activating photochemical reactions in the chloroplast (Hong et al. 2005; Mingyu et al. 2007). However, detrimental effects of the photosynthetic rate have also been reported due to a lack of stomatal regulation (Gao et al. 2013) and a negative impact on the structure and function of photosynthesis (Movafeghi et al. 2018). Similar results were obtained in the case of CuO NPs which affect the chlorophyll fluorescence increasing the dissipation of thermal energy and decrease in electron transport capacity of PS-II (Tighe-Neira et al. 2018). Increasing the dose of CuO NPs decreases photosynthetic rate as enzyme RuBP carboxylase gets inactivated; also changes in the rate of transpiration and photosynthetic efficiency of PS-II have higher impact on plant physiology (Nekrasova et al. 2011; Regier et al. 2015; Da Costa and Sharma 2016). Similarly, ZnO NPs have comparable effects on the photosynthetic process having more negative than positive effects, mainly due to functional impairments at higher application rates. In summary, the molecular and physiological responses of plants are linked to ENPs accumulation. Uptake of ENPs can trigger in vitro or in vivo plant responses subsequently, either alleviate the nano-toxicity or decrease the ENPs uptake via different pathways that needs to be studied in detail.

4.1.3 Effects of Engineered Nanoparticles on Molecular and Biochemical Properties of Plants

Molecular and biochemical effects of ENPs help in understanding the mechanism of plant responses. The most common biochemical effect is an upsurge in the reactive oxygen species (ROS) upon exposure to ENPs (Panda et al. 2011; Zhao et al. 2012; Speranza et al. 2013; Mukherjee et al. 2014). ROS plays a key role in signaling reactions in plants and can cause oxidative damage to plants (Mittler 2017). Various studies showed that excessive ROS results in alteration in phytohormones, as ROS plays vital role in hormone perception and transduction (Gechev et al. 2006; Zhang et al. 2017a, b). Syu et al. (2014) reported that ENPs exposure could influence gene expression which affects hormone signaling resulting in change in signaling transduction, and eventually imbalance in phytohormone's levels. Metal-based NPs cause oxidative stress in plants by generating ROS. These ENPs trigger the formation of ROS as they release ions which interact with different groups of proteins (Gorczyca et al. 2015).

Silver (Ag) NPs treated *O. sativa* showed excessive ROS generation which resulted in the upregulation of superoxide dismutases (SOD) genes viz. *FSD, MSD1,* and *CSD1* genes, *CAT* genes, and *APXa* and *APXb* genes (Nair and Chung 2014a). This abiotic stress causes an upsurge in ROS generation which eventually results in increased superoxide dismutase, peroxidases, and catalase enzymes activity (Rui et al. 2017). Proline was found to be accumulated which might be a protective mechanism in plants against excessive ROS to protect the cellular structures (Chiang and Dandekar 1995). Similarly, CeO_2 exposure (0, 100, 500, 1000, 2000 mg kg^{-1}) to *L. sativa* showed an increase in POD activity only in roots at 2000 mg kg^{-1}, while SOD activities were enhanced in both roots and shoots at 100 and 500 mg kg^{-1}, but decreased at 1000 and 2000 mg kg^{-1}.

A significant increase in MDA contents was observed at 1000 and 2000 mg kg^{-1} of CeO_2 owing to higher oxidative stress (Zhang et al. 2017a, b). Conversely, the foliar application of CeO_2 NPs on *S. bicolor* in drought conditions increases antioxidant enzyme activities causing lower lipid peroxidation (Djanaguiraman et al. 2018). CeO_2 NPs imitate SOD activity and efficiently convert O^{2-} to H_2O_2 than SOD (Heckert et al. 2008). The Ce^{4+} and Ce^{3+} oxidation states of CeO_2 NPs lead to redox reactions (Conesa 1995) that scavenge the ROS produced under drought.

ENP-treated plants exhibit an extensive regulation of gene and protein, which provide useful information for plant detoxification or tolerance. Exposure of TiO_2 NPs to *T. aestivum* showed no effect on SOD, catalase (CAT), and glutathione peroxidases (G-POX) activities but decreased the activities of ascorbate peroxidase (APX), tryptophan aminotransferase (TAA), and protein content and led to an alteration in GSH/GSSG ratio.

Graphene oxide NP which is highly used for various purposes and harms hydrophytes. They are highly mobile in water bodies such as lakes, rivers and cause negative impact if released. In addition, these molecules have a toxic effect on plant physiology such as oxidative stress in mitochondria (Miralles et al. 2012; Lv et al. 2019). In recent years, ENPs have achieved special attention as a potential agent for enhancing crop productivity. TiO_2 and SiO_2 are described to increase nitrate reductase activity, enhance absorption, and utilization of water, fertilizers, and antioxidant production (Peyrot et al. 2014; Xu et al. 2015).

5 Conclusion and Future Perspectives

The invention of different types of ENPs has revolutionized the field of science and technology. It has been used for various purposes and as the commercialization of ENPs are increasing, the risks of health and environmental contamination is an emerging challenge for scientific society. Today, researchers, industries, and companies are using a wide range of NPs for research and various other purposes. The increased production and volume of ENPs raise a potential concern for the environment and human health. Several studies reported a wide range of negative effects of ENPs on viruses, bacteria, plants, and animals. So, it is very crucial to find out a suitable solution to overcome the negative effects of NPs and their sustainable use. Thus, it is necessary to address these issues and open windows for the careful use of NPs for future applications. Further, there is a need to explore the knowledge on final fate and impact of NPs in every different type of contaminated environment and appropriate guidelines are required to be framed to avoid contamination. Overall, the future research should be focused on: (1) to determine the kinetics of ENPs interaction with soil and plants, (2) to explore the role of microorganisms on transformation of ENPs at interface of plants, (3) to unravel the interconnections with soil quality and climate change, (4) long-term studies should be conducted for impact assessment of ENPs on plants soil properties, and (5) majority of studies are basically laboratory scale studies in controlled environment which needs to be conducted in field.

References

Abu-Elsaad NI, Abdel Hameed RE (2019) Copper ferrite nanoparticles as nutritive supplement for cucumber plants grown under hydroponic system. J Plant Nutr 42(14):1645–1659

Adams J, Wright M, Wagner H, Valiente J, Britt D, Anderson A (2017) Cu from dissolution of CuO nanoparticles signals changes in root morphology. Plant Physiol Biochem 110:108–117

Arnon DI (1965) Ferredoxin and photosynthesis. Science 149 (3691):1460–1470

Arruda SCC, Silva ALD, Galazzi RM, Azevedo RA, Arruda MAZ (2015) Nanoparticles applied to plant science: a review. Talanta 131:693–705

Asadishad B, Chahal S, Cianciarelli V, Zhou K, Tufenkji N (2017) Effect of gold nanoparticles on extracellular nutrient-cycling enzyme activity and bacterial community in soil slurries: role of nanoparticle size and surface coating. Environ Sci Nano 4(4):907–918

Asli S, Neumann PM (2009) Colloidal suspensions of clay or titanium dioxide nanoparticles can inhibit leaf growth and transpiration via physical effects on root water transport. Plant Cell Environ 32 (5):577–584

Bao D, Oh ZG, Chen Z (2016) Characterization of silver nanoparticles internalized by Arabidopsis plants using single particle ICP-MS analysis. Front Plant Sci 7:32

Beddow J, Stolpe B, Cole P, Lead JR, Sapp M, Lyons BP, Colbeck I, Whitby C (2014) Effects of engineered silver nanoparticles on the growth and activity of ecologically important microbes. Environ Microbiol Rep 6(5):448–458

Bondarenko O, Juganson K, Ivask A, Kasemets K, Mortimer M, Kahru A (2013) Toxicity of Ag, CuO and ZnO nanoparticles to selected environmentally relevant test organisms and mammalian cells in vitro: a critical review. Arch Toxicol 87(7):1181–1200

Briat JF, Dubos C, Gaymard F (2015) Iron nutrition, biomass production, and plant product quality. Trends Plant Sci 20(1):33–40

Burns RG, DeForest JL, Marxsen J, Sinsabaugh RL, Stromberger ME, Wallenstein MD, Weintraub MN, Zoppini A (2013) Soil enzymes in a changing environment: current knowledge and future directions. Soil Biol Biochem 58:216–234

Cecchin I, Reddy KR, Thomé A, Tessaro EF, Schnaid F (2017) Nanobioremediation: Integration of nanoparticles and bioremediation for sustainable remediation of chlorinated organic contaminants in soils. Int Biodeterior Biodegrad 119:419–428

Chen G, Ma C, Mukherjee A, Musante C, Zhang J, White JC, Dhankher OP, Xing B (2016) Tannic acid alleviates bulk and nanoparticle Nd2O3 toxicity in pumpkin: a physiological and molecular response. Nanotoxicology 10(9):1243–1253

Chiang HH, Dandekar A (1995) Regulation of proline accumulation in Arabidopsis thaliana (L.) Heynh during development and in response to desiccation. Plant Cell Environ 18(11):1280–1290

Conesa J (1995) Computer modeling of surfaces and defects on cerium dioxide. Surf Sci 339(3):337–352

Conway JR, Beaulieu AL, Beaulieu NL, Mazer SJ, Keller AA (2015) Environmental stresses increase photosynthetic disruption by metal oxide nanomaterials in a soil-grown plant. ACS Nano 9(12):11737–11749

Courtois P, Rorat A, Lemiere S, Guyoneaud R, Attard E, Levard C, Vandenbulcke F (2019) Ecotoxicology of silver nanoparticles and their derivatives introduced in soil with or without sewage sludge: A review of effects on microorganisms, plants and animals. Environ Pollut

Da Costa M, Sharma P (2016) Effect of copper oxide nanoparticles on growth, morphology, photosynthesis, and antioxidant response in Oryza sativa. Photosynthetica 54(1):110–119

Darlington TK, Neigh AM, Spencer MT, Guyen OT, Oldenburg SJ (2009) Nanoparticle characteristics affecting environmental fate and transport through soil. Environ Toxicol Chem 28(6):1191–1199

Das CK, Srivastava G, Dubey A, Roy M, Jain S, Sethy NK, Saxena M, Harke S, Sarkar S, Misra K (2016) Nano-iron pyrite seed dressing: a sustainable intervention to reduce fertilizer consumption in vegetable (beetroot, carrot), spice (fenugreek), fodder (alfalfa), and oilseed (mustard, sesamum) crops. Nanotech Environ Eng 1(1):2

Debnath P, Mondal A, Sen K, Mishra D, Mondal NK (2020) Genotoxicity study of nano Al2O3, TiO2 and ZnO along with UV-B exposure: an Allium cepa root tip assay. Sci Total Environ 713:136592

DeRosa MC, Monreal C, Schnitzer M, Walsh R, Sultan Y (2010) Nanotechnology in fertilizers. Nat Nanotechnol 5(2):91

Dhankher OP, Li Y, Rosen BP, Shi J, Salt D, Senecoff JF, Sashti NA, Meagher RB (2002) Engineering tolerance and hyperaccumulation of arsenic in plants by combining arsenate reductase and γ-glutamylcysteine synthetase expression. Nat Biotechnol 20 (11):1140

Dietz K-J, Herth S (2011) Plant nanotoxicology. Trends Plant Sci 16 (11):582–589

Dimkpa CO, McLean JE, Latta DE, Manangón E, Britt DW, Johnson WP, Maxim IB, Anderson AJ (2012) CuO and ZnO nanoparticles: phytotoxicity, metal speciation, and induction of oxidative stress in sand-grown wheat. J Nanopart Res 14(9):1125

Dimkpa CO, Singh U, Bindraban PS, Elmer WH, Gardea-Torresdey JL, White JC (2018) Exposure to weathered and fresh nanoparticle and ionic Zn in soil promotes grain yield and modulates nutrient acquisition in wheat (Triticum aestivum L.). J Agric Food Chem 66 (37):9645–9656

Dimkpa CO, White JC, Elmer WH, Gardea-Torresdey J (2017) Nanoparticle and ionic Zn promote nutrient loading of sorghum grain under low NPK fertilization. J Agric Food Chem 65 (39):8552–8559

Djanaguiraman M, Nair R, Giraldo JP, Prasad PVV (2018) Cerium oxide nanoparticles decrease drought-induced oxidative damage in sorghum leading to higher photosynthesis and grain yield. ACS Omega 3(10):14406–14416

Du W, Gardea-Torresdey JL, Xie Y, Yin Y, Zhu J, Zhang X, Ji R, Gu K, Peralta-Videa JR, Guo H (2017) Elevated CO2 levels modify TiO2 nanoparticle effects on rice and soil microbial communities. Sci Total Environ 578:408–416

Du W, Yang J, Peng Q, Liang X, Mao H (2019) Comparison study of zinc nanoparticles and zinc sulphate on wheat growth: from toxicity and zinc biofortification. Chemosphere 227:109–116

Duran NdM, Savassa SM, Lima RGd, de Almeida E, Linhares FS, van Gestel CA, Pereira de Carvalho HW (2017) X-ray spectroscopy uncovering the effects of Cu based nanoparticle concentration and structure on Phaseolus vulgaris germination and seedling development. J Agric Food Chem 65(36):7874–7884

Feizi H, Moghaddam PR, Shahtahmassebi N, Fotovat A (2013) Assessment of concentrations of nano and bulk iron oxide particles on early growth of wheat (*Triticum aestivum* L.). Annu Res Rev 752–761

Fierer N, Jackson RB (2006) The diversity and biogeography of soil bacterial communities. Proc Natl Acad Sci 103(3):626–631

Foley JA, Ramankutty N, Brauman KA, Cassidy ES, Gerber JS, Johnston M, Mueller ND, O'Connell C, Ray DK, West PC (2011) Solutions for a cultivated planet. Nature 478(7369):337–342

Frenk S, Ben-Moshe T, Dror I, Berkowitz B, Minz D (2013) Effect of metal oxide nanoparticles on microbial community structure and function in two different soil types. PLoS One 8(12)

Gao J, Xu G, Qian H, Liu P, Zhao P, Hu Y (2013) Effects of nano-TiO_2 on photosynthetic characteristics of Ulmus elongata seedlings. Environ Pollut 176:63–70

García-Gómez C, Obrador A, González D, Babín M, Fernández MD (2017) Comparative effect of ZnO NPs, ZnO bulk and ZnSO4 in the antioxidant defences of two plant species growing in two agricultural soils under greenhouse conditions. Sci Total Environ 589:11–24

Gardea-Torresdey JL, Rico CM, White JC (2014) Trophic transfer, transformation, and impact of engineered nanomaterials in terrestrial environments. Environ Sci Technol 48(5):2526–2540

Gechev TS, Van Breusegem F, Stone JM, Denev I, Laloi C (2006) Reactive oxygen species as signals that modulate plant stress responses and programmed cell death. BioEssays 28(11):1091–1101

Ghoto K, Simon M, Shen Z-J, Gao G-F, Li P-F, Li H, Zheng H-L (2020) Physiological and root exudation response of maize seedlings to TiO_2 and SiO_2 nanoparticles exposure. Bio Nanosci 1–13

González Linares M, Jia Y, Sunahara GI, Whalen JK (2020) Barley (*Hordeum vulgare*) seedling growth declines with increasing exposure to silver nanoparticles in biosolid-amended soils. Can J Soil Sci 100:1–9

Gorczyca A, Pociecha E, Kasprowicz M, Niemiec M (2015) Effect of nanosilver in wheat seedlings and *Fusarium culmorum* culture systems. Eur J Plant Pathol 142(2):251–261

Gottschalk F, Sonderer T, Scholz RW, Nowack B (2009) Modeled environmental concentrations of engineered nanomaterials (TiO_2, ZnO, Ag, CNT, fullerenes) for different regions. Environ Sci Technol 43(24):9216–9222

Hao Y, Xu B, Ma C, Shang J, Gu W, Li W, Hou T, Xiang Y, Cao W, Xing B (2019) Synthesis of novel mesoporous carbon nanoparticles and their phytotoxicity to rice (*Oryza sativa* L.). J Saudi Chem Soc 23(1):75–82

Heckert EG, Karakoti AS, Seal S, Self WT (2008) The role of cerium redox state in the SOD mimetic activity of nanoceria. Biomaterials 29(18):2705–2709

Hendren CO, Mesnard X, Dröge J, Wiesner MR (2011) Estimating production data for five engineered nanomaterials as a basis for exposure assessment. ACS Publications

Holden PA, Schimel JP, Godwin HA (2014) Five reasons to use bacteria when assessing manufactured nanomaterial environmental hazards and fates. Curr Opin Biotechnol 27:73–78

Hong F, Zhou J, Liu C, Yang F, Wu C, Zheng L, Yang P (2005) Effect of nano-TiO_2 on photochemical reaction of chloroplasts of spinach. Biol Trace Elem Res 105(1–3):269–279

Huang Y, Zhao L, Keller AA (2017) Interactions, transformations, and bioavailability of nano-copper exposed to root exudates. Environ Sci Technol 51(17):9774–9783

Imada K, Sakai S, Kajihara H, Tanaka S, Ito S (2016) Magnesium oxide nanoparticles induce systemic resistance in tomato against bacterial wilt disease. Plant Pathol 65(4):551–560

Janvier C, Villeneuve F, Alabouvette C, Edel-Hermann V, Mateille T, Steinberg C (2007) Soil health through soil disease suppression: which strategy from descriptors to indicators? Soil Biol Biochem 39(1):1–23

Jini D, Sharmila S (2020) Green synthesis of silver nanoparticles from *Allium cepa* and its in vitro antidiabetic activity. Mater Today: Proceed 22:432–438

Jośko I, Oleszczuk P, Futa B (2014) The effect of inorganic nanoparticles (ZnO, Cr_2O_3, CuO and Ni) and their bulk counterparts on enzyme activities in different soils. Geoderma 232:528–537

Kandeler F, Kampichler C, Horak O (1996) Influence of heavy metals on the functional diversity of soil microbial communities. Biol Fertil Soils 23(3):299–306

Bhainsa KC, D'Souza SF (2006) Extracellular biosynthesis of silver nanoparticles using the fungus *Aspergillus fumigatus*. Colloids Surf B Biointerfaces 47(2):160–164

Keller AA, McFerran S, Lazareva A, Suh S (2013) Global life cycle releases of engineered nanomaterials. J Nanoparticle Res 15(6):1692

Kim S, Kim J, Lee I (2011) Effects of Zn and ZnO nanoparticles and Zn^{2+} on soil enzyme activity and bioaccumulation of Zn in *Cucumis sativus*. Chem Ecol 27(1):49–55

Konate A, He X, Zhang Z, Ma Y, Zhang P, Alugongo GM, Rui Y (2017) Magnetic (Fe_3O_4) nanoparticles reduce heavy metals uptake and mitigate their toxicity in wheat seedling. Sustainability 9(5):790

Kowsalya E, MosaChristas K, Balashanmugam P, Rani JC (2019) Biocompatible silver nanoparticles/poly (vinyl alcohol) electrospun nanofibers for potential antimicrobial food packaging applications. Food Packag Shelf Life 21:100379

Kranjc E, Mazej D, Regvar M, Drobne D, Remškar M (2018) Foliar surface free energy affects platinum nanoparticle adhesion, uptake, and translocation from leaves to roots in arugula and escarole. Environ Sci Nano 5(2):520–532

Kumar A, Verma JP (2018) Does plant—microbe interaction confer stress tolerance in plants: a review? Microbiol res 207:41–52

Kumar A, Singh S, Gaurav AK, Srivastava S, Verma JP (2020) Plant growth-promoting bacteria: Biological tools for the mitigation of salinity stress in plants. Front Microbiol 11

Kumar A, Singh S, Mukherjee A, Rastogi RP, Verma JP (2021) Salt-tolerant plant growth-promoting Bacillus pumilus strain JPVS11 to enhance plant growth attributes of rice and improve soil health under salinity stress. Microbiol Res 242:126616

Lahiani MH, Nima ZA, Villagarcia H, Biris AS, Khodakovskaya MV (2017) Assessment of effects of the long-term exposure of agricultural crops to carbon nanotubes. J Agric Food Chem 66(26):6654–6662

Lee W-M, Kwak JI, An Y-J (2012) Effect of silver nanoparticles in crop plants Phaseolus radiatus and Sorghum bicolor: media effect on phytotoxicity. Chem 86(5):491–499

Lofts S, Criel P, Janssen CR, Lock K, McGrath SP, Oorts K, Rooney CP, Smolders E, Spurgeon DJ, Svendsen C (2013) Modelling the effects of copper on soil organisms and processes using the free ion approach: towards a multi-species toxicity model. Environ Pollut 178:244–253

López-Moreno ML, Avilés LL, Pérez NG, Irizarry BÁ, Perales O, Cedeno-Mattei Y, Román F (2016) Effect of cobalt ferrite ($CoFe_2O_4$) nanoparticles on the growth and development of *Lycopersicon lycopersicum* (tomato plants). Sci Total Environ 550:45–52

Luo J, Song Y, Liang J, Li J, Islam E, Li T (2020) Elevated CO_2 mitigates the negative effect of CeO_2 and Cr_2O_3 nanoparticles on soil bacterial communities by alteration of microbial carbon use. Environ Pollut 263:114–456

Lv J, Christie P, Zhang S (2019) Uptake, translocation, and transformation of metal-based nanoparticles in plants: recent advances and methodological challenges. Environ Sci Nano 6(1):41–59

Lv J, Zhang S, Luo L, Zhang J, Yang K, Christie P (2015) Accumulation, speciation and uptake pathway of ZnO nanoparticles in maize. Environ Sci Nano 2(1):68–77

Ma C, White JC, Dhankher OP, Xing B (2015) Metal-based nanotoxicity and detoxification pathways in higher plants. Environ Sci Technol 49(12):7109–7122

Ma R, Cm L, Judy JD, Unrine JM, Durenkamp M, Martin B, Jefferson B, Lowry GV (2014) Fate of zinc oxide and silver nanoparticles in a pilot wastewater treatment plant and in processed biosolids. Environ Sci Technol 48(1):104–112

Ma X, Wang Q, Rossi L, Zhang W (2016) Cerium oxide nanoparticles and bulk cerium oxide leading to different physiological and biochemical responses in *Brassica rapa*. Environ Sci Technol 50 (13):6793–6802

Mahmoodzadeh H, Aghili R, Nabavi M (2013) Physiological effects of TiO_2 nanoparticles on wheat (*Triticum aestivum*). Tech J Eng Appl Sci 3:1365–1370

Matorin D, Todorenko D, Seifullina NK, Zayadan B, Rubin A (2013) Effect of silver nanoparticles on the parameters of chlorophyll fluorescence and P 700 reaction in the green alga *Chlamydomonas reinhardtii*. Microbiology 82(6):809–814

McCully M (1995) How do real roots work? (Some new views of root structure). Plant Physiol 109(1):1

Michels C, Yang Y, Moreira Soares H, Alvarez PJ (2015) Silver nanoparticles temporarily retard NO_2^--production without significantly affecting N2O release by *Nitrosomonas europaea*. Environ Toxicol Chem 34(10):2231–2235

Miller G, Pushnik J, Welkie G (1984) Iron chlorosis, a world wide problem, the relation of chlorophyll biosynthesis to iron. J Plant Nutr 7(1–5):1–22

Mingyu S, Fashui H, Chao L, Xiao W, Xiaoqing L, Liang C, Fengqing G, Fan Y, Zhongrui L (2007) Effects of nano-anatase TiO_2 on absorption, distribution of light, and photoreduction activities of chloroplast membrane of spinach. Biol Trace Elem Res 118(2):120–130

Miralles P, Church TL, Harris AT (2012) Toxicity, uptake, and translocation of engineered nanomaterials in vascular plants. Environ Sci Technol 46(17):9224–9239

Mittler R (2017) ROS are good. Trends Plant Sci 22(1):11–19

Movafeghi A, Khataee A, Abedi M, Tarrahi R, Dadpour M, Vafaei F (2018) Effects of TiO_2 nanoparticles on the aquatic plant *Spirodela polyrrhiza*: evaluation of growth parameters, pigment contents and antioxidant enzyme activities. JEnvS 64:130–138

Mueller ND, Gerber JS, Johnston M, Ray DK, Ramankutty N, Foley JA (2012) Closing yield gaps through nutrient and water management. Nature 490(7419):254–257

Mukherjee A, Peralta-Videa JR, Bandyopadhyay S, Rico CM, Zhao L, Gardea-Torresdey JL (2014) Physiological effects of nanoparticulate ZnO in green peas (*Pisum sativum* L.) cultivated in soil. Metallomics 6(1):132–138

Nair PMG, Chung IM (2014) Physiological and molecular level effects of silver nanoparticles exposure in rice (*Oryza sativa* L.) seedlings. Chemosphere 112:105–113

Nair PMG, Chung IM (2014b) A mechanistic study on the toxic effect of copper oxide nanoparticles in soybean (*Glycine max* L.) root development and lignification of root cells. Biol Trace Elem Res 162 (1–3):342–352, Nandanapalli KR, Mudusu D, Lee S (2019) Functionalization of graphene layers and advancements in device applications. Carbon

Nekrasova G, Ushakova O, Ermakov A, Uimin M, Byzov I (2011) Effects of copper (II) ions and copper oxide nanoparticles on Elodea densa Planch. Russ J Ecol 42(6):458

Noori A, White JC, Newman LA (2017) Mycorrhizal fungi influence on silver uptake and membrane protein gene expression following silver nanoparticle exposure. J Nanoparticle Res 19(2):66

Oades J (1993) The role of biology in the formation, stabilization and degradation of soil structure. In: Soil structure/soil biota interrelationships. Elsevier, pp 377–400

Panda KK, Achary VMM, Krishnaveni R, Padhi BK, Sarangi SN, Sahu SN, Panda BB (2011) In vitro biosynthesis and genotoxicity bioassay of silver nanoparticles using plants. Toxicol in Vitro 25 (5):1097–1105

Peng C, Duan D, Xu C, Chen Y, Sun L, Zhang H, Yuan X, Zheng L, Yang Y, Yang J (2015) Translocation and biotransformation of CuO nanoparticles in rice (*Oryza sativa* L.) plants. Environ Pollut 197:99–107

Peng C, Xu C, Liu Q, Sun L, Luo Y, Shi J (2017) Fate and transformation of CuO nanoparticles in the soil–rice system during the life cycle of rice plants. Environ Sci Technol 51(9):4907–4917

Perreault F, Samadani M, Dewez D (2014) Effect of soluble copper released from copper oxide nanoparticles solubilisation on growth and photosynthetic processes of *Lemna gibba* L. Nanotoxicology 8 (4):374–382

Peyrot C, Wilkinson KJ, Desrosiers M, Sauvé S (2014) Effects of silver nanoparticles on soil enzyme activities with and without added organic matter. Environ Toxicol Chem 33(1):115–125

Peyrot A (2015) Photodegradation of methyl orange using nanostructures synthesized by microwave irradiation: TiO_2 nanotubes and Ag NPs

Praveen A, Khan E, Perwez M, Sardar M, Gupta M (2018) Iron oxide nanoparticles as nano-adsorbents: a possible way to reduce arsenic phytotoxicity in Indian mustard plant (*Brassica juncea* L.). J Plant Growth Regul 37(2):612–624

Qu X, Alvarez PJ, Li Q (2013) Applications of nanotechnology in water and wastewater treatment. Water Res 47(12):3931–3946

Rawat S, Pullagurala VL, Hernandez-Molina M, Sun Y, Niu G, Hernandez-Viezcas JA, Peralta-Videa JR, Gardea-Torresdey JL (2018) Impacts of copper oxide nanoparticles on bell pepper (*Capsicum annum* L.) plants: a full life cycle study. Environ Sci Nano 5(1):83–95

Regier N, Cosio C, Von Moos N, Slaveykova VI (2015) Effects of copper-oxide nanoparticles, dissolved copper and ultraviolet radiation on copper bioaccumulation, photosynthesis and oxidative stress in the aquatic macrophyte *Elodea nuttallii*. Chemosphere 128:56–61

Rico CM, Johnson MG, Marcus MA, Andersen CP (2017) Intergenerational responses of wheat (*Triticum aestivum* L.) to cerium oxide nanoparticles exposure. Environ Sci Nano 4(3):700–711

Rossi L, Sharifan H, Zhang W, Schwab AP, Ma X (2018) Mutual effects and in planta accumulation of co-existing cerium oxide nanoparticles and cadmium in hydroponically grown soybean (*Glycine max* (L.) Merr.). Environ Sci Nano 5(1):150–157

Rui M, Ma C, Tang X, Yang J, Jiang F, Pan Y, Xiang Z, Hao Y, Rui Y, Cao W (2017) Phytotoxicity of silver nanoparticles to peanut (*Arachis hypogaea* L.): physiological responses and food safety. ACS Sustain Chem Eng 5(8):6557–6567

Salehi H, Chehregani A, Lucini L, Majd A, Gholami M (2018) Morphological, proteomic and metabolomic insight into the effect of cerium dioxide nanoparticles to *Phaseolus vulgaris* L. under soil or foliar application. Sci Tot Environ 616:1540–1551

Sarmast MK, Salehi H (2016) Silver nanoparticles: an influential element in plant nanobiotechnology. Mol Biotechnol 58(7):441–449

Schwab F, Zhai G, Kern M, Turner A, Schnoor JL, Wiesner MR (2016) Barriers, pathways and processes for uptake, translocation and accumulation of nanomaterials in plants–critical review. Nanotoxicology 10(3):257–278

Schymura S, Fricke T, Hildebrand H, Franke K (2017) Elucidating the role of dissolution in CeO_2 nanoparticle plant uptake by smart radiolabeling. Angew Chem Int Ed 56(26):7411–7414

Serag MF, Kaji N, Gaillard C, Okamoto Y, Terasaka K, Jabasini M, Tokeshi M, Mizukami H, Bianco A, Baba Y (2011) Trafficking and subcellular localization of multiwalled carbon nanotubes in plant cells. ACS Nano 5(1):493–499

Shah K, Sharma PK, Nandi I, Singh N (2014) Water sustainability: reforming water management in new global era of climate change. Environ Sci Pollut Res 21(19)11603–11604

Sharma PK, Raghubanshi A, Shah K (2020a) Examining dye degradation and antibacterial properties of organically induced α-MoO₃ nanoparticles, their uptake and phytotoxicity in rice seedlings. Environ Nanotechnol Monit Manag 100315

Sharma PK, Raghubanshi AS, Shah K (2020b) Examining the uptake and bioaccumulation of molybdenum nanoparticles and their effect on antioxidant activities in growing rice seedlings. Environ Sci Pollut Res 1–15

Shenashen M, Derbalah A, Hamza A, Mohamed A, El Safty S (2017) Antifungal activity of fabricated mesoporous alumina nanoparticles against root rot disease of tomato caused by *Fusarium oxysporium*. Pest Manag Sci 73(6):1121–1126

Shin YJ, Kwak JI, An YJ (2012) Evidence for the inhibitory effects of silver nanoparticles on the activities of soil exoenzymes. Chemosphere 88(4):524–529

Siddiqi KS, Husen A (2016) Fabrication of metal nanoparticles from fungi and metal salts: scope and application. Nanoscale Res Lett 11 (1):98

Silva S, de Oliveira JMPF, Dias MC, Silva AM, Santos C (2019) Antioxidant mechanisms to counteract TiO₂-nanoparticles toxicity in wheat leaves and roots are organ dependent. J Hazard Mater 380:120889

Simonin M, Richaume A (2015) Impact of engineered nanoparticles on the activity, abundance, and diversity of soil microbial communities: a review. Environ Sci Pollut Res 22(18):13710–13723

Sindhura KS, Prasad TNVKV, Selvam PP, Hussain OM (2014) Synthesis, characterization and evaluation of effect of phytogenic zinc nanoparticles on soil exo-enzymes. Appl Nanosci 4(7):819–827

Singh RP, Ramarao P (2012) Cellular uptake, intracellular trafficking and cytotoxicity of silver nanoparticles. Toxicol Lett 213(2):249–259

Speranza A, Crinelli R, Scoccianti V, Taddei AR, Iacobucci M, Bhattacharya P, Ke PC (2013) In vitro toxicity of silver nanoparticles to kiwifruit pollen exhibits peculiar traits beyond the cause of silver ion release. Environ Pollut 179:258–267

Srivastava A, Rao D (2014) Enhancement of seed germination and plant growth of wheat, maize, peanut and garlic using multiwalled carbon nanotubes. Eur Chem Bull 3(5):502–504

Suresh AK, Pelletier DA, Doktycz MJ (2013) Relating nanomaterial properties and microbial toxicity. Nanoscale 5(2):463–474

Syu Y-y, Hung J-H, Chen J-C, Chuang H-w (2014) Impacts of size and shape of silver nanoparticles on *Arabidopsis* plant growth and gene expression. Plant Physiol Biochem 83:57–64

Terry N, Low G (1982) Leaf chlorophyll content and its relation to the intracellular localization of iron. J Plant Nutr 5(4–7):301–310

Tighe-Neira R, Carmora E, Recio G, Nunes-Nesi A, Reyes-Diaz M, Alberdi M, Rengel Z, Inostroza-Blancheteau C (2018) Metallic nanoparticles influence the structure and function of the photosynthetic apparatus in plants. Plant Physiol Biochem 130:408–417

Tombuloglu H, Slimani Y, Tombuloglu G, Almessiere M, Baykal A, Ercan I, Sozeri H (2019) Tracking of NiFe2O4 nanoparticles in barley (*Hordeum vulgare* L.) and their impact on plant growth, biomass, pigmentation, catalase activity, and mineral uptake. Environ Nanotechnol Monit Manag 11:100223

Tripathi DK, Singh S, Singh S, Srivastava PK, Singh VP, Singh S, Prasad SM, Singh PK, Dubey NK, Pandey AC (2017a) Nitric oxide alleviates silver nanoparticles (AgNps)-induced phytotoxicity in *Pisum sativum* seedlings. Plant Physiol Biochem 110:167–177

Tripathi KM, Bhati A, Singh A, Sonker AK, Sarkar S, Sonkar SK (2017b) Sustainable changes in the contents of metallic micronutrients in first generation gram seeds imposed by carbon nano-onions: life cycle seed to seed study. ACS Sustain Chem Eng 5(4):2906–2916

WWAP (2012) Facts and figures from the United Nations World Water Development Report 4 (WWDR4). United Nations World Water Assessment Programme, UNESCO-WWAP

Vanti GL, Nargund VB, Vanarchi R, Kurjogi M, Mulla SI, Tubaki S, Patil RR (2019) Synthesis of *Gossypium hirsutum*-derived silver nanoparticles and their antibacterial efficacy against plant pathogens. Appl Organomet Chem 33(1):e4630

Wang S, Wang F, Gao S, Wang X (2016) Heavy metal accumulation in different rice cultivars as influenced by foliar application of nano-silicon. Water, Air, Soil Pollut 227(7):228

Wang Y, Wang S, Xu M, Xiao L, Dai Z, Li J (2019) The impacts of γ-Fe₂O₃ and Fe₂O₄ nanoparticles on the physiology and fruit quality of muskmelon (*Cucumis melo*) plants. Environ Pollut 249:1011–1018

Xiao L, Wang S, Yang D, Zou Z, Li J (2019) Physiological Effects of MgO and ZnO Nanoparticles on the *Citrus maxima*. J Wuhan Univ Technol Ed 34(1):243–253

Xiong T, Dumat C, Dappe V, Vezin H, Schreck E, Shahid M, Pierart A, Sobanska S (2017) Copper oxide nanoparticle foliar uptake, phytotoxicity, and consequences for sustainable urban agriculture. Environ Sci Technol 51(9):5242–5251

Xu C, Peng C, Sun L, Zhang S, Huang H, Chen Y, Shi J (2015) Distinctive effects of TiO₂ and CuO nanoparticles on soil microbes and their community structures in flooded paddy soil. Soil Biol Biochem 86:24–33

Xue W, Han Y, Tan J, Wang Y, Wang G, Wang H (2017) Effects of nanochitin on the enhancement of the grain yield and quality of winter wheat. J Agri F Chem 66(26):6637–6645

Yang J, Jiang F, Ma C, Rui Y, Rui M, Adeel M, Cao W, Xing B (2018) Alteration of crop yield and quality of wheat upon exposure to silver nanoparticles in a life cycle study. J Agri F Chem 66 (11):2589–2597

You T, Liu D, Chen J, Yang Z, Dou R, Gao X, Wang L (2018) Effects of metal oxide nanoparticles on soil enzyme activities and bacterial communities in two different soil types. J Soil sediment 18(1):211–221

Zhang H, Yue M, Zheng X, Xie C, Zhou H, Li L (2017a) Physiological effects of single-and multi-walled carbon nanotubes on rice seedlings. IEEE Trans NanoBiosci 16(7):563–570

Zhang P, Ma Y, Liu S, Wang G, Zhang J, He X, Zhang J, Rui Y, Zhang Z (2017b) Phytotoxicity, uptake and transformation of nano-CeO₂ in sand cultured *romaine lettuce*. Environ Pollut 220:1400–1408

Zhang Q, Uchaker E, Candelaria SL, Cao G (2013) Nanomaterials for energy conversion and storage. Chem Soc Rev 42(7):3127–3171

Zhang W-x (2003) Nanoscale iron particles for environmental remediation: an overview. J Nanopart Res 5(3–4):323–332

Zhao L, Peng B, Hernandez-Viezcas JA, Rico C, Sun Y, Peralta-Videa JR, Tang X, Niu G, Jin L, Varela-Ramirez A (2012) Stress response and tolerance of *Zea mays* to CeO₂ nanoparticles: cross talk among H₂O₂, heat shock protein, and lipid peroxidation. ACS Nano 6 (11):9615–9622

Zhao Q, Ma C, White JC, Dhankher OP, Zhang X, Zhang S, Xing B (2017) Quantitative evaluation of multi-wall carbon nanotube uptake by terrestrial plants. Carbon 114:661–670

Zhou D, Jin S, Li L, Wang Y, Weng N (2011) Quantifying the adsorption and uptake of CuO nanoparticles by wheat root based on chemical extractions. JEnvS 23(11):1852–1857

Printed in the United States
by Baker & Taylor Publisher Services